Matter and Molecules

Matter and Molecules

A Broader and Deeper View of Chemical Thermodynamics

First Edition

Roberto J. Fernandez-Prini, Ernesto Marceca, and Horacio Roberto Corti

cognella® | ACADEMIC PUBLISHING

Bassim Hamadeh, CEO and Publisher
Amy Smith, Project Editor
Berenice Quirino, Associate Production Editor
Emely Villavicencio, Senior Graphic Designer
Sara Schennum, Licensing Associate
Natalie Piccotti, Director of Marketing
Kassie Graves, Vice President of Editorial
Jamie Giganti, Director of Academic Publishing

This book was originally published in Spanish by The University Editorial of Buenos Aires (EUDEBA) as Materia y Moléculas written by Roberto Fernández-Prini, Horacio Corti, and Ernesto Marceca © 2004 by EUDEBA.

ISBN: 978-1-51657-266-3

cognella® | ACADEMIC PUBLISHING

Contents

2

4

Preface

Teaching natural sciences is facing a critical moment. While research increasingly demands interdisciplinary approaches, many of today's scientific textbooks still address the different disciplines in an isolated manner and concepts are discussed within unconnected and complex compartments. Moreover, since current research has shown that a more ample view is necessary to solve problems in science as well as in professional practice, specialists will need to comprehend the profound interplay arising among subdisciplines of a given area, as well as among various disciplines in natural sciences. Having these issues in mind, we propose a novel approach to several topics involving physical, chemical and/or biological phenomena. Instead of a rigid canonical view of fundamental physical chemistry, we tried to bridge the gap among the areas of interest connected with physical chemistry.

The book introduces the existing knowledge from a fresh angle, contributing new perspectives and points of view in the presentation and discussion of the different contents. It provides the tools that are necessary to study the properties of macroscopic systems, frequently referred to as chemical thermodynamics, including the science required to describe them.

The aim of the book is not replacing any of the various books dealing with physical chemistry existing nowadays, but to complement the explanation of subjects making use of an interdisciplinary approach, stressing their application to other areas of science that must be faced and understood by chemists, biochemists, physicists, geologists, biologists, pharmacists, engineers, and others. We are convinced this will be the near future role of physical chemistry. So, our book will encompasses the analysis of molecular interactions and effects of size in the temporal and spatial scales of the processes described.

Many of the aspects mentioned could, in principle, be developed without explicit reference to the molecular features of the systems being evaluated, as is typical of many classical and important books on thermodynamics. In this book we have considered it necessary to present the description of different subjects also emphasizing the atomic-molecular features of the systems being analyzed. As an example, this characteristic is very clear in the articles of physical chemistry published nowadays.

Along part of its content, this book also highlights the role of time in the correct description of the behavior of material systems, whenever it is relevant. Time has an explicit role in the description of non-equilibrium systems, phenomena that constitute a natural extension of the central material developed throughout the book.

Hence, a chapter is included dealing with linear irreversible phenomena, and a section is dedicated to metastable systems. Time, having an implicit role in equilibrium phenomena, acquires an explicit significance, which is of practical importance, when dealing with new scientific fields that are frequently discussed. This is similar to its role in chemical kinetics.

Rigor is maintained in the derivation of equations, whenever they are within the proposed general level we pretend. When this is not possible, we privilege to introduce the physical concepts involved instead of giving a formal mathematical derivation of equations that are frequently based on dubious or wrong concepts. In our opinion, this is the best way for the reader to tackle more complex processes. Different fields of physical knowledge can, in the future, get over what is discussed here for the purpose of getting a deeper insight into aspects dealt in the book, which may be required afterwards for scientific or professional practice. Moreover, we have used along the book comprehensive models to increase knowledge of the behavior of material systems constituted by atoms and molecules.

In almost all the chapters of the book, worked examples have been included to underline the application of the material contained in this volume and which are not usually covered in physical chemistry textbooks for undergraduates and graduate students.

A final chapter, "Future Development," was added to the 2010 edition in Spanish. This is a rather short chapter dedicated to applications of the toolbox built in the main part of the text and now related to describe processes in the small time and/or spacial scales. This, already, is an important scientific field, and its quick development suggests that it will remain so in the following years –for instance, to develop molecular biomimetic systems and molecular machines, as well as in many other basic sciences that do not usually take into account the molecular time and spacial scales.

NOTE: This book is, to a large extent, a translation of "Materia y Moléculas," Buenos Aires: EUDEBA, 2010. The English version has a few additions compared to the previous edition in Spanish, especially to the last chapter. For instance, two relatively simple systems have been chosen from new publications in the field of physical and natural sciences which seem to defy the second principle of thermodynamics. As it is clearly described in the two original papers, quantitative justifications of their rather surprising results have been given in those articles. For the present book, these two examples are very valuable because they illustrate the fact that many of the contents in the various chapters of the book have a relevant role in the explanations provided, showing the need to connect various fields of knowledge.

List of Figures

10

Chapter 1

Thermodynamic Background

1.1 Introduction

Matter that constitutes our world is formed by myriad atoms and molecules: The peculiar characteristics that matter exhibits, according to its specific nature and state, are consequences of the behavior of the atoms and molecules that form it–the relation between the atomic-molecular components and the macroscopic matter they form–and is the central point developed in this book. In the twentieth century, a notable development of the atomic-molecular knowledge took place; now a question arises, can this increased knowledge describe the behavior of macroscopic matter, of the different material objects we see every day? Even today, it is not possible to characterize the behavior of macroscopic material systems based on the behavior of the atoms and molecules that form a macroscopic material object. A discipline developed in the nineteenth century, thermodynamics is dedicated to describe the properties of macroscopic matter and relate them through proper equations. That is what we call a phenomenological description of the systems. Thermodynamics is a discipline in itself, having resources general enough with universal validity. Stephen Hawkins wrote *that physical theories are mathematical models of reality.* Thermodynamics is a mathematical universal discipline describing the behavior of material systems and the processes they undergo, hence it allows relating different properties of macroscopic material systems. In spite of its generality and that thermodynamics can be developed in an abstract frame, it has given practical results of undoubted importance; it has shown its capacity to express observable phenomena in understandable and easily interrelated properties, having the features of material systems, which are well-known. In order to relate the macroscopic behavior with molecular behavior, we need to use statistical mechanics which, except for very simple questions, is beyond the target of this book. Thus, on the basis of molecular properties, like their masses, their spectroscopic, electronic, and nuclear properties, it is, in principle, possible to describe the behavior of material systems. However, that discipline is much more complicated than thermodynamics.

A peculiarity that thermodynamics shares with other disciplines, for instance Euclidian geometry, is the fact that its fundamental formulation is *axiomatic* (i.e., it does not require demonstration of its validity because it is considered obvious). Which is the more unique feature of thermodynamics? It is not only the discipline that accounts for the energy of material systems and their transformation in different types of *work* that lead to changes in the external coordinates of a system as, for example, its volume, its charge distribution upon the application of an external electric field, its position in Earth's gravitational field, etc. There is a type of energy which, when changing, *does not* introduce variations in the external coordinates of the systems and is related to the movement of atoms and molecules. This is the thermal energy and the transformation it produces. This is one of the most important contributions of thermodynamics; thus, this discipline lets us ignore the individual coordinates of the molecules that compose a system and, nevertheless, describes matter in terms of the *collective* macroscopic consequences of changes that happen in the molecular coordinates.

Frequently, it is argued that thermodynamics describes *equilibrium states* which are considered states that are invariant in time. However, the notion of equilibrium states depends critically on the time scale where observations are done. This building, that tree, today, are very similar to what they were yesterday; in the interval of a single day, they may look invariant and they might suggest that they are in an equilibrium state. How long will the appearance of invariability prevail? With the notion of metastable states (i.e., those which are invariant for a time interval but are not the *most stable state*) it may be possible to measure precisely the properties of metastable systems if done in a time scale short enough to avoid a change of state. Thus, time plays a role in describing the equilibrium states of different systems.

Time and Thermodynamics

There are often, when discussing various aspects of thermodynamics, sentences like this: if after some time the system's properties do not change, it can be said that the system is in a state of equilibrium. How long is *some time*? This question appears to be facetious, but it deserves a short digression to help to capture the full value of this discipline. The concept of equilibrium, as applied in the previous question, is relative. If we put over a table an open beaker containing water and we observe it for, say, one hour, we may conclude that the system (the water) is in a state of equilibrium because its properties have not changed during the observation. But, if we leave the open beaker one or two days over the table, we will observe that there is less water inside the beaker because it has partially evaporated. Then, if the system is observed during a few days, we say that the water contained in the beaker was not initially at equilibrium. This fact is even more clear in the case of the geological evolution of our planet. For times of observation commensurable with the average human life, the lithosphere, or the Earth's surface appears to be in a state of equilibrium (with the exception of the occurrence of volcanic eruptions or earthquakes during that time). However, extending the period of observation to thousands, or even millions of years, it is not possible to consider that the system is in a state of equilibrium. It is easy to think of other examples, and we will confirm in every case that the system will be in equilibrium *within an appropriate* scale of time, and that by increasing the time of observation sufficiently it will be evident that the system was not at equilibrium initially. This is due to the fact that all material systems are immersed in the universe, which is not in a state of equilibrium.

To these features we must add its capacity to describe phenomena that change with the time of observation (i.e., processes that are not in an equilibrium state and consequently are irreversible). The methodology we developed to deal with the thermodynamics of states of equilibrium may be easily extended to linear irreversible processes.

Thermodynamics is a pragmatic discipline that provides relations between different properties and proposes various roads for their calculation or experimental determination; the user will have to choose the most convenient strategy that will be that more closely related with the problem to be solved. Also there are different but equivalent formulations of the principles or axioms of thermodynamics. They are all *equivalent*, but each coherent set of axioms are more closely related with the particular uses of thermodynamics. In this book we want to be able to use thermodynamics to deal with physicochemical processes and their connections with chemical engineering, physics and biology. So, we have adopted an axiomatic scheme close to that proposed by Guggenheim,[1] which is also very convenient to extend thermo-dynamics to irreversible phenomena. Thermodynamics provides general relations among different properties of materials systems and can be developed without re-stricting it to particular systems. However, as already mentioned, it would be very limited if the molecular aspects of matter were not considered in the systems being studied. This will be done as an illustration and also to provide a complementary molecular vision to clarify some points of the thermodynamic presentation.

1.2   The Principles

First, we need to agree about the meaning of some terms that will be used often in this book. A *system* is a portion of matter that we wish to study; it will be separated from the rest of the universe by a surface, which can be virtual, that surrounds the system completely. The rest of the universe that contains our system will be referred as the *environment*. *Extensive* properties are those that depend on the size of the system, such as the volume, the mass, and the energy. On the other hand, the *intensive* properties are independent of the system's size if it is homogeneous (e.g., density, temperature, pressure, and others). Systems are *homogeneous* when their characteristic intensive properties (density, color, temperature, etc.) have the same values in every element of volume of the system. The *heterogeneous* systems are composed by different parts, which are themselves homogeneous and are referred to as *phases*. Phases are separated from each other by surfaces of discontinuity, macroscopically defined, which are referred to as *interfaces*. As will be discussed in chapter 7, they are really inhomogeneous regions with a large change in the values of the intensive properties. Inhomogeneous phases are those that show a continuous (but rather abrupt) change of the intensive properties when we go from one homogeneous phase to an adjoining one.

A *process* takes place when, at two different times, changes are observed in some macroscopic properties of the system. When changes are not observed at different times, the system is considered at a state of *equilibrium*. An *isolated* system is one where it is not possible that external changes in the environment produce changes in the system; then, if no changes are observed in an isolated system, we say the system is in a state of internal equilibrium. If a system is surrounded by a ther-mally insulated surface, any process that takes place in it due to external effects–for instance, the displacement of the system's surface, a change in charge distribution when an external electric field is applies, etc.–is called *adiabatic*. If two systems in

---

[1]E. A. Guggenheim, "Thermodynamics. An advanced treatment for chemist and physicist", North Holland, Elsevier Science Publishers, 1967.

internal equilibrium are separated by a thermally conductive surface, when they are in contact, it is observed in general that they are not in equilibrium. That is, their state gradually adjusts until they are in thermal equilibrium. Once this state of internal equilibrium is reached, no more variations are observed.

Zero Principle

If two systems are in thermal equilibrium with a third one, they are in thermal equilibrium with each other. If we consider as reference a system with well-defined properties, all the systems that are in thermal equilibrium with it share with the reference system the same value of a property–the temperature–which will be the same for all and be identical to that of the reference system.

There are qualitative differences between the systems and the environment. We take an example given by Guggenheim (see suggested books for further reading) to illustrate this point. Consider a small and thin metal wire immersed in a large amount of water, the whole being isolated. If the metal wire is the system being studied, its temperature will be constant and equal to the water temperature, the latter acting as a thermostat (i.e., the massive environment determines the temperature of the system). On the other hand, if the liquid water is the actual system, temperature fluctuations in the wire will induce only a very small change in the water temperature (i.e., the small environment does not perturb the massive system). A measuring sensor, such as a thermometer, must behave as a small environment in contact with a massive system; such a sensor will be much more sensitive to temperature perturbations than the system itself.

First Principle

This principle establishes the conservation of energy that may be transformed but not created or destroyed. It is formulated by[2],

$$dU = \delta q + \delta w \qquad (1.1)$$

The variation of internal energy (dU) is due to the exchange of heat (δq) and to the work done upon the system (δw). The first principle implies that the variation of internal energy does not depend on the manner in which the change in the process is done; if it did depend on the way it is done, then it would be possible to generate or destroy energy choosing a different path to start the process and a different one to return to the initial state. All the functions with such a characteristic are called

---

[2]The symbol δ indicates a small variation having a value that will depend on the selected path followed by the system. It is not an exact differential.

state functions because the value of their variations depends only on the initial state and the final state, no matter how the process occurs. Thus, dU is a total or exact differential;[3] its integral does not depend on the particular way the process was carried out. However, in general, heat and work exchanged by the system will depend on the particular path followed by the process. The signs of δq and δw are clear if we analyze whether these changes lead to an increase or a decrease of U. Then, for a thermally isolated (adiabatic) system,

$$w = U(2) - U(1) = \Delta U$$

Thus, w is independent of the path followed if the process is adiabatic. This gives an opportunity to determine ΔU by the work done to take the system through an adiabatic path from the initial state to the final one. There are some processes that cannot be accomplished through an adiabatic path (cf. second principle); in that case the adiabatic inverse process (from final to initial state) can occur and, since we know that $U(2) - U(1) = -[U(1) - U(2)]$, it is always possible to determine ΔU following an adiabatic path. As a consequence, a cyclic path will imply that $w = -q$ because in a cycle the initial and final states are the same.

This is clearly shown by the classic experiment designed by Joule, used to determine the heat equivalent of work illustrated in figure 1.1. The experiment required a rigid and adiabatic vessel, which contained water, a thermometer, and a stirrer that was rotated by dropping some weights attached to it. The system studied was the water in an adiabatic container, and an stirrer doing work on it by letting a weight of mass m to fall a distance h by action of the Earth's gravity (g) (i.e., $w = mgh$). It is clear that work has been done on the system, so its energy must have increased according to

$$\Delta U = mgh > 0$$

The consequence of the mechanical work will be to increase the temperature of water, and there will be no heat exchanged with the environment because the container is adiabatic. Hence, the system has increased its thermal energy in a value equal to the mechanical work done on the system.

The processes to which we refer may be essentially physical; like Joule's experiment, or the expansion of a gas, they may involve a change of phase like the fusion of ice, or may refer to a chemical reaction. In all cases, the difference between the value of a system's property in the final state and the initial state value will always be denoted by the operator Δ. For a chemical reaction, ΔU denotes the difference in internal energy of the products minus the internal energy of the reactants.

---

[3]If the differential of a function $z = f(x,y)$, $dz = M(x,y)dx + N(x,y)dy$ is exact, it means that $M(x,y) = \partial z/\partial x$ and $N(x,y) = \partial z/\partial y$.

Thermometer

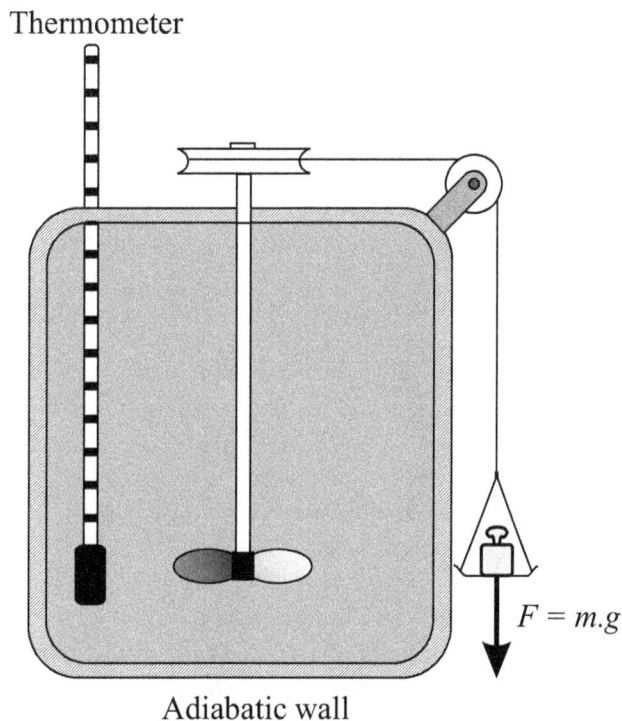

Adiabatic wall

Figure 1.1: Joule's experiment.

Let us assume that inside the adiabatic container, like that used in Joule's experiment, an exothermic chemical process takes place. The system is unable to exchange heat or work with the environment. The chemical energy liberated by the reaction will enhance the temperature of the system; that is, the reactants, which initially were at T_1, will be transformed into products at $T_2 > T_1$. Hence,

$$U(\text{products}, T_2) - U(\text{reactants}, T_1) = 0$$

This is a typical experience of adiabatic calorimetry, used to determine heats of reaction at constant volume, but we see that essentially it consists in determining the temperature change in the system. We start again from reactants at the initial state T_1, we introduce an electric resistor into the container, and we circulate an electric current I at a voltage V for a time t. Now, electric work was done on the system, $w_{\text{elec}} = VIt$, so that

$$U(\text{reactants}, T_2) - U(\text{reactants}, T_1) = VIt$$

Subtracting the last two expressions we have,

$$U(\text{products}, T_2) - U(\text{reactants}, T_2) = -VIt < 0$$

and this internal energy difference of reaction at T_2 and constant volume, is negative because it was supposed to be dealing with an exothermic reaction.

Second Principle

Conservation of energy allows us to balance the exchanges of energy between a system and its environment. The first principle establishes that the variation of energy of the system will be equal to the heat exchanged plus the work done upon the system. The conservation of energy is a particularly familiar axiom, but there is a feature of the behavior of material systems that is *evident*, albeit not so easily identifiable as the conservation of energy is. A more detailed analysis leads to a new principle or axiom. When the state of a system–which is not at equilibrium–changes with time, then we refer to this change as a spontaneous change. Spontaneous processes are those which take place naturally when starting with a system outside equilibrium; under those circumstances the inverse process *cannot* be observed. In order for these inverse processes to occur, it is necessary to do work on the system. For instance, if a beaker containing hot water was laid on a lab table, the liquid water would evaporate and the water would cool down until it reached the temperature of the lab. Nobody would expect that the water would heat up spontaneously until it started boiling at the expense of the heat in the laboratory (consequently cooling the lab temperature). Now, if we put in the container an aqueous solution of NaOH and add aqueous HCl, NaCl will be formed and the temperature of the system will increase. It is not to be expected that the NaCl solution would spontaneously turn into an aqueous mixture of NaOH and HCl and the container would cool itself. Finally, if a drop of black ink is poured in water, the dark drop would slowly turn into dark filaments that would, at the end, leave a gray or black color to the water, and it is not possible to turn the uniform blackish water into a drop of black ink in pure water.

These observations show the convenience to postulate the second principle of thermodynamics. A precise formulation of this principle states that there is an extensive property of the systems denominated entropy and whose variation for a system is given by

$$dS = d_{ex}S + d_iS \quad \text{and} \quad d_iS \geq 0 \tag{1.2}$$

where $d_{ex}S = \delta q/T_E$ is the change of entropy due to thermal energy transfer (δq) between system and environment. Such heat transfer occurs at the temperature of the part of the system in contact with environment; then it is equal to T_E. The change of entropy due to processes internal to the system (e.g., a chemical reaction or matter transfer between different phases within the system, the disappearance of Inhomogenities in a given phase of the system, will be contained in the term d_iS. If we want to apply eqn. (1.2) to open systems)[4] another term should be included in dS due to the flux of matter (cf. chapter 10). The term d_iS indicates if a process involving a change of entropy dS in the system occurs spontaneously or if it is in a state of equilibrium. Thus, the inequality in eqn. (1.2) implies that $d_iS = 0$ for a process that evolves not far away of successive states of equilibrium (reversible process), and is bigger than zero for spontaneous or irreversible processes. This expression of the second principle of thermodynamics is very convenient for the analysis of chemical systems and, fundamentally, very appropriate to extend thermodynamic concepts to systems and processes out of equilibrium, elaborated in chapter 10.

A good example to clarify the use of eqns. (1.2) is the transference of thermal energy from the environment to the system. Let us consider first the case of a system formed by two subsystems; the system as a whole is isolated from the environment and the subsystems are separated by a rigid wall that allows transfer of thermal energy, and the subsystems L and R are respectively at temperatures T_L and T_R and they exchange the quantity δq of heat between them. Since the entropy change of a system composed by P subsystems is $dS = \sum_P dS_P$, and the example corresponds to a composed system isolated from environment, $d_{ex}S = 0$, we have $dS = dS_L + dS_R = d_iS$. The changes of entropy in the two subsystems are functions of state and may be determined through the *reversible* transfer of thermal energy δq between them[5]

$$dS_P = \frac{\delta q_P}{T_P}$$

The heat received by subsystem L equals that lost by subsystem R, $\delta q_L = -\delta q_R$ and since the transfer of thermal energy is spontaneous, we get

$$dS = \delta q_L \left(\frac{1}{T_L} - \frac{1}{T_R} \right) = d_iS > 0$$

This expression indicates that if $\delta q_L < 0$, that is, if heat is transferred from L to R, then $T_L > T_R$; on the other hand if $\delta q_L > 0$, $T_L < T_R$ (i.e., heat always passes spontaneously from the hotter to the cooler subsystem).

---

[4] Open systems are those that can exchange matter with the environment.

[5] The inexact differential δq becomes exact when it is multiplied by the integration factor $(1/T)$.

As a second example, we consider a system at temperature T that is separated from the environment at temperature T_E by a diathermal wall that allows the transfer of an amount δq of heat. Under these conditions,

$$\mathrm{d}S = \frac{\delta q}{T_E} + \mathrm{d_i}S$$

The change of the system's entropy may be calculated considering that the process occurs reversibly, since S is a state function and its value is independent of the particular path followed from initial to final state. Then,

$$\frac{\delta q}{T} = \frac{\delta q}{T_E} + \mathrm{d_i}S$$

and,

$$\mathrm{d_i}S = \delta q \left(\frac{1}{T} - \frac{1}{T_E} \right) \geq 0$$

With the same reasoning used in the first example, we observe that the transfer of thermal energy proceeds from the higher to the lower temperature.

Heat and Entropy Transfer

Let us go back to the second example described on page 21, but now a finite quantity of heat is transferred Q between the environment at the temperature T_E and a system consisting in an ideal gas at an initial temperature T_i, and at the end of the process, which occurs at constant volume, the system ends up with a temperature T_f (see figure 1.2). In the initial and final states, the system is at internal equilibrium. The entropy change of the system will be

$$\Delta S = \frac{Q}{T_E} + \Delta_i S \tag{1.3}$$

ΔS is calculated by a reversible change of state (i.e., as if the heat transference took place at the system's temperature). Hence, defining the heat capacity at constant volume C_v as $\delta q = C_v \mathrm{d}T$, the change of entropy in the system when the transfer of thermal energy is reversible will be

$$\mathrm{d}S = \frac{\delta q}{T} = C_v \mathrm{d}\ln T \tag{1.4}$$

By integration of this equation we get

$$\Delta S = \int_{T_i}^{T_f} \frac{\delta q}{T} \mathrm{d}T = C_v \ln \frac{T_f}{T_i}$$

On the other hand, the total heat transferred is

$$Q = C_v(T_f - T_i)$$

Replacing these expressions in eqn. (1.3), the internal change of entropy may be written as

$$\Delta_i S = C_v \ln \frac{T_f}{T_i} - \frac{C_v}{T_{\mathrm{E}}}(T_f - T_i)$$

Making $\Delta T = T_f - T_i$, $T_{\mathrm{E}} = T_f$, and developing this equation in series, we obtain the expression

$$\Delta_i S = C_v \left(\frac{\Delta T}{T_i}\right)^2 \left[\frac{1}{2} - \frac{2}{3}\frac{\Delta T}{T_i} + \frac{3}{4}\left(\frac{\Delta T}{T_i}\right)^2 - \frac{4}{5}\left(\frac{\Delta T}{T_i}\right)^3 \cdots\right]$$

If $(\Delta T/T_i) \ll 1$, only the first term contributes on the right-hand side, which also leads to $\Delta_i S > 0$.

This example is very useful to see from which quantities the change of internal entropy between two states depends and to appreciate also how the process of thermal energy transfer is modified by following a different path for the whole process. Let us imagine now that instead of the process illustrated–transfer of Q from environment at T_f to the system initially at T_i, so that at the end both system and environment have the same temperature–Q is transferred in two stages. In the first stage, Q_1 is transferred from the environment at T_l and then Q_2 is transferred from the environment at T_f, until final thermal equilibrium is attained, so that $Q = Q_1 + Q_2$. Using the same assumptions as with the first example, the difference of $\Delta_i S$ between the process $T_i \to T_f$ and $T_i \to T_l \to T_f$ will be

$$\Delta(\Delta_i S) = C_v \frac{\Delta T_1 \Delta T_2}{T_i^2} + \ldots > 0$$

where $\Delta T_1 = T_l - T_i$ and $\Delta T_2 = T_f - T_l$. The internal entropy generated by the spontaneous process, when the heat is transferred between two more separated temperatures, is larger than that observed when the system goes from the initial to the final temperature through an intermediate temperature. It may be said that for more spontaneous processes

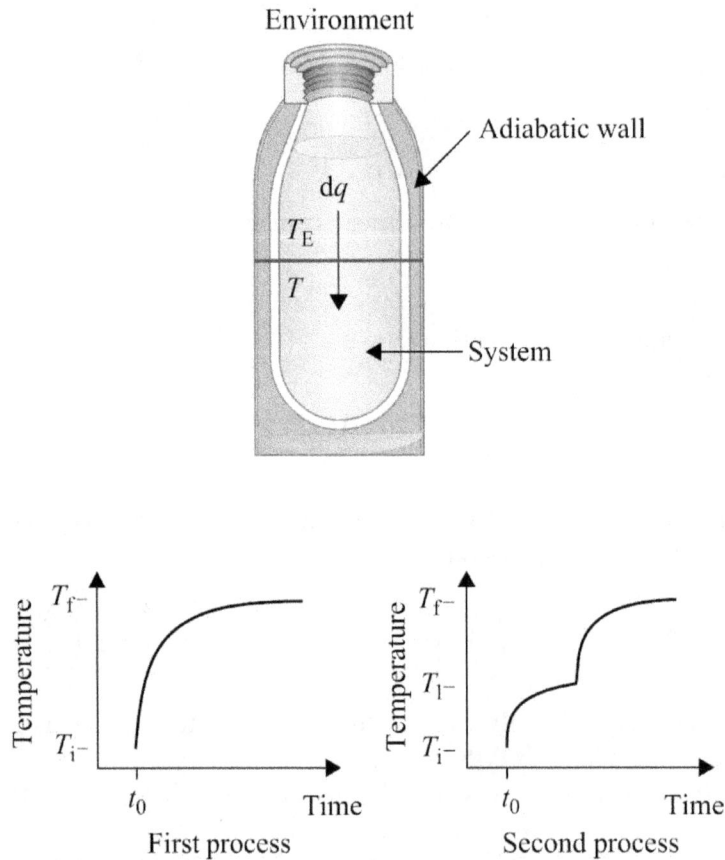

Figure 1.2: Heat transfer between the system and the environment.

(in the preceding example, when the temperature difference is bigger) the internal entropy is larger. This result is important, it explains thermodynamically why many natural processes occur through intermediate stages that diminish the change of internal entropy; that is, the loss of energy is minimized, and the entropy dissipation decreases.

This analysis can also be used to make evident what happens when a system receiving thermal energy cannot homogenize quickly enough. In this case, there will be in the system a region close to the heat-conducting wall with a temperature bigger than the rest of the system. In order to discuss how it homogenizes and which are the changes of internal entropy while homogenization occurs, it can be assumed that this situation corresponds to a system composed of several subsystems, with a temperature that progressively goes from T_i to T_f.

1.3   Other Thermodynamic State Functions

Using the equations for the first and the second principles of thermodynamics,[6] one obtains

$$dU = T_{\text{ex}}dS - p_{\text{ex}}dV + \sum Y_{\text{ex}}dX - T_{\text{ex}}d_iS \tag{1.5}$$

where Y_{ex} is an intensive variable and X is its corresponding extensive variable, so that the products $Y_{\text{ex}}dX$ refer to terms contributing to the total work made on the process, excepting the volume work or mechanical work, which is separated from other works for traditional and methodological reasons. The quantity Y_{ex} refers to the value of variable Y in the system's exterior, so that it is also denoted as Y_{E}. Now, considering that the environment has the same intensive variables T, p, and Y as the system, we can write

$$dU \leq TdS - pdV + \sum YdX \tag{1.6}$$

and also,

$$dS \geq \frac{1}{T}dU + \frac{p}{T}dV - \sum \frac{Y}{T}dX \tag{1.7}$$

These two inequalities arise because the sign of the term Td_iS and the meaning given when explaining eqn. (1.2).

The systems for which we have derived the expression 1.6 are closed systems having the same values of pressure and temperature as the environment surrounding them. Then, it is convenient to reflect over which spontaneous processes can take place in those systems to which the inequality eqn. (1.6) applies. They are processes that occur in systems that are not in internal equilibrium, that is, systems where chemical reactions, mass transfer between different phases, or Inhomogeneous phases become homogeneous, can take place. Open systems, or those with differences of pressure or temperature with the environment, will be discussed in chapter 10.

---

[6]In this section, some important mathematical relations are used. For further details concerning mathematical aspects, refer to appendix A.

From the equations 1.6 and 1.7 we get, in the absence of any work excepting mechanical work ($dX = 0$),

$$dU_{S,V} \leq 0 \quad \text{and} \quad dS_{U,V} \geq 0 \tag{1.8}$$

For isolated systems, the change of entropy is larger than zero if processes are spontaneous, and equal to zero for reversible processes occurring through very near equilibrium states. The change dU is negative or equal to zero when the processes are isoentropic (adiabatic), occurring at constant volume. These relations that indicate the reversibility or spontaneity of a process are not very valuable for describing processes that exchange thermal energy and work in the laboratory because they refer to conditions where U and V, or S and V, are constant. For many processes, it is more convenient to define other thermodynamic functions; really, it is possible to define different fundamental relations that are more adequate for a simpler and more direct descriptions of the various processes. Thus, it is convenient to define three new fundamental relations. It should be always remembered that an important feature of thermodynamics is the capacity it has to yield very different ways of representing the processes and the properties of the systems involved; it is always better to choose the expressions more directly related to those features we want to underline.

The functions enthalpy, $H = U + pV$; Helmholtz energy, $A = U - TS$; and Gibbs energy, $G = U + pV - TS$ are the ones we will more frequently use. Writing the differentials of these functions and assuming there is only mechanical work, and replacing dU by eqn. (1.6), we have

$$dH \leq TdS + Vdp \tag{1.9}$$

or $dH_{S,p} \leq 0$ for processes involving only volume work.

Likewise,

$$dA \leq -pdV - SdT \tag{1.10}$$

or $dA_{T,V} \leq 0$ for processes where there is only heat transfer.

And,

$$dG \leq Vdp - SdT \tag{1.11}$$

or $dG_{T,p} \leq 0$ for processes where there is volume work and heat transference.

The last three inequalities clearly show that thermodynamic functions are either maxima or minima, when the systems are at equilibrium (where the equal sign applies) and the process takes place under conditions that are indicated by the subscripts' identifying quantities that have been held constant during the process. For instance, the entropy at constant V and U (isolated system) will be a maximum at an equilibrium state. Helmholtz energy, at V and T constants, will be a minimum. And so on for the other thermodynamic functions.

The thermodynamic functions have exact differentials; for the thermodynamic function $z = f(x, y)$,

$$dz = \frac{\partial z}{\partial x}dx + \frac{\partial z}{\partial y}dy$$

That is, the coefficients multiplying the differentials of the variables are the partial derivatives of the function with respect to those variables. The integral of an exact differential is *independent* of the integration path and that is why the thermodynamic functions are denoted states functions. On the other hand, the previous expression allows obtaining the following general relation:

$$\frac{\partial}{\partial y}\left(\frac{\partial z}{\partial x}\right) = \frac{\partial}{\partial x}\left(\frac{\partial z}{\partial y}\right)$$

These relations are known in thermodynamics as Maxwell's relations.

In this way, a very large number of relations that link different system properties are obtained and shown in appendix A. This is a very important feature of thermodynamics, as it allows us to find the most convenient relationships to describe a given system or process. As an example we will calculate the Maxwell's equations related to eqn. (1.11), which gives dG; easily we can identify the following equations,

$$\left(\frac{\partial G}{\partial T}\right)_p = -S \qquad \left(\frac{\partial G}{\partial p}\right)_T = V \qquad (1.12)$$

Hence, G necessarily increases when there is an increase in pressure because $V > 0$. At this point, it is convenient to state that in spite of the fact the partial derivative with respect to a variable, has a clear mathematical meaning–all the other independent variables are constant–this is normally indicated explicitly in thermodynamics because the way in which a thermodynamic function depends on a variable will be different according to which other variables are kept constant. As an example, we calculate $(\partial G/\partial p)_V$ and compare it with $(\partial G/\partial p)_T$, given in eqn. (1.12); the expression is obtained by deriving eqn. (1.11) with respect to pressure at constant volume. Thus, we get

$$\left(\frac{\partial G}{\partial p}\right)_V = V - S\left(\frac{\partial T}{\partial p}\right)_V = \left(\frac{\partial G}{\partial p}\right)_T - S\left(\frac{\partial T}{\partial p}\right)_V$$

To understand the difference between these two quantities, we must remember that when p is varied at constant V, temperature will also vary. Thus, in this case it is necessary to add to the term $(\partial G/\partial p)_T = V$, another one accounting for the change of G due to the modification of temperature.

Many important relations are called fundamental relations or fundamental functions. These relationships have the *the same thermodynamic information* as eqn. (1.5). Equations obtained from derivatives of fundamental relations are called equations of state and do *not* contain the same information as the fundamental relations. This is due to the process of differentiation, where information is lost (e.g., eliminating the contributions of all the constant quantities). Thus, if a given quantity is written as a function of a different set of independent variables, information will be lost. For instance, $S = f(V, T)$ is not a fundamental function, although $S = f(V, U)$ is, because U has been replaced by the independent variable $T = (\partial S / \partial U)_V$. In other words, different information will be available depending on the independent variables chosen for a given quantity.

When the processes we are interested in are chemical reactions, the difference of a property $\Delta_R X$ will be

$$\Delta_R X = X(\text{products}) - X(\text{reactants})$$

Here, the subscript R identifies the process as a chemical reaction.

For instance, for a mole of oxygen,

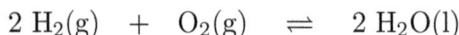

$$2\,H_2(g) \quad + \quad O_2(g) \quad \rightleftharpoons \quad 2\,H_2O(l)$$

and if $X = G$ we have

$$\Delta_R G = 2G(H_2O, l) - 2G(H_2, g) - G(O_2, g)$$

Also in these cases, expressions (1.8–1.11) are applied and, in particular, it turns out that $(\Delta_R G)_{p,T} = 0$ for a reaction that is in equilibrium at p and T constant.

1.4 Qualitative Relationships Between Thermodynamic and Molecular Properties

Let us analyze the contributions to the change of internal energy of a system. As already established, the fundamental eqn. (1.6) tells us that $U = f(S, V, X)$. Despite that this equation contains all thermodynamic information, it is very useful to know how U depends on T and V. The following example is developed for the case where there is only mechanical work; shortly we will extend it.

The change of U due to the change of temperature and volume can be written by a differential equation consisting of two terms,

$$dU = \left(\frac{\partial U}{\partial V}\right)_T dV + \left(\frac{\partial U}{\partial T}\right)_V dT \tag{1.13}$$

On the right-hand side of this equation, the first term $(\partial U/\partial V)_T$ has information about how the energy of the system changes when its microscopic configuration changes. It essentially describes the contribution of the potential energy–due to intermolecular interactions–to the total system's internal energy, which depends on the average distance between molecules (density). For instance, when the system's volume changes, the average distance between two of its molecules will also change, and that involves a change of the average potential energy of the molecules and, hence, also to a change of the internal energy U (a system's macroscopic property). This is a direct consequence to the fact that intermolecular forces change with the average separation among the molecules, which will be discussed in detail in the following chapter.

At constant temperature, the change in the value of U when the volume is altered can be expressed by the following equation of state

$$\left(\frac{\partial U}{\partial V}\right)_T = T\left(\frac{\partial p}{\partial T}\right)_V - p \tag{1.14}$$

In order to derive that equation, we start using eqn. (1.6), which can be written as

$$dU = \left(\frac{\partial U}{\partial V}\right)_S dV + \left(\frac{\partial U}{\partial S}\right)_V dS$$

By differentiating the previous equation, with respect to volume at constant temperature, it yields,

$$\left(\frac{\partial U}{\partial V}\right)_T = \left(\frac{\partial U}{\partial V}\right)_S + \left(\frac{\partial U}{\partial S}\right)_V \left(\frac{\partial S}{\partial V}\right)_T = -p + T\left(\frac{\partial S}{\partial V}\right)_T$$

And, using a Maxwell relation $(\partial S/\partial V)_T = (\partial p/\partial T)_V$, expression (1.14) is obtained. It is important to note that there is another thermodynamic equation of state based on the change of enthalpy with pressure at constant temperature,

$$\left(\frac{\partial H}{\partial p}\right)_T = V - T\left(\frac{\partial V}{\partial T}\right)_p \tag{1.15}$$

It is easy to verify that in ideal systems without intermolecular interactions, as in the case of ideal gases, $(\partial U/\partial V)_T$ and $(\partial H/\partial p)_T$ are zero. Hence, for systems with no intermolecular interactions, internal energy and enthalpy do not vary with volume or pressure, respectively, at constant temperature.

On the other hand, the amount of variation of internal energy due to thermal effects, as represented by the quantity $(\partial U/\partial T)_V$ in eqn. (1.13), takes account of the contribution of the kinetic energy, associated to particle translation and *internal modes*,[7] to the internal energy. In a gas, an important amount of the internal energy of the system will be associated to the random translational movements of its molecules. Moreover, in addition to translational modes, molecules can store energy in internal modes associated with molecular vibrations and rotations, or to the excitation of electronic or nuclear modes.

In a similar way, it is possible to derive equations of state for systems able to exchange other types of energy; e.g., elastomers can be subjected to elastic work, although their volume does not change appreciably (and hence $-pdV$ is negligible). The equation of state derived in this case is analogous to eqn. (1.14), also its microscopic interpretation; internal energy of elastomers have a component due to interactions between elastomer molecules, and another related to the internal modes of their, usually long, chains of atomic groups. This case will be developed with detail in chapter 3 where the discussion will be more profitable. Finally, the changes of U and H with temperature will be added. The change of the system's internal energy with temperature at constant volume is known as the system's *heat capacity at constant volume*, C_V, defined as

$$C_V = \left(\frac{\partial U}{\partial T}\right)_V \qquad (1.16)$$

This relation implies that when energy is given to a system, its temperature will increase inversely proportional to its molar heat capacity. That is, if a system has a small heat capacity, for example, because vibrational modes are not active at the actual temperature, the molecular energy can only be stored as kinetic energy, leading to a stronger increase in temperature than for larger values of C_V. The first principle, see eqn. (1.1), establishes that changes of internal energy at constant volume are due exclusively to the exchange of thermal energy, that is, to the flow of heat, leading to

$$dU = \delta q_V \qquad \text{and} \qquad C_V = \frac{\partial q_V}{\partial T}$$

so that C_V is the heat absorbed by the system to increase 1 K, its temperature.

Using the definition of enthalpy in a similar way, the *heat capacity at constant pressure*, C_p, is defined as

$$C_p = \left(\frac{\partial H}{\partial T}\right)_p \qquad (1.17)$$

---

[7]Different modes through which molecules store internal energy, as vibrations and rotation.

and also

$$\mathrm{d}H = \delta q_p \quad \text{and} \quad C_p = \frac{\partial q_p}{\partial T}$$

Microorganisms Growth

It is interesting to apply the thermodynamic quantities already presented in order to describe the growth of microorganisms. At the beginning of the twentieth century there was a tendency to consider that the tools employed to describe the behavior of material systems could not be applied to biological systems. Nowadays, it is accepted that the principles of thermodynamics and their consequences are also applicable to biological systems, and it is worth to analyze them.

All the relationships between thermodynamic quantities that will be developed in the first nine chapters of this book refer to states of equilibrium; this with the exception that the second principle explicitly adds the condition that the quantity $\mathrm{d_i}S$ is positive for spontaneous processes, which naturally occurs out of equilibrium. Obviously, the growth of microorganisms and, in general, the life of biological organisms, cannot be considered a process that occurs under equilibrium conditions (if it were so, life would be a reversible path!), hence thermodynamic equilibrium is not apt to explain the behavior of biological systems. Moreover, when trying to describe the growth of organisms, the systems have to be considered *open*: Nutrients enter them and other substances exit them; these are the products of cellular metabolism and of their reproduction

The first difficulty that arises is how to deal with irreversible systems using the tools that we have already developed. This is, to a large extent, accomplished realizing that it is possible to assume that microorganisms are in a *steady state*. To say that a system is in stationary state implies that the thermodynamic properties in every volume element of the system do not vary with time; nevertheless, the system will generate flows of energy and/or matter, a point that will be dealt in chapter 10. For instance, imagine a metallic bar that receives heat in one extreme, while on the other extreme it is kept at room temperature. Along the bar a temperature profile will be established that will not change with time, and will produce a flow of heat. Another case is a chemical continuous reactor; reactants enter it and products come out of it, and the concentrations profiles of all the substances inside the reactor will be independent of time, but a flow of matter will exist. These are two irreversible processes that will produce internal entropy d_iS, but the condition of a steady, or stationary, state, requires that the entropy produced be a minimum that is compatible with the process–there is an *irreversibility economy*, which leads to a minimum dissipation of entropy, as shown in the section 8.5. Under these conditions, the usual thermodynamic relations can be validly applied.

The following illustration is based on an article by von Stockar and Liu[8] which starts with a well-known statement of Erwin Schrödinger: *An organism feeds on negative entropy. Or, putting it in a less paradoxical sentence, what is essential in the metabolism is that the organism can be able to liberate itself of all the entropy that it will irremediably produce while living.* This is how it looks because the molecules it produces by biosynthesis are complex, like proteins or polynucleotides, having lower entropy than that of the simpler molecules that intervened in their synthesis. On the other hand, the global irreversible process of growth must lead to an increase of the entropy of the universe. Taking into account these two contributions, it seems logical to think that the organisms must feed on substances of low entropy so that the final balance will show and excess of entropy as a characteristic of the spontaneous process. von Stockar and Liu analyze four examples of microbe growth and evaluate, using calorimetric data, the contributions leading to global production of entropy. To make the analysis, it is necessary to characterize microorganisms' growth through various conditions. Since the system is in a steady state, the concentration of substances inside the cells must be constant, hence all the substances that enter to the cell as nutrients must leave in the form of products and/or of the generated biomass. Obviously, inside the organisms many chemical reactions will occur, which are responsible for the cellular metabolism and for the synthesis of biomass. Thus, the following expression is obtained

$$d_iS = \sum_j -\frac{\Delta_{\mathrm{R}j}G}{T}dn_{\mathrm{R}j} = \sum_j \left(\Delta_{\mathrm{R}j}S - \frac{\Delta_{\mathrm{R}j}H}{T}\right)dn_{\mathrm{R}j}$$

where $n_{\mathrm{R}j}$ is equal to the moles of reaction j that has taken place.

---

[8]U. von Stockar, and J.-S. Liu, "Does Microbial Life Always Feed on Negative Entropy? Thermodynamic Analysis of Microbial Growth", *Biochim. Biophys. Acta*, 1412 (1999):191-211.

Then, what is found is that the process' spontaneity can be evaluated as a function of the thermodynamic of equilibrium corresponding to chemical processes occurring in the microorganisms, even in the presence of diffusion and flow of matter. That is why the analysis can be carried out based on the calorimetric data obtained measuring the heat exchanged with environment. Also, it is necessary to use data for enthalpies of formation and combustion of the substances involved.

If Schrödinger's statement was correct, $\Delta_R S = \sum_j \Delta_{Rj} S$ should be positive and this would be the origin of the thermodynamic contribution to the global production of entropy.

Oxidative growth of yeast:

As an illustration of the procedure, the authors analyze the case of *Kluyveromyces fragilis*, using glucose as nutrient, where the reaction can be written

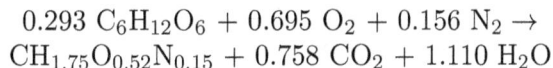

$$0.293 \ C_6H_{12}O_6 + 0.695 \ O_2 + 0.156 \ N_2 \rightarrow$$
$$CH_{1.75}O_{0.52}N_{0.15} + 0.758 \ CO_2 + 1.110 \ H_2O$$

If glucose is the nutrient, it is observed that *K. fragilis* has a $\Delta_R G = -345.1$ kJ/(C-mole) (where one C-mole refers to one mole of carbon atoms), close to $\Delta_R H = -302$ kJ/(C-mole); hence, the change of reaction entropy is very small. It can be concluded that the microorganisms growing oxidatively employing organic substances as nutrients do *not* feed on negative entropy. The production of entropy that necessarily will occur during the growth process is due almost entirely to the exchange with the environment of the heat generated by the reaction.

Anaerobic growth of archaeobacters:

An equivalent reaction for the anaerobic growth of *Methanobacterium thermoautotrophicum*, has an enthalpy $\Delta_R H = -3730$ kJ/(C-mole), that is, much more negative than the Gibbs energy of reaction, $\Delta_R G = -798.1$ kJ/(C-mole). In this case, there is a negative entropy of reaction because the process of growth implies that one mole of CO_2 and four moles of H_2 are converted into one mole of CH_4 and two moles of H_2O. It may be said that in order for the growth of *M. thermoautotrophicum* to occur spontaneously, $\Delta_R H$ must contribute to $\Delta_R G < 0$ (driving force that produces growth) and also compensate the fact that $\Delta_R S < 0$. That is, the large exothermicity of the process is not only used to generate the necessary production of entropy of the microorganism, it also must compensate the unfavorable $\Delta_R S$. It is evident that in this case the process does not correspond to Schrödinger's statement.

Anaerobic fermentation of yeasts to produce ethanol:
Saccharomyces cerevisae is a yeast feeding on glucose to produce ethanol. In this case we have $\Delta_R H = -101$ kJ/(C-mole) and $\Delta_R G = -310$kJ/(C-mole). There is an important contribution of $\Delta_R S$ to the spontaneous growth process that can be thought as due to the reaction of converting glucose molecules to CO_2 and ethanol, accompanied by a much smaller production of biomass.

Methanogenesis using acetate as nutrient:
Methanobacterium soehngenii feeds with acetate and produces CH_4. In this case, the enthalpy of reaction is positive, and therefore the growth process is sustained by a very large and positive $\Delta_R S$; it may be said that the microorganism feeds on negative entropy.

   As seen, there is a great variety of situations to explain the spontaneity of the growth of microorganisms.

_____

1.5   Multicomponent Systems

Systems consisting of various phases containing more than one substance, at least in one of the phases, are denoted as multicomponent systems. A system will be considered closed if it cannot exchange mass with the environment. Now, for a closed system, is it possible to know if it is composed of more than one substance from measurements of p, T, and V? The answer is that it will not be possible; externally, the system does not show that it is formed by more than one substance. A clear example is air, formed by two main substances with large concentrations and a series of other substances in minor concentration. There are tables that record values of the thermodynamic properties of air, as if it were a simple pure substance. An example, even more clear than air, is water containing isotopologues;[9] however, in most circumstances that we report or refer to the properties of water, it

_____

[9]Isotopologues are molecules that only differ in their isotopic composition. For example, the isotopologues of methane are: CH_4, CH_3D, CH_2D_2, CHD_3, and CD_4.

is not considered a multicomponent system and the carefully built equations for its thermodynamic properties, as established by the International Association for the Properties of Water and Steam (IAPWS) are known as the general and scientific formulation. A feature of thermodynamics, that will be expatiated below, is that it describes systems with details according to the available information and the need to take account of minor differences in its composition.

An example where the composition of the system changes from one place to another is sea water. Its composition may change appreciably in different geographical points; so, it has been necessary to define a standard or Vienna Standard Mean Ocean Water (VSMOW) in order to carry representative studies of this medium.

Whenever a multicomponent system has a fixed or relatively constant composition, its thermodynamic behavior will not permit us to establish whether it is formed for more than one substance. When a system's composition changes (changing the number of moles of one or more substances that form it) one cannot ignore that the quantities of matter of each component are thermodynamic variables and that it is then necessary to know precisely their values to characterize the state of the system. Multicomponent systems have more thermodynamic degrees of freedom than those formed only by one pure substance. For closed systems, the changes of quantities of mass of each component (measured, for instance, by their number of moles n_i of each substance) may occur by chemical reactions or by transfer of one or more components between the phases that form the given system. That is how eqn. (1.6) is extended to take account of multicomponent systems in equilibrium; starting from the change of internal energy for a system of c components, we have,

$$\mathrm{d}U = T\mathrm{d}S - p\mathrm{d}V + \sum Y\mathrm{d}X + \sum_{i=1}^{c} \mu_i \mathrm{d}n_i \qquad (1.18)$$

where μ_i is known as the chemical potential of component i. This is a really important and central thermodynamic magnitude to describe multicomponent systems. Eqn. (1.18) is obtained considering first that the environment has the same intensive variable T, p, and Y–the system is under equilibrium with the environment. Then, the term $-T\mathrm{d}_i S$ is replaced in eqn. (1.5), taking into account the effect produced by changes in the amount of substances involved in the internal processes, represented by the last term in eqn. (1.18),

$$-T\mathrm{d}_i S = \sum_{i=1}^{c} \mu_i \mathrm{d}n_i \qquad (1.19)$$

As mentioned above when discussing eqn. (1.2), $\mathrm{d}_i S$ it is the change of entropy due to the internal processes in the system, like chemical reactions or the transference of matter between phases of the system.

A simple but clear illustration of the relation between d_iS and the processes occurring in the system can be obtained from the thermochemical processes carried out inside adiabatic calorimeters, like the generic example of chemical reactions discussed on page 19. There, the method used to determine the variation of observed temperature in a constant volume adiabatic calorimeter was analyzed. In this case, $(\Delta U)_{SV}$ or $(\Delta S)_{UV}$ determine the spontaneity of the process, which being adiabatic, implies $d_{ex}S = 0$. Thus,

$$dS = d_iS$$

The value of the system's entropy may be determined following a reversible path because it is a function of state. Then, it results that

$$\Delta S = \int_{T_1}^{T_2} C_p(\text{products}) d\ln T + \Delta_R S(T_2)$$

Since it is a spontaneous process, $\Delta_i S = \Delta S > 0$. The same analysis can be used when the internal process is a transfer of matter between phases or the system is homogenized.

Using the definitions of H, A, and G, and limiting for simplicity to the case where only mechanical work is done, eqns. (1.9–1.11) can be extended to cover equilibrium systems having more than one component,

$$dH = TdS + Vdp + \sum_{i=1}^{c} \mu_i dn_i \tag{1.20}$$

$$dA = -pdV - SdT + \sum_{i=1}^{c} \mu_i dn_i \tag{1.21}$$

$$dG = Vdp - SdT + \sum_{i=1}^{c} \mu_i dn_i \tag{1.22}$$

Recalling that thermodynamic functions have exact differentials, one concludes from eqns. (1.18) and (1.20-1.22) that

$$\mu_i = \left(\frac{\partial U}{\partial n_i}\right)_{S,V,n_{j\neq i}} = \left(\frac{\partial H}{\partial n_i}\right)_{S,p,n_{j\neq i}} = \tag{1.23}$$
$$= \left(\frac{\partial A}{\partial n_i}\right)_{T,V,n_{j\neq i}} = \left(\frac{\partial G}{\partial n_i}\right)_{T,p,n_{j\neq i}}$$

It may be seen that chemical potential is the same as the change of the fundamental functions when there is a change in the amount of matter of any component, keeping constant all other natural variables of the function.

From expression (1.19), it may be concluded that, for spontaneous processes, the variation in the quantities for the components it holds is

$$\sum_{i=1}^{c} \mu_i \mathrm{d}n_i \leq 0 \qquad (1.24)$$

The equality applies to equilibrium processes and the inequality to spontaneous processes, but the relation is applicable *for all thermodynamic paths.* This derivation took into account that the total system is closed, so the amount of matter of every component in the system will not vary by exchange with the environment; however, chemical reactions can occur in the system. That is why the inequalities (1.6) and (1.9–1.11) can be used to establish the thermodynamic equations in equilibrium processes as well as to spontaneous ones. Also, eqn. (1.24), together with eqns.(1.18) and (1.20-1.22), allow establishing for these types of processes, involving changes in the amount of matter of the components, conditions for which the sign that takes the change of the thermodynamic functions $\mathrm{d}U$, $\mathrm{d}H$, $\mathrm{d}A$, and $\mathrm{d}G$ can be employed as a criterion of spontaneity. Similar to the case of one component-systems, eqns. (1.8-1.11), the spontaneity of processes in multicomponent systems can be associated with the following inequalities

$$\mathrm{d}U_{S,V,n} \leq 0 \quad , \quad \mathrm{d}H_{S,p,n} \leq 0$$

$$\mathrm{d}A_{T,V,n} \leq 0 \quad \text{and} \quad \mathrm{d}G_{T,p,n} \leq 0$$

A final consideration will help understanding how one goes from equations that describe the behavior of multicomponent systems to systems formed by only one substance. According to eqn. (1.22), if there is only one component in one-phase system, $\mathrm{d}G$ for that phase is given by

$$\mathrm{d}G = V\mathrm{d}p - S\mathrm{d}T + \mu\mathrm{d}n$$

hence, $\mu = G/n$. The last term in the previous expression takes account of how much G varies when the amount of matter changes; it is the term appearing when the system is open. This point will be dealt with again in chapter 3 when analyzing the method of integration of Euler.

1.6   Membrane or Osmotic Equilibrium

According to the previous section, all processes of matter transference between phases in a system, and every chemical reaction, lead to changes of composition in the phases; consequently, they will be subjected to the condition given by eqn.

(1.24). When a closed system is formed by different phases γ, its Gibbs energy will change according with

$$dG = \sum_{\gamma} dG^{\gamma} = \sum_{\gamma} \left(V^{\gamma} dp - S^{\gamma} dT + \sum_{i=1}^{c} \mu_i^{\gamma} dn_i^{\gamma} \right) \qquad (1.25)$$

If there are only two phases, α and β, and there is only transfer of substance from α to β, keeping p and T constant, the previous expression will reduce to

$$dG)_{p,T} = \mu_l^{\alpha} dn_l^{\alpha} + \mu_l^{\beta} dn_l^{\beta} \leq 0 \qquad (1.26)$$

Since the system is closed, $dn_l = dn_l^{\alpha} + dn_l^{\beta} = 0$, i.e. $dn_l^{\alpha} = -dn_l^{\beta}$. Using this relation in eqn. (1.26), we have

$$(\mu_l^{\beta} - \mu_l^{\alpha}) dn_l^{\beta} \leq 0 \qquad (1.27)$$

Since dn_l^{β} is arbitrary, but positive because it was assumed that component l goes from phase α to phase β, it is necessary that

$$\mu_l^{\alpha} \geq \mu_l^{\beta}$$

in order to satisfy eqns. (1.27). That is, when the system is at equilibrium, the chemical potential of i must be the same in both coexisting phases. When the mass transfer is spontaneous, matter will flow from the phase where it has larger chemical potential to that having lower chemical potential. A logical result, in order that the system attains a state of equilibrium (chemical potential identical in both phases), it will be that

$$\left(\frac{\partial \mu_i}{\partial n_i} \right)_{p,T,n_{j \neq i}} > 0 \qquad (1.28)$$

Transport will increase μ_i^{β} because n_i^{β} has increased and μ_i^{α} will therefore decrease by the reduction of n_i^{α}.

If eqn. (1.24) is used to analyze the transference of matter of the c components between the same two phases at constant T and p, we have

$$\sum_{\gamma=\alpha,\beta} \sum_{i}^{c} \mu_i^{\gamma} dn_i^{\gamma} \leq 0 \qquad (1.29)$$

and, being the composed system a closed one, that is, $dn_i^{\alpha} = -dn_i^{\beta}$, one gets

$$\sum_{i}^{c} (\mu_i^{\beta} - \mu_i^{\alpha}) dn_i^{\beta} \leq 0 \qquad (1.30)$$

When there are several substances being transferred between two phases, what characterizes a spontaneous process is that it should comply with eqn. (1.30). The total balance of matter transference has to follow that relation, but it may happen that the spontaneous transference may produce *a rise of chemical potential of a few components in one phase*. This is a very important situation in complex systems and, in particular, in biological systems.

These considerations also show that the condition of equilibrium for matter transfer between two or more phases requires *that the components are able to pass from one phase to the other* and it's not sufficient that those component exist in the phases. That is why this situation is denoted as osmotic or membrane equilibrium.

Transport Against Concentration Gradient or Active Transport

Cussler et al.[10] described the following sodium pump (see figures 1.3A and B). Aqueous solutions in the two compartments are well stirred and are separated from each other by a membrane containing solutions of the antibiotic monesine dissolved in hexanol. The experiment starts with both compartments having the same amount of Na^+ ion, but having different pH–one containing an alkaline solution and the other an acid solution. The difference of sodium ion concentration was determined as a function of time, and the curve in figure 1.3A was obtained. This implies the concentration of Na^+ ion increases in one compartment and decreases in the other; that is, diffusion of sodium through the membrane takes place *against* the Na^+ concentration gradient. The process observed is the transport of sodium ion from one compartment to the other, despite that $\mu_i(Na^+)$ is equal in both phases, according to eqn. (1.27). The spontaneous process does not maintain the equality of the chemical potential of sodium ion in both phases; on the contrary, it creates a difference of the chemical potential of the ion in both compartment separated by a membrane.

This experiment could appear as a paradox; however, taking into account eqn. (1.29), the paradox disappears because at the same time as the sodium ion is transported from one side to the other; there is a H^+ transport in the opposite direction. The mechanism of this sodium

---

[10]E. L. Cussler, D. F. Fennell Evans, and S. M. A. Matesich, "Theoretical and Experimental Basis for a Specific Countertransport System in Membranes", *Science*, 172 (1971): 377-379.

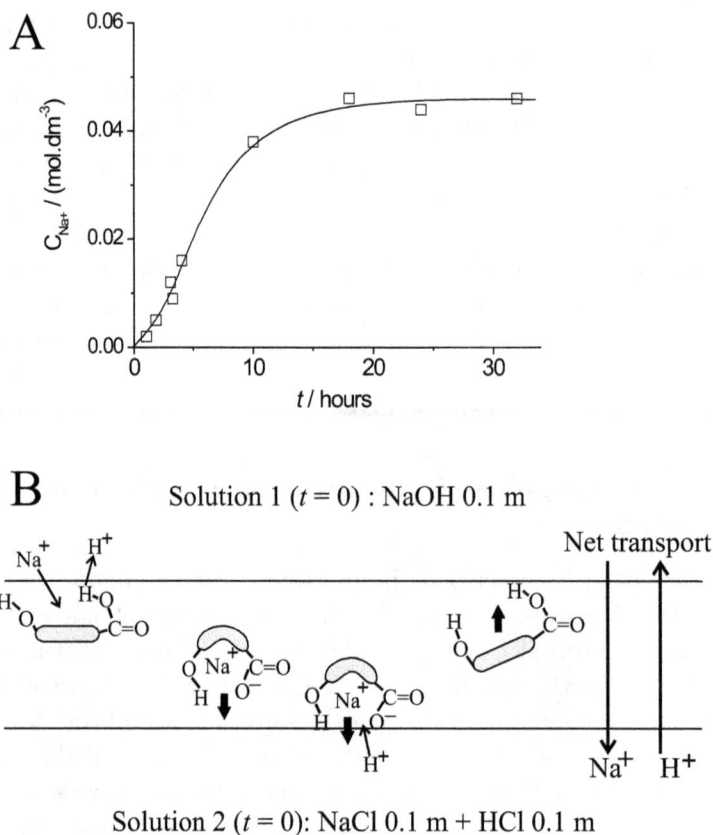

Figure 1.3: Transport against gradient, or active transport.

pump is schematized in figure 1.3B. It is clear that in the compartment with alkaline pH the antibiotic donates a proton and changes its molecular conformation. In this new conformation, it can embed a sodium ion, forming a new *cage-like* structure, and, in this way it is transported through the membrane. When arriving to the other side of the interface, the antibiotic molecules that transport Na^+ will be neutralized by the acid on the other compartment. This produces another change in the antibiotic structure, which returns to its original conformation, thus liberating the transported sodium ion into the acid compartment. Hence, this sodium pump is coupled to the diffusion of protons which will be neutralized by OH^- ions. The Gibbs energy of neutralization provides the energy necessary for the sodium pump to work.

1.7 Digression About the Second Principle

Eqn. (1.18) suggests that the term including the chemical potential is another form or work–chemical or osmotic work–which contains an intensive term (which depends on the type of work) and a differential of an extensive variable. This is another form of energy and, when there is transport of matter, the *force* is the gradient of chemical potentials, as shown in chapter 10. It must be emphasized that chemical work not only refers to the transfer of matter, it also includes transformation of matter by chemical reactions. Thus, the fact that it is not possible to obtain work from an isolated system at equilibrium can be generalized if chemical equilibrium is included.

We have not yet exploited a consequence of the second principle. If fundamental functions go through extreme points when in equilibrium state, it means that their derivatives are zero in those states, and their second derivatives have a definite sign. U, H, A, and G go through a minimum at equilibrium, meaning that their second derivatives must have a positive value; meanwhile, S is a maximum at equilibrium and its second derivative will be negative. These are *strong* conditions that guarantee the stability of the equilibrium states of macroscopic systems.

On the other hand, a small, but not a negligible, change of internal energy around an equilibrium state, due to changes in the variables on which the state depends, can be expressed by a series around the given equilibrium state. The series is given by the following expression,

$$\delta U = \mathrm{d}U_{\mathrm{eq}} + \mathrm{d}^2 U_{\mathrm{eq}} + \ldots$$

The first term $\mathrm{d}U_{\mathrm{eq}}$ is zero due to the equilibrium condition, so that

$$\delta U = \mathrm{d}^2 U_{\mathrm{eq}} + \ldots$$

and $\mathrm{d}^2 U_{\mathrm{eq}}$ for a one-component system can be expressed in terms of the second derivatives of U, with respect to its variables S and V, taken at the state of equilibrium. Moreover, we know that being U, a minimum in the state of equilibrium, $\mathrm{d}^2 U > 0$. Hence,

$$(\mathrm{d}^2 U)_{\mathrm{eq}} = \frac{1}{2}\left[\left(\frac{\partial^2 U}{\partial S^2}\right)_{\mathrm{eq}}\delta^2 S + 2\left(\frac{\partial^2 U}{\partial S \partial V}\right)_{\mathrm{eq}}\delta S \delta V + \left(\frac{\partial^2 U}{\partial V^2}\delta^2 V\right)_{\mathrm{eq}}\right] > 0$$

This inequality requires, at least,

$$\left(\frac{\partial^2 U}{\partial S^2}\right) = \frac{T}{C_V} > 0$$

and

$$\left(\frac{\partial^2 U}{\partial V^2}\right) = -\left(\frac{\partial p}{\partial V}\right)_S > 0$$

so that when either δS or δV are zero, the inequality sign must be maintained, considering that the differential of the variables are completely arbitrary.

This derivation shows that at least C_V and $-(\partial p/\partial V)_S = V \kappa_S$, where κ_S is the adiabatic compressibility and must be positive. These two thermodynamic quantities are, respectively, response functions of the system when it is perturbed thermally at constant volume, and also mechanically at constant entropy. Their finite sign implies stability of the equilibrium state. Also, C_p and κ_T, the isothermic compressibility defined by

$$\kappa_T - \frac{1}{V}\left(\frac{\partial p}{\partial V}\right)_T$$

are response functions for the systems.

In appendix A, the following relation between both types of compressibility and the two heat capacities,

$$\frac{\kappa_S}{\kappa_T} = \frac{C_V}{C_p}$$

is demonstrated. Therefore, positive values of C_p and κ_T also imply conditions of thermodynamic stability.

Although we will not derive it here, in multicomponent systems, it is necessary that the following conditions applies to guarantee material stability (eqn. (1.28)).

Problems

Problem 1

An ideal gas occupies 1.5 dm^3 at a pressure of 2.50 bar.

a) Calculate the work that is required to compress it to 0.8 dm^3 at a constant pressure of 4.7 bar, and then compress it to 0.4 dm^3 at a constant pressure of 9.4 bar.

b) Calculate the isothermal reversible work required to take it from the initial to the final states. Compare both results.

Answers: a) $w = 705$ J; b) $w = 496$ J

Problem 2

One mole of ideal monoatomic gas is contained in a cylinder having a piston and adiabatic walls. The piston is kept initially fixed, so the gas is contained in a volume of 5 dm^3 at 400 K. The external pressure p_{ex} is zero.

a) For an expansion against vacuum that carries the system to a final volume of 10 dm^3, calculate p and T at the final state, the heat exchanged, the work, ΔU, and ΔS. Verify that $\Delta U < T\Delta S - p_{ex}\Delta V$.

b) Starting at the same initial state, an adiabatic reversible expansion is carried out (system's pressure equal to that of the environment) until the system occupies a volume of 10 dm^3, and then another reversible isocoric process is made to take the system to the same final state, as in item (a). Calculate ΔU and ΔS for this reversible process.

In both cases, draw the graphs of pressure as a function of volume describing the processes.

Answers: a) $T = 400$ K, $p = 3.325$ bar, $w = 0$, $q = 0$, $\Delta U = 0$, $\Delta S = 5.76$ J/(mole K); b) $\Delta U = 0$, $\Delta S = 5.76$ J/(mole K)

Problem 3

In a cylinder with a mobile piston surrounded by a heating mantle that is at 100 ^{o}C, 0.1 mole of benzene is heated from 25 ^{o}C to 100 ^{o}C at an external pressure of 1 bar. By considering that the vapor is ideal calculate q, w, ΔH and ΔU for benzene.

Data: $\Delta_v H(80\ ^o\text{C}) = 30.75$ kJ/mol; $T_b = 80\ ^o$C; $C_p(\text{l}) = 136.1$ J/(mole K); $C_p(\text{v}) = 111.9$ J/(mole K)

Answers: $q = 3.991$ kJ; $w = 3.103$ kJ/mol; $\Delta H = 3.991$ kJ/mol; $\Delta U = 888$ kJ/mol

Problem 4

Find an expression for $(\partial T/\partial p)_S$ as a function of C_p and of the isobaric expansivity $\alpha = (\partial V/\partial T)_p/V$.

a) When water is adiabatically compressed starting at $T = 275$ K, will its temperature rise of decrease?

b) The isobaric expansivity of benzene at a 298 K is 1.257×10^{-3} K^{-1}, its molar volume is 88.9 cm^3/mole, and its $C_p = 132.5$ J/(mole K). Calculate the temperature variation when the liquid is adiabatically compressed from 0.1 MPa to 10.0 MPa.

Answers: a) it diminishes; b) $\Delta T = 2.45$ K

Problem 5

Calculate $(C_p - C_V)$ for metallic aluminum at 330 K.

Data: $(1/V)(\partial V/\partial T)_p = 2.31 \times 10^{-5}$ K^{-1}; $-(1/V)(\partial V/\partial p)_T = 1.53 \times 10^{-5}$ bar^{-1}; $\rho(\text{Al}) = 2.70$ g cm^{-3}

Answer: $(C_p - C_V) = 1.17 \times 10^{-2}$ J/(mole K)

Problem 6

The heat capacities of the substances involved in the reaction

$$CO(g) + 2\,H_2(g) \rightleftharpoons CH_3OH(g)$$

follow the expression

$$C_p = a + bT + cT^2$$

where the coefficients a, b and c are listed in the following table.

Compound	a/R	$10^3 \cdot b/R$ K^{-1}	$10^6 \cdot c/R$ K^{-2}	$\Delta_f H(298\text{ K})$ kJ/mole	$S(298\text{ K})$ J/(mole K)
CO(g)	3.191	0.924	0.1410	-110.54	197.66
H$_2$(g)	3.496	-0.1007	0.2420	0	130.68
CH$_3$OH(g)	2.213	12.215	3.449	-201.25	239.80

Calculate $\Delta_R H^{\ominus}(420\text{ K})$, $\Delta_R S^{\ominus}(420\text{ K})$, and $\Delta_R G^{\ominus}(420\text{ K})$ for the gases in the standard state.

Answers: $\Delta_R H^{\ominus}(420\text{ K}) = -95.15$ kJ/mol, $\Delta_R S^{\ominus}(420\text{ K}) = -218.91$ kJ/(mole K), and $\Delta_R G^{\ominus}(420\text{ K}) = -3.21$ kJ/mol

Chapter 2

Molecules, Statistics, and Matter

2.1 Properties of Matter

Any portion of macroscopic matter is constituted by an enormous quantity of particles animated by constant movement–this occurs by intra- and intermolecular displacements. In the same manner as the thermodynamic degrees of freedom are independent variables that allow fixing the state of a system, the molecular degrees of freedom–vibrations, rotations, and translations– determine the energetic and dynamic state of atoms and molecules. In principle, it looks reasonable that atomic movements of vibration and rotation can be separated from those that involve the displacement of the center of mass of molecules, which affect the kinetic energy and the relative distance between molecules. This separation of molecular movements is possible due to the usually very different magnitude of the involved energies. Intermolecular displacements are related to the potential and kinetic energies of molecules; the kinetic energy is responsible for the displacement of the centers of mass of molecules, and, in the case of gases and liquids, they lead to what is denoted as Brownian motion; the potential energy, on the other hand, involves interactions between molecules that mainly depend on the average distance among them. It may be considered that intramolecular movements have relation with changes in the individual molecules (i.e., that the proximity of other molecules do not affect them); on the other hand, intermolecular displacements are a manifestation of the collective effect of the system's behavior. This distinction is important because it implies a quite different way of dealing with each type of molecular degrees of freedom.

The properties of matter are manifestations of their atomic-molecular behavior, but it is clear that these properties do not depend on $6N$ degrees of freedom, the number of independent variables that characterizes the dynamic state of a system constituted by N atoms[1] The macroscopic properties represent the *average* behavior of all the particles in the system. That is why it is necessary to use statistical concepts to relate the atomic-molecular degrees of freedom (behavior of atoms and molecules) with the thermodynamic degrees of freedom, which fix the state of the system.

When determining, experimentally, the value of a thermodynamic variable or function $\mathcal{L}$ for a system in state of equilibrium, what is measured is its value during a time interval, and the average value of this measurement, $\bar{\mathcal{L}}$, is the value finally reported. Since the particles constituting the systems are in constant movement, the thermodynamic state will undergo fluctuations; that is, variables that characterize an equilibrium state will fluctuate around their average values. The basis of the statistical treatment of molecules that form a system consists in assuming that the temporal average of $\mathcal{L}$ is identical to the average value of the variable or function taken over a very large number M of instantaneous snapshots of the system. Each of those snapshots records a microscopic configuration of the system different to the others. Thus, the basis of the statistical description can be written as

$$\bar{\mathcal{L}} = \lim_{\tau \to \infty} \frac{1}{\tau} \int_0^\tau \mathcal{L}(t)\mathrm{d}t \equiv \lim_{M \to \infty} \frac{1}{M} \sum_{s=1}^{M} \mathcal{L}_s \qquad (2.1)$$

This implies that during a *sufficiently long* interval ($\tau \to \infty$), the real system goes over all possible microscopic configurations s, corresponding to the respective thermodynamic state. Eqn. (2.1) again becomes a very important expression playing a central role and that is why the adverb *sufficiently* has a central value. This point will be discussed in other parts of this book.

System's microconfigurations are distributed according with the relation between its energies E_i and the mean thermal energy. For macroscopic systems in most of the usual conditions, the Boltzmann's statistics allows us to express the probability $\mathcal{P}_s$ of a energy state E_s as

$$\mathcal{P}_s = \frac{\exp(-\beta E_s)}{\int_\Omega \exp(-\beta E_s)\mathrm{d}\omega} \qquad (2.2)$$

This expression correctly represents the distribution of energies. In the equation $\beta = 1/kT$, k is Boltzmann's constant, and $\mathrm{d}\omega$ is a volume element of the phase

---

[1]Each particle has three coordinates that determine its position and three components of the velocity, which determine its movement; in total, six dynamic degrees of freedom for each particle.

space Ω.[2] This distribution is related to the fundamental thermodynamic functions, taking into account that the mean energy value of the system, $\bar{E}$, is equal to the internal energy of the system, U. Then,

$$\bar{E} = \int_\Omega E_s \mathcal{P}_s \mathrm{d}\omega = \frac{\int_\Omega E_s \exp(-\beta E_s)\mathrm{d}\omega}{\int_\Omega \exp(-\beta E_s)\mathrm{d}\omega} \equiv U \tag{2.3}$$

Using the thermodynamic relation between U and A (cf. appendix A),

$$U = \left(\frac{\partial(A/T)}{\partial(1/T)}\right)_V = \left(\frac{\partial(\beta A)}{\partial \beta}\right)_V \tag{2.4}$$

and from eqn. (2.3) we get

$$\left(\frac{\partial(\beta A)}{\partial \beta}\right)_V = \frac{\int_\Omega E_s \exp(-\beta E_s)\mathrm{d}\omega}{\int_\Omega \exp(-\beta E_s)\mathrm{d}\omega} = -\frac{\frac{\partial}{\partial \beta}\left[\int_\Omega \exp(-\beta E_s)\mathrm{d}\omega\right]}{\int_\Omega \exp(-\beta E_s)\mathrm{d}\omega}$$

an expression that can be written

$$\left(\frac{\partial(\beta A)}{\partial \beta}\right)_V = -\frac{\partial}{\partial \beta}\left(\ln \int_\Omega \exp(-\beta E_s)\mathrm{d}\omega\right)$$

Finally, integrating, we have

$$\beta A = -\ln \int_\Omega \exp(-\beta E_s)\mathrm{d}\omega = -\ln Q_N \tag{2.5}$$

where Q_N is the partition function of a system having N particles; it is a summary of the statistical weight of all microconfigurations the system can have in that thermodynamic state. Rigorously, the partition function is a summation over all the system's microstates,

$$Q_N = \sum_{s=1}^{M} \exp(-\beta E_s) \tag{2.6}$$

which very frequently is replaced by an integral. The partition function is the base of the connection between atomic-molecular and macroscopic properties of the systems. This expression provides a link to the thermodynamic behavior of any system. However, the calculation of the integral is not simple, although it is possible to advance with a simpler description if we remember that it is possible to separate the intramolecular or internal degrees of freedom from those corresponding to intermolecular interactions.

---

[2] Ω denotes the phase space consisting of all the dynamic coordinates or dynamic degrees of freedom for a system containing N particles.

2.2   The Partition Function

The energy E_s, characterizing a microconfiguration of a given material system, contains contributions having different origins, and each of them can be analyzed separately because generally their magnitudes are very different. There are nuclear and electronic degrees of freedom, as well as vibrational and rotational degrees of freedom, having smaller energies than the two previous ones, that correspond to intramolecular or internal modes. Finally the translational degrees of freedom are intermolecular. As already mentioned, all the contributions to the energy of the internal modes of atoms and molecules only depend on their nature; hence, they are practically the same when we deal with a single particle or with a mole of them. Only the potential energy depends on the average density, which also reflects the type of interactions being dealt with and the particular interaction among the particles; they can be considered intermolecular modes.

The nuclear and electronic contributions to the energy of atoms and molecules, in most cases treated in this book, do not vary in simple chemical processes; hence, they are assumed constant. Vibrations and rotations are separable modes in the majority of systems, although there are situations in which their clear separation into purely vibrational or purely rotational modes may be difficult. Nonetheless, they will be assumed separable in all the cases that interest us. On the other hand, vibrational energies are larger than those corresponding to rotations; hence, the summation that defines the partition function (eqn. (2.6)) cannot be approximated by an integral because vibrational energy levels are appreciable separated one from the another. If temperature is not very high (i.e., around room temperature), rotational energy levels are considered sufficiently close to each other to enable replacing the summation by an integral. Then, for internal degrees of freedom, the contribution to the partition function are

the electronic,

$$(q_\text{e})_i = \sum_m g_m \exp(-\beta \epsilon_{m,i}) \tag{2.7}$$

the vibrational,

$$(q_\text{v})_i = \sum_m \frac{\exp(-\beta h \nu_{m,i}/2)}{1 - \exp(-\beta h \nu_{m,i})} \tag{2.8}$$

and the rotational,

$$(q_\text{r})_i = \frac{8\pi^2 I_i}{\beta \sigma_i h^2} \tag{2.9}$$

In the previous expressions, g_m is the number of degenerate electronic levels m of molecule i; in the majority of the molecules we will deal with, only the fundamental electronic level contributes appreciably. In the vibrational partition function, $\nu_{m,i}$ is the normal mode vibrational frequency m of molecule i. The expression for the rotational partition function refers to linear molecules having only a single moment of inertia I_i; σ_i is the number or factor of symmetry of molecule i. For diatomic homonuclear molecules $\sigma = 2$, and for diatomic heteronuclear ones $\sigma = 1$. The expression for q_r has been integrated and this is valid, assuming that the temperature is high enough to consider that the various rotational levels are very close.

Regarding the translational levels, they are so close from each other that the integration is valid for almost all the systems in which we are interested. This integration is made for two types of dynamic variables, the momenta (product of mass times velocity), and the coordinates of the center of mass of the molecules. Both variables are separable: The integration over the momenta leads to a result that *does not* depend on the relative position of the molecules and, for fluids, it becomes the Maxwell-Boltzmann description of the kinetic theory of gases. Thus, the translational contribution is

$$\sum_s \exp[-\beta(E_t)_s] = \left(\frac{2\pi m_i}{\beta h^2}\right)^{3N/2} \int_V \exp[-\beta \sum u_{ab}(r_{ab})]\mathrm{d}\mathbf{r_{ab}} =$$

$$= (q_{tr})_i^N \int_V \exp[-\beta \sum u_{ab}(r_{ab})]\mathrm{d}\mathbf{r_{ab}} = (q_{tr})_i^N Z_N \tag{2.10}$$

where $\sum u_{ab}(r_{ab})$ is the total potential energy of the system considered equal to the sum of potential energy between pairs (ab) of molecules. In this expression, $(q_{tr})_i = (2\pi m_i/\beta h^2)^{3/2}$. Hence, the system's partition function, when it has N particles, can be written

$$Q_N = \frac{(q_{int})_i^N}{N!} Z_N \tag{2.11}$$

where $(q_{int})_i = (q_e\, q_v\, q_r\, q_{tr})_i$ is the product of the partition functions corresponding to the various degree of freedom *that only depend on a single molecule i* and will be denoted collectively as internal partition function.[3] Z_N is the configurational integral; it summarizes all the intermolecular interactions, that is, the energy dependence of the system on the relative position of its molecules–the potential energy. For systems in which the molecules do not interact with each other (*v. gr.* ideal gases), $u_{ab} = 0$, and $Z_N = V^N$. The term $N!$ in the denominator of eqn. (2.11) is due to the elimination of irrelevant microconfigurations for the description of the system's state. Effectively for fluids, liquids or gases, all those microconfigurations due

---

[3]In this book, part of the translational contribution is included inside the internal partition function because, albeit that it does not refer to intramolecular movements, its contribution involves properties of a single molecule.

to permutation of indistinct particles, as the permutation of molecules of the same substance, are not relevant because all are identical; that is why the total number of microstates must be divided by the number of permutations among indistinguishable molecules.

2.3   Heat Capacities and Equipartition of Energy

The energy equipartition theorem states that the average energy of the different dynamic molecular modes (i.e., translations, rotations, vibrations, etc.) is, at high temperature, equal to $(kT/2)$ for each contribution to the energy,[4] the total energy of the system being the sum of the contributions from every mode. Thus, at high temperature, the translational energy of a molecules will average $3(kT/2)$, indicating the effect of the three modes or translational degrees of freedom existing for each independent particle.

In a similar way, the rotational energy of a molecule will be in the high temperature limit $(kT/2)$ for each molecular axis of rotation or rotational degree of freedom. For instance, the limiting rotational energy of CH_4 is $3(kT/2)$, but for HCl it is $2(kT/2)$ because this molecule is linear and has only two degrees of rotational freedom.

The principle of energy equipartition also establishes that the mean vibrational energy of each normal vibrational mode will be (kT), because the oscillatory movement associated with each vibration has one contribution from kinetic energy $(kT/2)$ and a second one due to the potential energy which also contributes $(kT/2)$. The number of vibrational degrees of freedom will be equal to the total number of degrees of freedom of the system having N atoms reduced by the number of translational and rotational modes, that is $(3N - 5)$ for linear molecules and $(3N - 6)$ for non-linear molecules.

From the values of energy, it is possible to calculate the limiting contribution made by each molecular mode to the heat capacity at constant volume. The total heat capacity will be the sum of the contributions of all the excited modes. Thus, it is predicted that in the classic limit, the heat capacities will be independent of temperature. Nevertheless, frequently the temperature will be such that the limiting values are not attained, especially for the vibrational modes, because the energy of the vibrations is relatively large and the requirement that kT be bigger than the system's temperature (i.e., it must be hundreds or even thousands of K). Since

---

[4] *High temperature* refers to temperatures high enough so that the molecular modes have reached their classical statistic limit. For a given mode to attain such a limit, the energy of its fundamental mode must be much smaller than kT.

the majority of simple molecules have no vibrational modes excited, they cannot participate in the storage of energy as established by classic physics. Thus, in general, only at high temperatures the heat capacity and the energy will have the limiting values. Generally, the electronic and nuclear modes are not excited because a very high temperature is required and the molecules will be decomposed at those temperatures (thermolysis).

According to eqns. (1.16), (2.4), and (2.5), the contribution of mode s to the heat capacity at constant volume will be,

$$(C_V)_s = Nk\beta^2 \left(\frac{\partial^2 \ln q_s}{\partial \beta^2} \right)_V \tag{2.12}$$

The expressions for q_s are given by eqns. (2.7), (2.8), (2.9), and (2.11).

2.4 Heat Capacity of Solids

The energy equipartition principle predicts that the solid's heat capacity at constant volume of an elemental solid formed by N atoms is equal to $3R$ per mole of atoms, independently of the solid's nature. This statement, known as the law of Dulong and Petit, is based on the assumption that every atom forming the crystalline lattice of the solid vibrates with respect to the position of the neighboring atoms in three direction, which are independent and perpendicular to each other. This law has been applied with some success to metallic solids; however, it fails when used to describe the behavior of polyatomic solids, because in this case it is necessary to use a model that also takes into account the contribution of internal modes of the molecules to store energy.

In terms of classical physics, the formulation of Dulong and Petit would imply that a solid sample behaves as a set of N independent oscillators, which determine exclusively the internal energy of the system $U = 3NkT$ at temperature T. In spite of its partial practical success, a careful comparison of the predicted values for C_V with the experimental ones, shows that the agreement is not satisfactory for the majority of solids. The situation worsens when the comparison is done at low temperatures, and it fails completely when temperature is close to 0 K (Nernst's heat theorem, see page 70).

The classic description suggests that even at very low temperatures the amplitude of the vibrations, which represents the solid's internal energy, will be proportional to T, so that its derivative with temperature, C_V, is constant. Experimental observation shows that C_V, decreases strongly until it becomes zero at 0 K. This discrepancy can only be overcome if we abandon the classical description of atomic vibrations in the solid and use the concept that energy is a quantal magnitude.

Einstein's model to calculate C_V for solids was the first able to reproduce, with some success, the behavior of that quantity down to very low temperatures. This model considers that atoms in the solid's atomic crystalline lattice are linked to their equilibrium positions by a harmonic atomic force. Two oscillators can vibrate around the equilibrium positions in the three Cartesian axes at a natural frequency ν_E. The energy of these oscillators is equal to a quantum $h\nu_E$ if they are excited. Really, the vibration of each atom in the crystal can be considered to be coupled harmonically to the neighboring atoms, instead of assuming the latter particles are fixed in the lattice. This coupling gives $3N$ modes of collective vibrations, denoted as *normal modes*. The frequencies distribution of these modes is very wide, although the majority of the vibrational modes of a solid are distributed in a narrow range of high frequencies. Einstein's model describes the frequency spectrum of a solid in the simplest possible way, considering a unique frequency of vibration ν_E, which will be close to the most probable frequency in their distribution. Figure 2.1 shows the experimental vibrations spectrum of a solid, also indicating the vibration frequency ν_E for an Einstein solid at the same temperature.

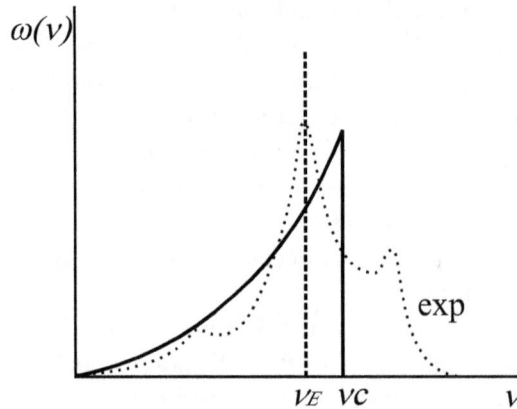

Figure 2.1: Vibration frequency distribution of solids at low temperature: $\cdots\cdots$, experimental; $---$, Einstein model and ———, Debye model.

The partition function of the model system proposed by Einstein can then be written as a set of $3N$ equivalent and independent subsystems, each one representing a one-dimensional harmonic oscillator of frequency ν_E,

$$Q = \exp(-\epsilon_0/kT)\, q^{3N} \tag{2.13}$$

where ϵ_0 represents the zero of energy for atoms infinitely separated and fixed. We can represent the partition function of each subsystem q in the following way

$$q = \sum_0^\infty \exp(-\epsilon_n/kT) \qquad \epsilon_n = (n + \frac{1}{2})h\nu_E \qquad n = 0, 1, 2, ...$$

where n represents the vibrational quantum number and ϵ_n the energy levels of each one-dimensional oscillator.

Defining the characteristic temperature $\theta_E = h\nu_E/k$, also denoted as Einstein's temperature, and assuming that $\exp(-\theta_E/T) \ll 1$, q is expressed by

$$q \approx \frac{\exp(-\theta_E/2T)}{1 - \exp(-\theta_E/T)} \tag{2.14}$$

based on eqns. (2.12), (2.13), and (2.14), the final expression for the heat capacity for an Einstein solid as a function of temperature is

$$C_V = 3Nk \left(\frac{x \exp(x/2)}{1 - \exp(x)} \right)^2 \tag{2.15}$$

where $x = \theta_E/T$. The characteristic constant θ_E is the model's parameter, and its value is adjusted to the solid's behavior.

Raising the temperature, the energy received by an Einstein solid can be distributed in a larger number of oscillators and, consequently, the heat capacity will increase. At very high temperatures $x = (\theta_E/T) \ll 1$; hence, it can be expressed as a series of exponentials of the previous equation, ignoring the higher order terms (i.e., if $\exp(x) \approx 1 + x$), and Dulong and Petit's expression for C_V is obtained,

$$C_V = 3Nk \ (1 + x/2 + ...)^2 \approx 3Nk$$

Under these conditions, the thermal perturbation exceeds the capacity of energy absorption by the oscillators. Thus, the majority of oscillators will be excited, populating even high vibrational states.

If energy is transferred to a solid at low temperature, behaving according to Einstein's model, only a small fraction of oscillators will be activated. This is due to the fact that each quantic oscillator has a single quantum of energy $h\nu_E$, and because $h\nu_E \gg kT$ at low temperatures and only a few oscillators will be activated under these conditions. Consequently, the heat capacity measured for an Einstein solid will be low compared with a real solid at temperatures close to 0 K. Under these conditions, eqn. (2.15) is reduced to

$$C_V = 3R \ [x \exp(-x/2)]^2$$

If the limit of the previous expression is analyzed when temperature approaches 0 K, it is observed that the heat capacity of an Einstein solid decreases exponentially as temperature decreases. The linear factor in eqn. (2.15), which diverges to $+\infty$ as T gets closer to 0 K, cannot compensate the increase of the exponential factor.

A more elaborated model for the solid's heat capacity was proposed by Debye. In this model, atomic vibrations are not considered to be independent from each other. Debye took into account high frequency collective vibrations (having large force constants) generated by atomic displacements between nearby nodes in the crystal lattice, including also low frequency ones (having small force constants), produced by collective atomic displacements (phonons). Figure 2.1 shows that the vibrational frequency spectrum of a solid is better represented by the Debye model than by the model of isoenergetic oscillators proposed by Einstein.

Assuming that the set of normal frequencies of vibrations is known $\{\nu_i\}$, the expression for the partition function of the system is

$$Q = \exp(-\epsilon_0/kT) \prod_{i=1}^{3N} q(\nu_i) \qquad (2.16)$$

where ν_E in eqn. (2.13) is replaced by ν_i. On the other hand, eqn. (2.14) is still valid to represent the partition function of each subsystem $q(\nu_i)$, substituting Einstein's characteristic temperature θ_E by that of Debye $\theta_D = h\nu_c/k$, where ν_c is a model's characteristic frequency.

Due to the large number of normal vibration frequencies ($3N$), it is possible to assume a continuous function of vibrational excitations, $\omega(\nu)$, fixing the total number of modes in the solid, as

$$\int_0^\infty \omega(\nu)d\nu = 3N \qquad (2.17)$$

From eqns. (2.14) and (2.16), an expression is obtained to enable calculation of the total partition function, once the frequency distribution function of the solid is known,

$$-\ln Q = \exp(-\epsilon_0/kT) + \int_0^\infty \left[\ln[1 - \exp(-\theta/T)] + \frac{\theta}{2T} \right] \omega(\nu)d\nu \qquad (2.18)$$

For instance, if a frequency distribution $\omega(\nu)$ is considered, which is infinitely or extremely sharp centered at frequency ν_E, the previous equation becomes that corresponding to Einstein's model.

Using classical physics tools, it is possible to demonstrate that the frequency distribution in a continuous tridimensional solid is given by $\omega(\nu) \propto \nu^2$. Debye's model uses this classical distribution function when applied to monoatomic solids over a range of frequencies, ranging from $\nu = 0$ up to a frequency equal to the model's characteristic frequency ν_c. It is necessary to assume a discontinuity in the function $\omega(\nu)$, at $\nu = \nu_c$, to guarantee that the total number of oscillators is conserved, as indicated by eqn. (2.17), leading to

$$\int_0^{\nu_c} \omega(\nu)\mathrm{d}\nu = \int_0^{\nu_c} \omega_0 \nu^2 \mathrm{d}\nu = \frac{\omega_0 \nu_c^3}{3} = 3N$$

From this expression, the analytic form of $\omega(\nu)$ can be obtained, leading to

$$\omega(\nu) = \begin{cases} 9N\nu^2/\nu_c^3 & 0 \le \nu \le \nu_c \\ 0 & \nu > \nu_c \end{cases} \tag{2.19}$$

Figure 2.1 illustrates the function $\omega(\nu)$ employed by Debye's model. The classical frequency distribution used by this model is an approximation, particularly for high frequencies. This is due to the fact that the description of these modes cannot be done employing models for continuous systems and omitting the discrete nature of an atomic lattice.

The calculation of C_V, employing Debye's model, can be done with eqns. (2.12), (2.18), and (2.19). Defining $x = (h\nu/kT)$, the final expression for the solid's heat capacity as a function of temperature is

$$C_V = 3Nk\, 3 \left(\frac{1}{x_0}\right)^3 \int_0^{x_0} \left(\frac{x^2}{1 - \exp(x)}\right)^2 \mathrm{d}x \tag{2.20}$$

where $x_0 = \theta_D/T$. Thus, the calculated heat capacity depends on the parameter θ_D, which is known as the Debye temperature. At a given temperature, the model predicts low heat capacity for those substances having higher Debye temperatures. At temperatures close to 0 K, $T \ll \theta_D$, eqn. (2.20), corresponding to the Debye model, can be reduced to

$$\lim_{T \to 0} C_V = \frac{12\pi^4 Nk}{5\theta_D^3} T^3 \tag{2.21}$$

The previous equation predicts that the heat capacity of a solid must be proportional to $(T/\theta_D)^3$. This behavior has been confirmed for numerous solids and is used to elaborate thermodynamic cycles of entropy, the calculation of residual entropies, etc.

Figure 2.2 shows the dependence of C_V on temperature for Einstein and Debye solids. The fact that Debye's model accounts for the possible excitation of low frequency oscillators that increases the value of the calculated heat capacity over all the ranges of temperatures, making this value closer to the experimental one.

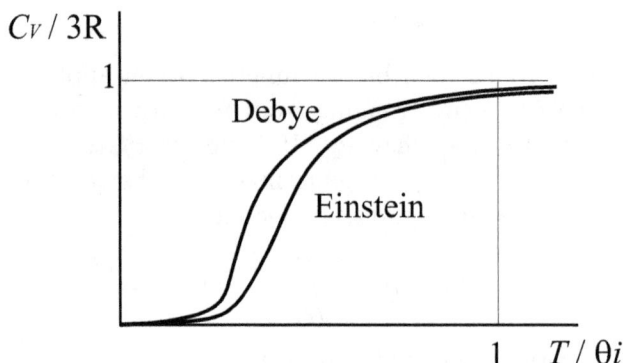

Figure 2.2: Einstein and Debye heat capacities at constant volume of solids, as a function of temperature.

2.5   Molecular Interactions: Their Origin

The ideal gas model proposes that there are no interactions among the gases' molecules. It is known that such a model is good to describe the behavior of gases at very low pressures, but as the gas density increases (by increase of pressure or number of molecules at constant temperature), gaseous systems deviate from the behavior predicted by the ideal gas equation of state. This deviation is progressive and can be attributed to the fact that as density increases, the average distance between its molecules is shorter and, therefore, intermolecular interactions become significant.

The magnitude of these interactions, which are denoted as intermolecular, can be estimated by the magnitude of the enthalpies of vaporization or sublimation—that is, the enthalpy difference among molecules very far from each other, implying negligible intermolecular interactions, and the enthalpy of molecules in a condensed phase, liquid or solid, where molecules are on average very close. The magnitude of the interaction energies is, for small molecules, between 20 and 40 kJ/mole. This value is much lower than that for chemical bonds; however, their magnitude is not negligible to describe the behavior of gaseous systems.

Interactions between atoms and molecules can lead to interactions of molecules as a whole or to chemical reactions. The latter situation implies a modification of the chemical identity of the particles involved. This will lead to formation or breaking of chemical bonds. On the other hand, the electronic and nuclear distributions in a reacting molecule changes appreciably, while molecular interactions do not alter the nuclear distribution and only, in some cases, slightly alter the electronic distribution of the molecules.

Intermolecular interactions are of electrostatic nature and are due to the fact that different atoms existing in molecules have different electronegativities, and the positive charges in the nuclei will not have the same center as the negative charges of electrons. Even in diatomic homonuclear diatomic nuclear molecules, like N_2, there is a charge distribution that produces electrostatic interactions between particles that are close to each other. Figure 2.3 shows the charge distribution of a few molecules.

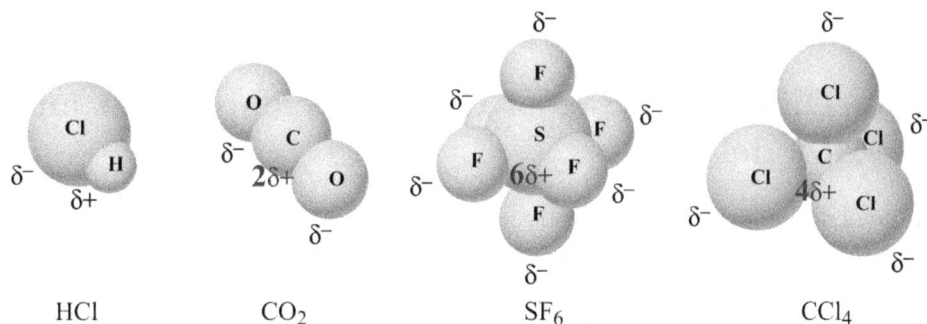

Figure 2.3: Charge distribution in different molecules.

The main models used to calculate the potential energy among molecules are based on the charge distribution in the molecules in vacuum. Recently, a few studies have tried to take into account the change induced on the electronic distribution of a molecule in the vicinity of another molecule; the estimated effect is very small. The main effect is due to the molecular polarizability.

When particles have a net charge, they are called ions. Nevertheless, it is quite common the case of neutral molecules that present a difference between the position of the center of positive charges and that of the negative ones. This is very obvious in heteronuclear molecules like HCl in the gaseous phase–depicted in Figure 2.3; in that case, the molecule can be represented as a dipole with a dipole moment equal to $\mu = q \cdot l$ (i.e., a positive charge q separated by a distance l from a negative charge $-q$, having both the same absolute value). These molecules are called dipolar, or simply polar.

Molecules possessing atoms of different electronegativity will form polar chemical bonds; albeit the molecular symmetry may result such that these molecules have a zero dipolar moment, and then they are non-polar molecules. An example illustrated in figure 2.3 is the CO_2 molecule having a null dipole moment; however, the intramolecular charge distribution is capable of producing intermolecular interactions between two close molecules. That charge distribution, as depicted in the figure, is due to a quadrupolar moment that is not zero. Other molecules having more atoms can have a greater symmetry that leads also to a zero quadrupolar moment. For instance, in figure 2.3 is exemplified by CCl_4 and SF_6; the first has a zero quadrupole, but an octopolar moment different from zero. Other molecules, like SF_6, having higher symmetry may lead to quadrupolar and octopolar moments equal to zero, but with finite hexadecupolar moment. All these charge distributions have effective intermolecular interactions when another molecule approaches them; however, the higher the order of the first nonzero dipolar moment, the weaker the intermolecular interaction.

2.6   Different Types of Intermolecular Forces and Models

When two molecules are separated by an infinite distance, they do not interact with each other; when the molecules approach each other, the electrostatic interaction arising from their electronic distribution will contribute to the potential energy of both molecules. At such distances, the interaction between molecules is attractive.[5] If the two molecules are very close, the potential energy will become positive and also the force between them becomes repulsive. This universal phenomenon implies that matter cannot interpenetrate, which at atomic-molecular level is due to the exclusion principle of Pauli–two electrons cannot be in the same energy level. As molecules approach, the same occurs with their electrons; the interlap energy giving the probability that electrons superpose will rapidly become smaller until it vanishes to zero.

Figure 2.4 shows a typical curve for $u_{ij}(r)$, where the repulsive and attractive energies are identified, and also the total force. For condensed matter having a large density, the repulsive interactions determine the final disposition of molecules, so they determine the microscopic structure of the phase. When one deals with gases or expanded fluids, the predominant interaction will be attractive.

The repulsive energies, due to the impossibility of having more than one electron in the same quantum level, vary exponentially with distance; i.e.,

$$(u_{ij})_{\text{rep}} \propto \exp(-br)$$

As shortly will be seen, attractive energies depend on negative potentials varying with their distance, according to

$$(u_{ij})_{\text{atr}} \propto -r^{-m}$$

---

[5]The force F_{ij} between two particles is minus the derivative of the intermolecular energy, u_{ij}, with respect to the distance r separating them (i.e., $F_{ij} = -du_{ij}/dr$).

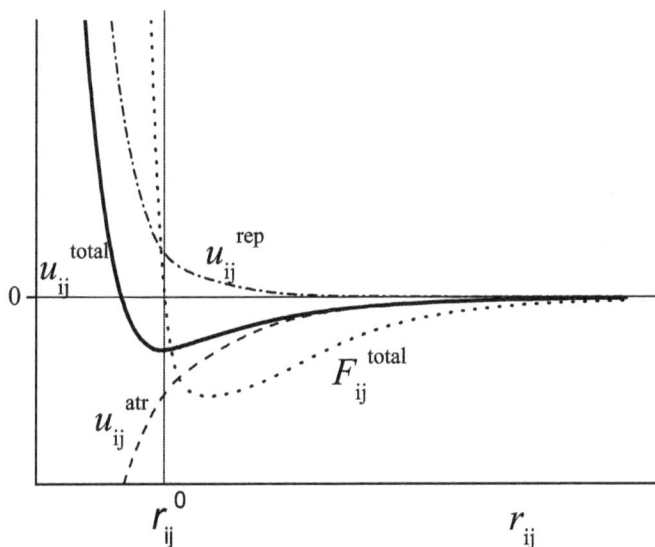

Figure 2.4: Intermolelcular energy (———) as a function of the distance between two molecules. Repulsive component, $-.-.$; attractive component, $---$; intermolecular force, ... Distance r_{ij}^o corresponds to zero force at the minimum distance of approach.

The value of exponent m will depend on the charge distribution of the molecule. Often, the repulsive energy is represented by means of a negative power, n, of the distance of separation of the two molecules; then, the potential energy is expressed as

$$u_{ij} = \frac{C_{\text{rep}}}{r^n} - \frac{C_{\text{atr}}}{r^m}$$

2.7 Electrostatic Interactions in Vacuum

The interaction between two molecules that can be represented by charge distributions, is given by Coulomb's law, applied to every pair of charges that can be formed. Thus, for molecules i and j having charges $q_{i\alpha}$ and $q_{j\beta}$, we have,

$$u_{ij} = \sum_{\alpha,\beta} \frac{q_{i\alpha}q_{j\beta}}{4\pi\epsilon_0 r_{\alpha\beta}}$$

where $r_{\alpha\beta}$ is the distance between the sites $(i\alpha)$ and $(j\beta)$, and ϵ_0 is vacuum's permitivity, a universal constant having the value $\epsilon_0 = 8.854185 \times 10^{-12}$ C/(V m). The total energy of interaction of the molecules will be the sum of all the energies u_{ij} of each pair of molecules. Whenever $q_{i\alpha}$ has the same sign as $q_{j\beta}$, the resultant energy between those two sites will be positive (repulsive), as it is the case of the force acting between those two sites. The opposite occurs when both charges have different signs. The simplest case is the electrostatic interaction between ions having a single charge; this will vary as r^{-1}, with the sign given by the product of both charges. For ionic interactions we use

$$u_{i,j}(r) = -\frac{q_{i\alpha}\,q_{j\beta}}{4\pi\epsilon_0 r} + \text{repulsion} \tag{2.22}$$

The potential energy existing between two molecules of HCl can be modeled assuming that it is equal to the interaction of two dipoles, that is between two charges (one positive and the other negative on each molecule). See the following scheme, where there are four interactions for each pair of HCl molecules: $(q_{1+}\,q_{2+})$, $(q_{1+}\,q_{2-})$, $(q_{1-}\,q_{2+})$ and $(q_{1-}\,q_{2-})$; two of these interactions are repulsive and two attractive. This case is illustrated in figure 2.5: The forces at a fixed distance between the centers of mass of each molecule are vectorial, tensorial, etc.–quantities that depend on their relative orientation. The models used normally give the average energy of interaction. Thus, the dipoles in the orientation 1, as shown in figure 2.5, have a repulsive energy given by

$$u_{ij} = \frac{\mu_i\mu_j}{4\pi\epsilon_0 r^3}$$

When the dipolar orientation is that of panel 4 in figure 2.5, the energy is

$$u_{ij} = -\frac{\mu_i\mu_j}{4\pi\epsilon_0 r^3}$$

and the force is, in this case, attractive. Averaging over all the possible orientations, we get that there is an attractive interaction varying as r^{-3}.

So far, we have considered that the electron distribution in a molecule is not modified by the presence of other molecules close to it. However, the molecular electric field can induce a charge separation (i.e., generate a dipole on another molecule close to it). This phenomenon can be described through the molecular polarizability α; the induced dipole is proportional to the electric field E that induces it. Thus,

$$\mu_{\text{ind}} = \alpha E$$

Polarizability is related with the molecular deformation of the electronic orbitals in one molecule, and the average energy of interaction between a permanent dipole and an induced one is proportional to r^{-6}.

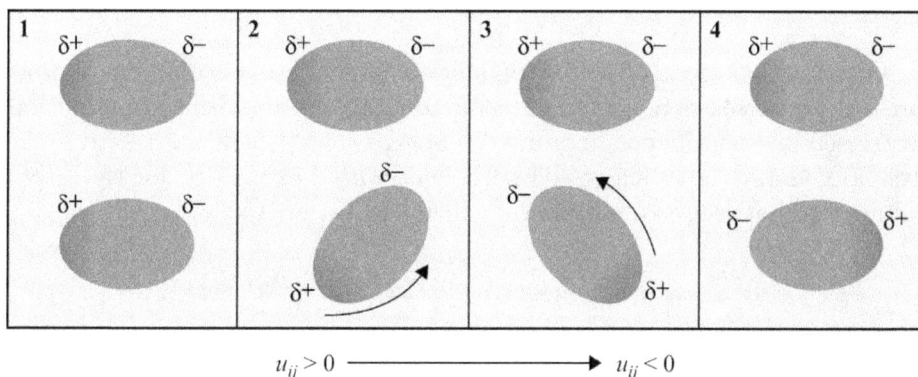

Figure 2.5: Effect of relative orientation on the interaction between two dipolar molecules.

2.8   Dispersion Forces, Lennard-Jones Potential

Even monoatomic molecules, like the noble gases, interact with each other. This is evident because otherwise they could not exist as liquids because that implies intermolecular interactions. Nevertheless, the noble gas atoms have an average charge distribution equal to zero. In this case, the only interaction is due to the movement of the electron clouds in the atoms; this generates an oscillating electric field which, in turn, induces dipoles in neighboring atoms. Induced dipoles on the neighboring atoms will follow the oscillation of the electrons on the central particle, so that the movement of the induced dipoles and that of the electric field of the central atom will be in phase; this implies that there will always produce an attractive interaction. This interaction force is known as a dispersion force.

Polyatomic molecules also are affected by dispersion forces and, in spite of having charge distribution, often the dispersion forces predominate. F. London deduced from the model of oscillating electrons a formula that expressed the attractive interaction as $C\,r^{-6}$; the constant C obtained by London was

$$C = -\frac{3I_iI_j}{2(I_i + I_j)}\frac{\alpha_i\alpha_j}{(4\pi\epsilon_0)^2}$$

where I_k is the ionization potential of the molecule k. It may be seen that the more deformable the electronic density of the interacting molecules (larger α_k) and the lower their electron binding energies (larger I_k), the larger will be C. For example, organic molecules with conjugated bonds or aromatic molecules, will exhibit relatively larger dispersion forces. These forces exist in all the molecules and their effect adds to those due to the charge distribution of the interacting molecules.

Often, it has been observed that attractive intermolecular interactions are expressed by terms depending on r^{-6} and that the repulsive contribution may be expressed by terms depending on r^{-n}, being the exponent n a relatively large number. J. Lennard-Jones suggested a frequently used expression for the intermolecular interaction between two molecules, which can be written as

$$u_{i,j}(r) = 4\epsilon_{i,j}\left[\left(\frac{\sigma_{i,j}}{r}\right)^{12} - \left(\frac{\sigma_{i,j}}{r}\right)^{6}\right] \tag{2.23}$$

where $\epsilon_{i,j}$ and $\sigma_{i,j}$ are model parameters.

Figure 2.6 shows, qualitatively, the curves corresponding to interactions between ions (eqn. (2.22)) and for dispersion forces. It is clear that the most notable difference is the range of action of the two types of interactions. Lennard-Jones interactions are short ranged and at three molecular diameters they are negligible, while the Coulombic attraction of opposite-charged ions is long ranged, being strong even at an appreciable distance of separation.

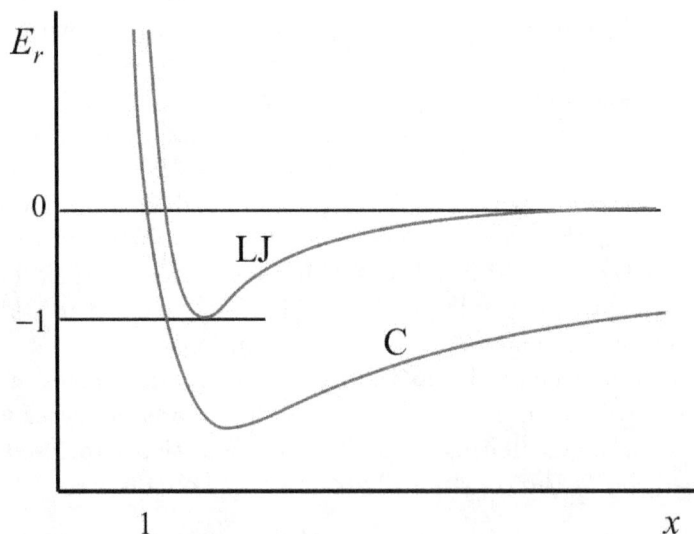

Figure 2.6: Reduced intermolecular energy, $E_r = (u_{i,j}/\epsilon_{i,j})$, as a function of the reduced distance, $x = (r/\sigma_{i,j})$, for Lennard-Jones (LJ) and Coulombic (C) interactions. A notable difference of range for the attractive interaction is observed for both models due to the different value of exponent n in the attractive term, as shown by eqns. (2.22) and (2.23). The force acting between two moleucles is given by the slope of the curve with a change in sign.

Eqn. (2.23) can be written in a dimensionless form as

$$E_r(x) \equiv \frac{u_{i,j}(r)}{\epsilon_{i,j}} = 4\left(x^{-12} - x^{-6}\right) \tag{2.24}$$

$E_r(x)$ is a universal function of the reduced intermolecular distance $x = (r/\sigma_{ij})$ (i.e., a function that is independent of the particular molecule when the intermolecular potential is given by the Lennard-Jones expression). This observation lets us establish that the properties of all substances well described by that intermolecular potential show a universal behavior in terms of adimensional variables–it is the principle of corresponding states.

The hydrogen bond is another type of intermolecular interactions, generally stronger than those due to the distribution of molecular charges and the dispersion forces. However, in this case, there is a geometrical or configurational condition between the two molecules forming a hydrogen bond, that they must be close to a linear configuration.

The interaction between molecules having permanent dipoles, hydrogen bonds, etc,. cannot be expressed by eqn. (2.23). The corresponding state principle cannot be applied to molecules having an interaction that requires more than two parameters to be represented. Nonetheless, frequently in those cases one can use the universality of function $E_r(x)$ as a first approximation.

2.9   The Second Virial Coefficient

Even today, the description of the effect of intermolecular interaction is done to a large extent on the basis of what occurs in gaseous systems; this means systems having low molecular density. Despite that this is a big model simplification of real situations prevailing in denser systems, it is important to analyze at an introductory level the effect produced by interaction between two gas molecules.

According to eqns. (2.5) and (2.11) we have,

$$\beta A = \beta A^\circ + N \ln \frac{V}{N} + \ln \frac{Z_N}{V^N} \tag{2.25}$$

and from the relation between pressure and Helmholtz energy,

$$-\beta p = \left(\frac{\partial \beta A}{\partial V}\right)_T = -\frac{N}{V} + \left(\frac{\partial \ln(Z_N/V^N)}{\partial V}\right)_T \tag{2.26}$$

The total intermolecular energy U_N may be expressed as the sum of interactions between pairs of molecules (this is a good approximation and only in a few special cases it cannot be applied); then using the Mayer function, defined by $f_{ij} = \exp(-\beta u_{ij}) - 1$, we have

$$U_N = \sum_{i,j>i} u_{ij}$$

$$\exp(-\beta U_N) = \prod_{i,j>i} \exp(-\beta u_{ij}) = \prod_{i,j>i} (1 + f_{ij}) = \left(1 + \sum_{i,j>i} f_{ij} + ...\right)$$

For the case of a low density gas,

$$\frac{Z_N}{V^N} = 1 + \frac{N(N-1)}{2V} \int_V f_{12} dV \simeq 1 + \frac{N^2}{2V} \int_V f_{12} dV$$

$$\left(\frac{\partial \ln(Z_N/V^N)}{\partial V}\right)_T = \frac{\partial}{\partial V}\left(\ln\left[1 + \frac{N^2}{2V} \int_V f_{12} dV\right]\right) \simeq$$

$$\simeq \frac{\partial}{\partial V}\left(\frac{N^2}{2V} \int_V f_{12} dV\right)_T$$

Then, replacing $(\partial \ln Z_N/\partial V)_T$ in eqn. (2.26), we get

$$\beta p = \frac{N}{V} - \frac{N^2}{2V^2} \int_V f_{12} dV$$

which is the ideal gas equation with a correction for non-ideality denominated second virial coefficient, $B_2(T)$. That is,

$$B_2(T) = -\frac{1}{2} \int_V f_{12} dV \tag{2.27}$$

As illustrated by figure 2.7 for molecular interactions given by eqn. (2.23), f_{ij} will be different from zero only when the two particles (i, j) are very close to each other.

Mayer's function then allows the description of interaction as a function of a series of integrals that involve an increasing number of interacting molecules, whose contributions become more important as the gas density increases. This is the typical technique used to solve the configuration integral (eqn. (2.10))–to reduce the multiple integral to integral functions that involve the positions of a few particles. This strategy applies to systems having molecules with short range interactions; thus, it describes the non-ideal behavior by a succession of contributions corresponding to a progressively increasing number of interacting molecules (second, third, and fourth virial coefficients). For long-range interactions, Mayer's method must be modified, thus allowing the use of a similar type of description; this will be important for ionic

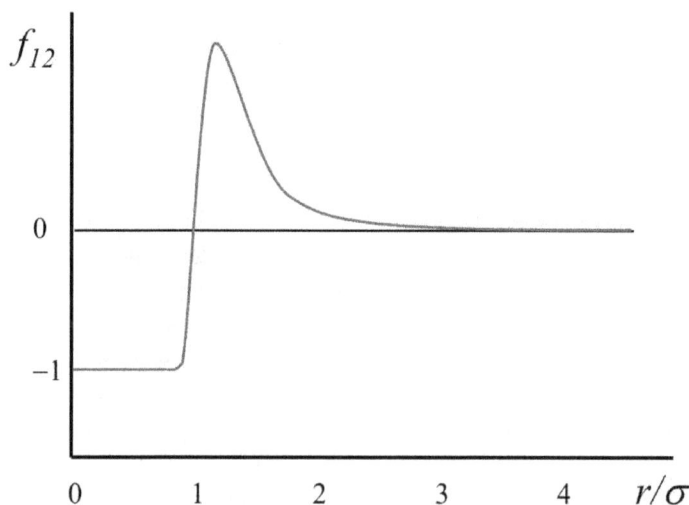

Figure 2.7: Mayer's function for two molecules interacting with a Lennard-Jones potential. f_{12} has very short range (i.e., it becomes zero at very short distance from $(r/\sigma) = 1$).

systems. It is possible to give a universal expression for the second virial coefficient, reducing the variables as indicated in the previous section. Based on eqns. (2.24) and (2.27), and the definition of Mayer's function, we have

$$B_2^* \equiv \frac{B_2}{\sigma_{ij}^3} = -2\pi \int_0^\infty \exp[-\beta E_r(x) - 1]x^2 \mathrm{d}x \qquad (2.28)$$

Hence, the second virial coefficient divided by the molecular diameter at the third is a universal function of the reduced energy E_r/kT, and the same is valid for the nth virial coefficient B_n if it is divided by $(\sigma_{ij}^3)^n$.

The thermodynamic states of systems following the same adimensional universal expression are denoted as corresponding states and their properties, expressed in reduced variables, are identical in these states. The derivation of eqn. (2.28) shows the applicability of the principle of corresponding states and its limitations to the case of molecules with the same type of intermolecular interactions, as mentioned on page 63.

2.10   Interactions Between Molecular Aggregates

The main interest of this book is the description, by means of thermodynamic relations, of the behavior of material systems on the basis of the molecules that constitute them. In this chapter, some effective intermolecular potentials have been presented and are often used to calculate properties of several systems by describing the interactions between two isolated molecules in vacuo. However, there are many practical situations where molecules interact with aggregates of atoms or molecules. These aggregates may form microscopic particles containing many atomic-molecular units, that is the case of colloids, emulsions, aerosols, micelles, etc. The microscopic particle can also be a macromolecule. In all cases, the particles are of size ranging from 0.1 to 100 μm, that is containing between 10^8 and 10^{15} atoms or small molecules. How are the interactions described in these cases? Let's focus in the interaction energy between isolated molecules with the aggregates, and also to interactions between two aggregates. In principle, the interactions between individual molecules of a portion of matter are the same as that between two isolated molecules, but the fact that they may agglomerate and form larger particles implies that many molecules are interacting simultaneously and this fact, in general, alters the dependence of the interparticle energy with the distance between the two unities. Qualitatively, it is clear that the origin of the change in the distance dependence between two entities is the fact that the number of *effective* interacting molecules will change with the distance separating the two entities, thus adding a new effect to the mere approach or separation of the interacting units.

In this section, some distinguishable aspects are presented, which originate in the interaction between aggregates of simple atoms and molecules. It will be shown that in all the presented cases, despite that interactions between isolated nonpolar molecules are described by the Lennard-Jones effective intermolecular potential, the attractive energy between two aggregates has a longer range than that corresponding to a potential energy varying with r^{-6}, as would correspond to isolated nonpolar molecules. This effect is due to the *collective action* of the molecules that form the aggregates and from the particular symmetry of the situation being studied (this is developed following Israelachvili's book, cf. books for consultation).

Now, we analyze one example to illustrate the origin of this collective effect. Let a nonpolar molecule, denoted probe, interacts with an infinite solid plane also formed by atoms or nonpolar molecules, and all the interactions between the probe and each of the molecules in the planar solid are well described by the Lennard-Jones potential. If the probe were in vacuo close to an isolated molecule of the solid, its interaction would be correctly describe by the Lennard-Jones potential, but in the case of an infinite planar solid, it is necessary to add the contribution of all the molecules of the solid that interacts simultaneously with the probe.

Figure 2.8 illustrates this calculation. The probe is at a distance R from the surface of the solid and its interaction with a solid's cylindrical ring, of width $\mathrm{d}x$ and depth $\mathrm{d}z$, which will depend on the number of molecules within the solid's ring, given by $2\pi\rho x\mathrm{d}x\mathrm{d}z$, where ρ is the molecular density of the solid. Considering only the attractive contribution of the Lennard-Jones between the probe and every molecule in the solid at distance r from it, the potential energy will be

$$\phi(r) = -\frac{A}{r^6} \tag{2.29}$$

where $r^2 = x^2 + z^2$, and $z = R$, being the coordinate perpendicular to the the surface (cf. figure 2.8).

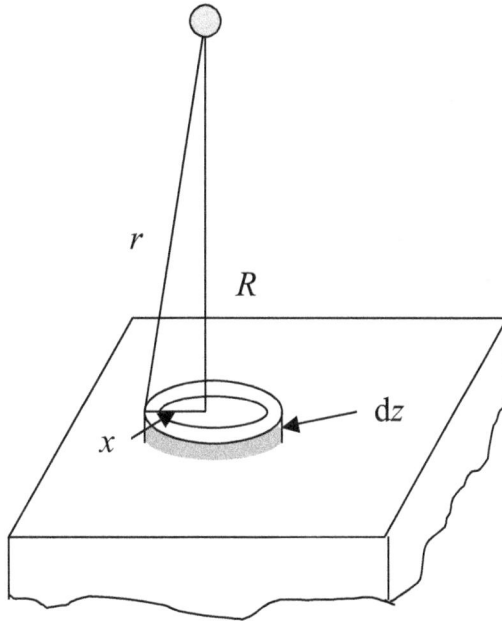

Figure 2.8: Probe's position with respect to the surface.

Then,

$$w(R) = \pi\rho \int_R^\infty \mathrm{d}z \int_0^\infty \mathrm{d}x^2 \left(-\frac{A}{r^6}\right)$$

and, as in the second integral z is constant (that is, $\mathrm{d}x^2 = \mathrm{d}r^2$), and considering that $r = z$ when $x = 0$, we have

$$w(R) = 2\pi\rho \int_R^\infty \mathrm{d}z \int_z^\infty -\frac{A}{r^5}\mathrm{d}r \qquad (2.30)$$

and finally

$$w(R) = -\frac{\pi\rho}{2}A \int_R^\infty \mathrm{d}z \left(\frac{1}{z^4}\right) = -\frac{\pi\rho A}{6R^3} = -\frac{B}{R^3} \qquad (2.31)$$

The previous equations show that the potential energy between the molecular probe and the solid matter changes with distance between them at the power -3, that is, it has a longer range than the interaction between two isolated molecules in vacuo. The potential energy also will depend on the molecular density of the solid, according to eqn. (2.31).

The same calculation can be used for other probes and aggregates and it will be verified that the dependence of the interaction energy on the distance separating both objects changes with the particular situation. Figure 2.9 summarizes the dependence on R of the interactions between probes and atomic-molecular aggregates, or between two aggregates. Some cases are illustrated in figure 2.9, which yield different dependences of the interaction on the distance between the two objects.

These examples give evidence that, despite that the attractive interaction between two isolated nonpolar molecules will change with r^{-6}, for one molecule interacting with a surface the dependence goes as R^{-3}, for two plane surfaces is R^{-2}, and they can even have an R^{-1} dependence, as in cases c and e in figure 2.9, as if they were Coulombic interactions between two isolated ions. The value of the constant B depends on the particular situation.

The different relations given in this section are very important and have great practical implications; some of them will be analyzed in chapter 9, for the case of colloids.

2.11   Entropy and Information

Another very important relationship between macroscopic properties and molecular properties was proposed by L. Boltzmann; he concluded that entropy was related to the logarithm of the thermodynamic multiplicity of states, W, thus

$$S = k \ln W \qquad (2.32)$$

This expression was formulated independently from that derived on the basis of the partition functions, but it has the same fundamentals. The quantity W is much

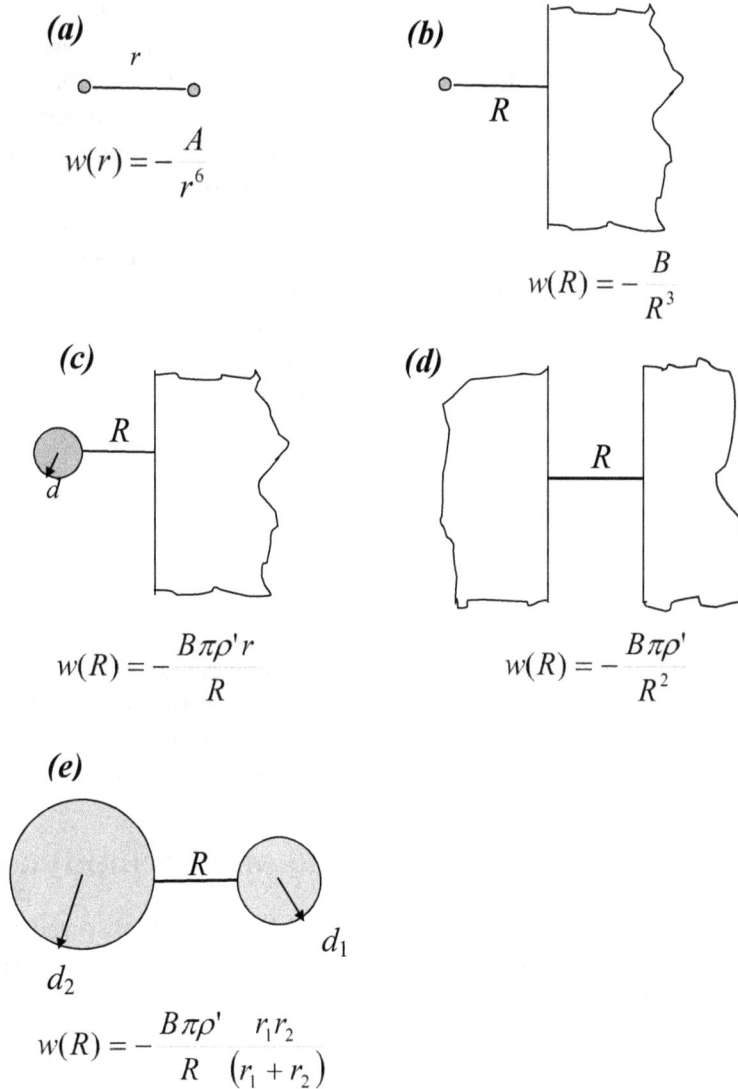

Figure 2.9: Interactions between probes and aggregates. ρ' is the molecular density of the second solid.

greater than unity, and is related to the number of microstates (all of them equally probable) that exist in a thermodynamic state. In some cases, this quantity is identical to the system's state degeneracy. That is, W is related to the volume of the phase space that determines the state of the system; really, it is a large number of the order of Avogadro's number. Hence, it is valid to say that the quantity $1/W$ is the probability to find the system in a given microstate.

Eqn. (2.32) gives the relation existing between information and the reciprocal of entropy–a system being in a microscopic state clearly determined has minimum entropy; for instance, if the system has a unique state available, its entropy will be zero. When the number of possible microstates is large, the system's instantaneous microstate will change continuously; hence, there is no certainty of the position and energy of every molecule and the entropy of the system will be large.

It is obvious that entropy and heat capacity are related. According to eqn. (1.7), whenever there is only mechanical work (change of volume), we obtain

$$\left(\frac{\partial S}{\partial T}\right)_V = \frac{1}{T}\left(\frac{\partial U}{\partial T}\right)_V = \frac{C_V}{T} \tag{2.33}$$

and analogously from eqn. (1.9),

$$\left(\frac{\partial S}{\partial T}\right)_p = \frac{1}{T}\left(\frac{\partial H}{\partial T}\right)_p = \frac{C_p}{T} \tag{2.34}$$

These expressions let us determine calorimetrically the entropic difference between a molecule in a given configuration and that of the same molecule having a random configuration. For example, in the case of chiral compounds, the molar heat capacity of the racemic mixture will be larger than those of each pure enantiomer.

2.12   The Third Principle of Thermodynamics

At the beginning of twentieth century, W. Nernst observed that for reactions between solids, the change of Gibbs energy was quite similar to the change of reaction enthalpy as the temperature approached absolute zero. Thus, it appeared that

$$\lim_{T\to0}\frac{\Delta G - \Delta H}{T} = 0 \tag{2.35}$$

then,

$$\lim_{T\to0}\Delta S = 0 \tag{2.36}$$

This limiting expression, also called the *heat theorem*, is the basis of the third principle of thermodynamics. It is an axiom because its demonstration is outside the realm of thermodynamics and is not a consequence of previous principles. However, statistical mechanics explains its origin. The Boltzmann relation between thermodynamic probability of an equilibrium state and its entropy allows seeing how this principle arises at the molecular level.

The present formulation of the third principle implies that chemical reactions between *perfect* crystals will have a change of entropy equal to zero at 0 K. This principle can also be formulated as the impossibility to reach a zero absolute temperature. Its relation with eqn. (2.32) is clear if we consider that perfect solids (having no defects) at 0 K will have a single microscopic state describing the macroscopic equilibrium state. In other words, all the molecules are in states of minimum energy, forming a unique microscopic configuration, hence the system has a thermodynamic probability of unity in such a state. As a consequence, the information about such a state is a maximum; knowing the position of a single particle, we can have the complete description of the microconfiguration. Then, the following expression for the third principle can be accepted.

$$S(\text{cr}, 0\text{ K}) = 0 \tag{2.37}$$

This principle has been verified experimentally with great care because it is the basis for the calculation of thermodynamic properties from calorimetric measurements of heat capacity, or from determination of the sound velocity in the systems and their molar volumes.

Much research was performed to verify the third principle were based, as mentioned already, on the knowledge of the heat capacity for two different crystalline phases of a given element (allotropes), for example, of Sn or of S. In these cases, often one of the two phases is a metastable state, albeit this limitation, it has been possible to determine the properties with great precision.

If the solid at 0 K results an amorphous glass or a solid solution, there will be more than one microconfiguration corresponding to the thermodynamic equilibrium state. Hence, the entropy of this solid will not be zero at 0 K. There are many other cases involving pure crystalline solids where it seems as if the third principle is not valid. Solid CO and also solid NO have positive values of entropy at 0 K when the experimental calorimetric result is compared with the spectroscopic statistical values obtained for the corresponding molecule. For CO, $S(298\text{ K})$ has the following values:

Calorimetric value = 193.3 J/(mole K)
Spectroscopic value = 197.5 J/(mole K)

This difference of 4.2 J/(mole K) can be explained by taking into account that the CO molecule can be oriented in the direction C=O or O=C, along an arbitrary axis, on each site of the crystalline lattice, meaning there are two degenerate states, hence,

$$S(298 \text{ K}) = R \ln 2 = 5.86 \text{ J/(mole K)}$$

and this result agrees quite closely with the experimental value.

For NO, the difference observed is 12.6 J/(mole K) and a similar explanation is given; crystalline NO exists as a dimer which may adopt two molecular configurations,

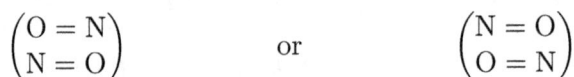

$$\begin{pmatrix} O = N \\ N = O \end{pmatrix} \qquad \text{or} \qquad \begin{pmatrix} N = O \\ O = N \end{pmatrix}$$

The entropy remaining in crystalline systems at 0 K is known as residual entropy.

An important case is that of water (ice) at 0 K, where the excess entropy at 0 K is 3.35 J/(mole K). Linus Pauling explained this value on the basis of the undefinition in the position of the hydrogen atoms, which can be bound to the oxygen atoms in two different ways–through covalent bonds with oxygen or forming hydrogen bonds with them. Pauling calculated a thermodynamic probability of (3/2), which implies an entropy of 3.37 J/(mole K) in very good agreement with the experimental value. As a consequence of this observation, it results that all crystalline hydrates will have nonzero residual entropies, because in their crystals there is uncertainty over which type of bond exists between the oxygens and the hydrogens. A final example, in some way apparently opposing the case of water, is the ion $(FHF)^-$. The compounds having that ion gave zero for the residual entropy of the crystals, thus one may conclude that the H atom is situated symmetrically between two F atoms and its position is *not* uncertain.

The Case of Water

N water molecules have $2N$ hydrogen atoms and each of them can be situated in two positions between two neighboring oxygens: One, closer to the oxygen of a given molecule to which it is bound covalently, and the another more separated, corresponding to a hydrogen bond. In total, there will be 2^{2N} possible configurations. In ice, a given molecule has its oxygen atom surrounded by four hydrogens; this corresponds to sixteen different configurations, but only six of them have two H atoms bound covalently and two hydrogen bonded to the O atom. The probability of these configurations is $(6/16)$, hence the number of degenerate microconfigurations when there are N molecules will be

$$W = 2^{2N} \left(\frac{3}{8}\right)^N = \left(\frac{3}{2}\right)^N$$

The Third Principle and Internal Molecular Rotations

In the previous example, it has been shown how the existence of more than one microconfigurations in crystalline solids leads to residual entropy at 0 K, a fact that does not invalidate the application of the third principle of thermodynamics. Now, another aspect will be discussed: The relation between the statistical and the phenomenological behavior of molecular systems.

When in a molecule there are two atoms, that we will call central, bound chemically to each other, and having other atoms bound to them, it is possible to observe three types of distinguishable situations that will be illustrated with the analysis of the behavior of simple and substituted hydrocarbons.

i) The type of chemical bond between the two central atoms does not allow rotation of one of them with respect to the other; in this case, the motion of the central atoms is a torsion that can be described as an oscillation or vibration. Examples of this case are hydrocarbons with the central C atoms involved in a double or triple bond and C atoms forming aromatic bonds, that is, $H_2C=CH_2$ and pairs of adjacent C atoms in the benzene ring, etc.

ii) There is free rotation of a central atom around the other–an internal free rotation having a partition function equal to that given by eqn. (2.9), but for a single degree of rotational freedom (rotation of central atoms around the chemical bond joining them). Eqn. (2.9) gives $(q_r)_i$ for diatomic molecules which, being linear, have two rotational degrees of freedom, while for free internal rotation a molecule has a single rotational degree of freedom. It should be noted in this case that the symmetry factor σ can have values greater than 2; this factor eliminates microconfigurations that are irrelevant, that is those that do not give a different configuration when rotating the molecule. For instance, the rotation of cyclopropane on the molecular plane has $\sigma = 3$ because each time that the molecule rotates at an angle of $(2\pi/3)$, the molecular configuration is repeated. For benzene, rotation in the molecular plane has $\sigma = 6$, because the same configuration appears when the angle changes $2\pi/6$.

Using the relation between Helmholtz energy and the partition function (eqn. (2.5)) values of the thermodynamic functions may be calculated and, thus, we can know how much the contribution of the internal rotation of molecules is. Denoted by $(q_{r,\text{lib}})_i$ the contribution of a free rotation around the chemical bond, we will have

$$(q_{r,\text{lib}})_i = \frac{1}{\sigma}\left(\frac{8\pi^2 I_i kT}{h^2}\right)^{1/2}$$

Then, each molecule contributes to the thermodynamic functions[6] according to

$$A_{r,\text{lib}} = -kT\ln(q_{r,\text{lib}}); \quad U_{r,\text{lib}} = \frac{kT}{2};$$

$$(C_V)_{r,\text{lib}} = \frac{k}{2}; \quad S_{r,\text{lib}} = k\left(\frac{1}{2} + \ln(q_{r,\text{lib}})\right) \tag{2.38}$$

iii) There are rotations between central atoms chemically bound, but they are not completely free. Figure 2.10 illustrates the case of ethane (CH_3—CH_3); it is evident that when hydrogen atoms bound to one central C superpose with those bound to the other central C atom, there will be some repulsion between them that will limit the rotation to some extent, and hence the rotation will require some energy to overcome that repulsive barrier.

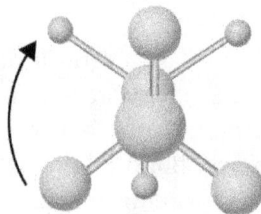

Figure 2.10: Internal rotation of ethane's C—C bond.

---

[6]A good exercise for the reader is to deduce these simple expressions

Li[7] determined experimentally by calorimetry the entropy of dimethyl cadmium (CH_3—Cd—CH_3). Table 2.1 reports the contributions to en-

Table 2.1: Internal rotation and entropy of dimethylcadmium at 298.2 K[a]

Origin of the value	Type of contribution to the entropy	$S(298.2\ K)$ J/(mole K)
Statistical calculation	Vibrations	36.64
	Rotations and Translations	253.8
	Total	290.4
Experimental		302.9 ± 0.8
Difference		12.5
	Free internal rotation	12.24

[a] $I = 2.70 \times 10^{-47}$kg m^2; $\sigma = 3$.

tropy for dimethyl cadmium calculated with spectroscopic data and the experimental value. The difference between them can be explained adequately if the internal free rotation of the two methyl groups is taken into account; they are free rotations because the two methyl groups are separated by a relatively large distance, thus not hampering sterically the rotation. The coincidence of the statically calculated value and the calorimetrically measured one strongly supports the validity of the third principle.

The cases of ethane or of 1,1,1-trichloroethane correspond to internal rotations restricted by the repulsion occurring when atoms bound to different C atoms are superposed. For ethane, it is difficult to determine spectroscopically the value of the repulsive barriers of that torsional mode because it produces neither IR nor Raman absorption due to the particular selection rules applying to this mode. In the case of 1,1,1-trichloroethane Pitzer and Hollenberg[8] used the position of a combination band present in its IR spectrum to estimate the height of the

---

[7]J. C. M. Li, "The Thermodynamic Properties of Cadmium Dimethyl: Heat Capacities from 14 to 291 K., Heats of Transition, Fusion and Vaporization, Vapor Pressure up to 296 K. and the Entropy of Ideal Gas", *J. Am. Chem. Soc.*, 78 (1956): 1081-1083.

[8]K. S. Pitzer, and J. L. Hollenberg, "Methylchloroform: The Infrared Spectrum from 130-430 cm^{-1}, the Energy Levels and Potential for Internal Rotation and the Thermodynamic Properties", *J. Am. Chem. Soc.*, 75 (1953): 2219-2221.

steric barrier.

In table 2.2, it may be observed that the internal restricted rotation contributes 9.0 J/(mole K) to the molecular entropy. The repulsive barrier restricting the internal rotation resulted equal to 12.41 kJ/mole. Also, in this case, the agreement between the calorimetric and statistical results are very satisfactory.

Table 2.2: Internal rotation and entropy of 1,1,1-trichloroethane at 286.5 K

Origin of value	Type of contribution to entropy	$S(286.5)$ J/(mole K)
Statistical calculation	Vibrations	36.7
	Rotations and Translations	276.1
	Restricted internal rotation [a]	9.0
	Total	318.8
Experimental		318.9 ± 0.8

[a] For a repulsive barrier of 12.41 kJ/mole.

Barriers restricting internal rotation will result in that molecule at low temperature behaving as if it had one more degree of freedom corresponding to a torsional vibration. At high temperature, internal rotation will be free. Figure 2.11 depicts schematically a calculation performed by T. Hill (cf. suggested books for further reading) for ethane's C_V.

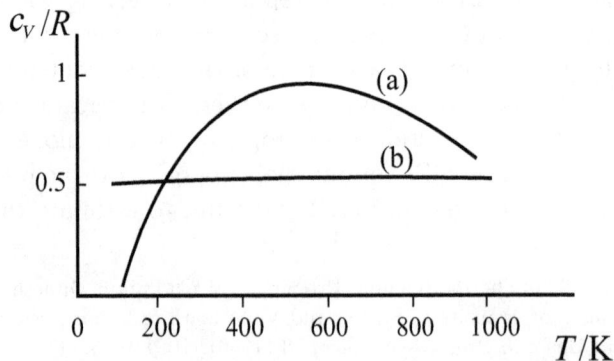

Figure 2.11: Ethane's heat capacity. a) experimental, b) free internal rotation.

Curve (a) up to 200 K is very similar to that for a crystal having a vibration in a single direction, as calculated with Einstein's model. Curve (b) corresponds to a free internal rotation. Adjusting curve (a) to the experimental behavior, Hill obtained a value of 12.97 kJ/mole for the repulsive barrier that restricts the inter-rotation of ethane.

Problems

Problem 1

At room temperature there are three solid allotropes of phosphine, β–PH_3 is stable up to 49.43 K, and γ–PH_3 is stable up to 30.29 K; at those temperatures, each of the crystalline phases is transformed into α–PH_3. From calorimetric measurements of C_p, and careful extrapolation to 0 K, it is known that $S^\beta(49.43 \text{ K}) = 18.32$ J/(mole K) and that $S^\gamma(30.29 \text{ K}) = 11.21$ J/(mole K). Moreover, it was established that $S^\alpha(49.43 \text{ K}) = S^\alpha(30.29 \text{ K}) + 20.08$ J/(mole K).

The enthalpies of transition of the two allotropes to the α phase at $T = 49.43$ K are $\Delta_{tr}H(\beta \to \alpha) = 777.0$ J/mole, and $\Delta_{tr}H(\gamma \to \alpha) = 82.0$ J/mole.

a) Calculate the entropy of α–PH_3 a 49.43 K i)using the data for β–PH_3, and ii) using data for γ–PH_3.

b) Discuss the values obtained in terms of the third principle of thermodynamics.

Answers: a) i) 34.04 J/(mole K), ii) 34.00 J/(mole K); b) both crystalline allotropes have the same entropy at 0 K

Problem 2

Using eqn. (2.4), calculate the expressions for U and C_V of a harmonic oscillator; plot them as a function of $(kT/h\nu)$. For HBr, considered as a harmonic oscillator having $\nu = 2649.7$ cm^{-1}, estimate the minimum temperature at which (a) U, and (b) C_V, differ less than 10 percent from the values corresponding to the classical limit.

Answers: a) 2500 K; b) 3600 K

Problem 3

Frequently the function $-[G^\theta - H^\theta(0\,\text{K})]/T \simeq -[A^\theta - U^\theta(0\,\text{K})]/T$ is used to calculate thermodynamic quantities from the partition function; in these cases, $U^\theta(0\ \text{K}) \simeq H^\theta(0\ \text{K})$ includes the zero point vibration. Calculate this function for NO at 1800 K assuming that the molecule is a harmonic oscillator and a rigid rotor having $\nu^* = 1904.0$ cm^{-1} and $r(\text{NO}) = 0.115$ nm.

Answers: Rotational contribution to the function: -54.84 J/(mole K), vibrational contribution: -2.04 J/(mole K)

Problem 4

Two Ar atoms move one against the other with a kinetic energy equal to kT. Calculate the distance d at which the total kinetic energy is transformed into potential energy at 300 K for:

a) Only repulsive intermolecular energy acts, given by

$$U(r) = 1.403 \times 10^{-15} \exp\left(-\frac{2r}{0.53}\right) \text{J}$$

b) Lennard-Jones equation describes properly the intermolecular interactions with $(\epsilon/k) = 121.94$ K and $\sigma = 0.343$ nm.

Answers: a) $d = 0.339$ nm; b) $d = 0.343$ nm

Problem 5

Calculate the second virial coefficient of a gas of hard spheres having a diameter $\sigma = 0.40$ nm, and the obtained results, for:

a) $T = 440$ K

b) $T = 650$ K

Answer: $B_2 = 80.7$ cm^3/mole at both temperatures

Problem 6

Calculate the molar standard entropy of HCN(g) at 1000 K. This molecule is linear, with $I = 1.882 \times 10^{-46}$ kg·m^2, and has two vibrational stretching modes with $\nu_1^* = 2096.7$ cm^{-1}, and $\nu_2^* = 713.5$cm^{-1}, and two bending vibrational modes which are degenerate ($\nu_3^* = 3311.5$ cm^{-1}).

Answer: $S(1000K) = 253.6$ J/(mole K). Compare this value with the experimental one 253.7 J/(mole K)

Problem 7

The translational entropy of a monoatomic ideal gas is given by the equation of Sackur-Tetrode,

$$S_{tr} = R \ln \left[\frac{(2\pi mkT)^{3/2}}{h^3} \frac{kT}{p} e^{5/2} \right]$$

Calculate S_{tr} for Kr at 350 K and a pressure of 1.5 bar.

Answer: $S_{tr} = 20.66$ J/(mole K)

Problem 8

In an atomic force microscope (AFM), a tip, approached as an nanosphere made of SiN with radius d = 10 nm, is attached to a very small metallic arm (cf. figure).

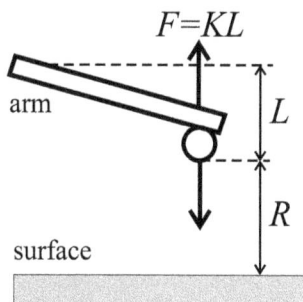

Figure 2.12: AFM cantilever.

The arm can be considered as a spring having a force constant of $K = 0.1$ N/m. A surface approaches the point very slowly, produced by a piezoelectric device. The attractive interaction energy between the tip and the planar surface is $B = 1.48 \times 10^{-20}$ J. Assuming that the repulsion of the sphere occurs at a distance of 0.2 mm:

a) Calculate the critical distance, R, at which the point jumps spontaneously over the surface (i.e., when the arm's deflection distance L is bigger than R)

b) Calculate the attractive force at R and compare it with the adhesion force between the sphere and the surface, calculated at a distance $R = 0.2$ nm

Answers: a) $R = 1.14$ nm; b) $F(1.14 \text{ nm}) = 1.14 \times 10^{-10}$ N, and $F(0.2 \text{ nm}) = 3.70 \times 10^{-10}$ N

Chapter 3

Partial Molar Properties and Phase Transitions

3.1 Euler's Integration Method for Extensive Properties

It has been shown that multicomponent systems cannot be distinguished from the single component ones if their compositions are not modified in the processes in which they participate. Compositions are thermodynamically relevant variables when they change along a given process, and the chemical potential of the species present, μ_i, is the state function accounting for such changes. According to eqns. (1.23), μ_i is equal to the change of the system's Gibbs energy when one mole of component i is added at constant temperature and pressure, keeping the amount of matter of the other components. This way of *expressing* the chemical potential was used with generality for other extensive properties of multicompenent systems, like volume, enthalpy, heat capacity, etc. This generalization is very useful because, in practice, many processes occur at constant (p, T) and under these conditions the changes in extensive properties, as in the case of the function G, are quite convenient for expressing the contribution due to one component of the mixture to the extensive property being considered.[1] The definition of the partial molar properties, X_i, of a

---

[1]It is frequent the application of thermodynamic concepts to dilute solutions, where the solvent is close to its critical point; in these cases, many partial molar quantities are not so useful because they tend to diverge at the critical point of the solvent.

system is

$$X_i = \left(\frac{\partial X}{\partial n_i}\right)_{p,T,n_{j\neq i}} \tag{3.1}$$

As will be shown, the values of partial molar quantities depend on the system's composition. Thus, this expression must be interpreted in the following way: At constant (p, T) one mole of component i is added to a large quantity of the multi-component system, so that it is possible to neglect any changes in its concentration, and the variation observed in the quantity X can be determined. Alternatively, partial molar quantities may be thought as the addition of one molecule of component i to a macroscopic amount of the solvent, a virtual process that does not change the composition of the system, and the variation of X has to be multiplied by Avogadro's number. The state functions are extensive properties that can depend on intensive variables (u, v), like p and T, and also of extensive variables (x, y), like S, V, and n_i. Being an extensive property, a state function f, represented by $f(u, v; x, y)$, must comply with the following relation,

$$kf(u, v; x, y) = f(u, v; kx, ky) \tag{3.2}$$

This equation indicates that if all extensive variables have been multiplied by a factor k, function f will also be k times larger. That is, if extensive variables of a system, volume, number of moles, etc. increase (or decrease) by a factor k without varying the values of the intensive variables, temperature, density, etc., all the system's thermodynamic properties will be modified by a factor k. This process may be conceived as an increase (or decrease) of the system's *size*. Differentiation of the previous equation with respect to factor k yields

$$f(u, v; x, y) = \left[\frac{\partial f(u, v; kx, ky)}{\partial kx}\right]_{u,v,y} \frac{\partial kx}{\partial k} + \tag{3.3}$$
$$+ \left[\frac{\partial f(u, v; kx, ky)}{\partial ky}\right]_{u,v,x} \frac{\partial ky}{\partial k}$$

and since $(\partial kx/\partial k) = x$ and $(\partial ky/\partial k) = y$, we have

$$f(u, v; x, y) = x \left[\frac{\partial f(u, v; kx, ky)}{\partial kx}\right]_{u,v,y} + y \left[\frac{\partial f(u, v; kx, ky)}{\partial ky}\right]_{u,v,x} \tag{3.4}$$

It is possible to make $k = 1$, then

$$f(u, v; x, y) = x \left[\frac{\partial f(u, v; x, y)}{\partial x}\right]_{u,v,y} + y \left[\frac{\partial f(u, v; x, y)}{\partial y}\right]_{u,v,x} \tag{3.5}$$

Eqn. (3.5) summarizes Euler's integral. As examples, we write the expressions for the case that f is equal to G, V, or A. For the two first thermodynamic functions, the extensive variables are the amounts of matter of the components n_i,

$$G = f(p, T, \sum_i n_i) = \sum_i n_i \mu_i = \sum_i n_i \left(\frac{\partial G}{\partial n_i}\right)_{T, p, n_{j \neq i}}$$

$$V = f(p, T, \sum_i n_i) = \sum_i n_i V_i = \sum_i n_i \left(\frac{\partial V}{\partial n_i}\right)_{T, p, n_{j \neq i}}$$

For the case of A, which also depends on the extensive variable V, we have

$$\begin{aligned}
A &= f(V, T, \sum_i n_i) = \sum_i n_i \left(\frac{\partial A}{\partial n_i}\right)_{T, V, n_{j \neq 1}} + V \left(\frac{\partial A}{\partial V}\right)_{T, n_i} = \\
&= \sum_i n_i \mu_i - pV
\end{aligned}$$

If (u, v) are (p, T) and (x, y) are (n_1, n_2), the derivatives of $f(u, v; x, y)$ with respect to the extensive variables, are the molar partial properties of the system and, since they do not depend on k, have to be intensive properties. This can be seen using eqn. (3.2) and differentiating both members with respect to x. In this case, it yields,

$$f_x(u, v; x, y) = \frac{\partial f(u, v; x, y)}{\partial x} = f_x(u, v; kx, ky) \tag{3.6}$$

Making $k = (1/y)$, the result is

$$f_x(u, v; x, y) = f_x(u, v; \frac{x}{y}) \tag{3.7}$$

That is, x and y are not two independent variables of function f_x; the quotient x/y becomes the independent variable. Hence, for a two-component system with $x = n_1$ and $y = n_2$, eqn. (3.7) indicates that the partial molar properties, which will correspond to f_x, depend on the relation (n_1/n_2) (i.e., on the concentration and not on the number of moles of each individual component).

Due to chemical potential's bearing for the rest of the subjects dealt within this book, it is convenient to use eqn. (3.7) to write clearly the dependence of chemical potential on the thermodynamic variables p, T, and the mole fractions x_i of the mixture.

$$d\mu_i = \left(\frac{\partial \mu_i}{\partial p}\right)_{T, x_l} dp + \left(\frac{\partial \mu_i}{\partial T}\right)_{p, x_l} dT + \sum_{j=1}^{C-1} \left(\frac{\partial \mu_i}{\partial x_j}\right)_{T, p, x_{l \neq j}} dx_j \tag{3.8}$$

In the previous equation, there are $(C - 1)$ independent molar fractions because $\sum_i x_i = 1$, so that the molar fraction of one of the components may be expressed as a function of all the others. Since

$$\mu_i = \left(\frac{\partial G}{\partial n_i} \right)_{T,p,n_{j \neq i}}$$

is possible to use Maxwell's relations to write the differential quotients in eqn. (3.8) in terms of entropy and partial molar volumes,

$$\mathrm{d}\mu_i = V_i \mathrm{d}p - S_i \mathrm{d}T + \sum_{j=1}^{C-1} \left(\frac{\partial \mu_i}{\partial x_j} \right)_{T,p,x_{l \neq j}} \mathrm{d}x_j = V_i \mathrm{d}p - S_i \mathrm{d}T + \mathrm{D}\mu_i \qquad (3.9)$$

where the quantity $\mathrm{D}\mu_i$, suggested by Guggenheim, represents the term that contains the sum in eqn. (3.9).

3.2   Gibbs-Duhem's Relation and the Phase Rule

By means of Euler's integral (eqn. (3.5)), one obtains for the Gibbs energy the following expression

$$G = \sum_i \mu_i n_i$$

Then, differentiating this equation, and remembering eqn. (1.22), one gets the following expression

$$\mathrm{d}G = V \mathrm{d}p - S \mathrm{d}T + \sum_i \mu_i \mathrm{d}n_i = \sum_i \mu_i \mathrm{d}n_i + \sum_i n_i \mathrm{d}\mu_i$$

Thus, we get the extremely important relation

$$S \mathrm{d}T - V \mathrm{d}p + \sum_i n_i \mathrm{d}\mu_i = 0 \qquad (3.10)$$

This expression is universally known as the Gibbs-Duhem equation, essentially of central importance for the treatment of homogeneous multicomponent phases in equilibrium.

In a similar fashion for volume, considered as function of pressure, temperature, and moles of the various components, we have,

$$\mathrm{d}V)_{p,T} = \sum_i \left(\frac{\partial V}{\partial n_i} \right)_{p,T,n_{j \neq i}} \mathrm{d}n_i = \sum_i V_i \mathrm{d}n_i + \sum_i n_i \mathrm{d}V_i$$

Hence,

$$\sum_i n_i dV_i = 0$$

The meaning of this relation may be seen better if it is written for a system of two components. In this case,

$$dV_2 = -\frac{n_1}{n_2}dV_1$$

It is seen that the changes of partial molar volumes are interdependent. Figure 3.1 illustrated the case of a mixture of propanone and trichloromethane; it is clear that when the partial molar volume of propanone goes through a maximum, that of trichloromethane goes through a minimum. On the other hand, the figure shows that when $n_1 = 0, dV_2 = 0$; this means that $V_2 = V_2^*$ and the curve's tangent representing V_2 is horizontal when $n_1 \to 0$. The supraindex indicates pure component, thus V_2^* is the molar volume of pure component 2.

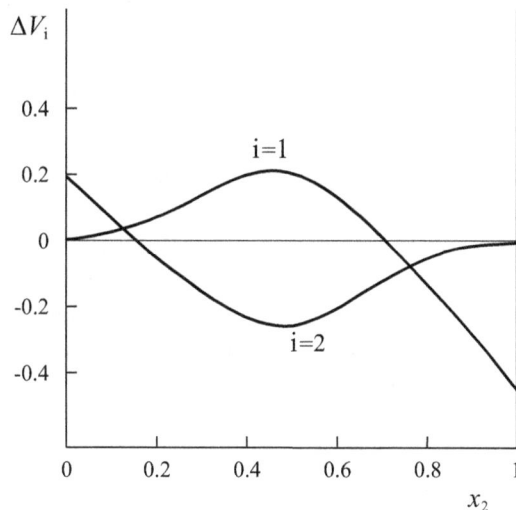

Figure 3.1: Partial molar volumes of the mixture components: Propanone (1) and trichloromethane (2), at 298.2 K. $\Delta V_i = V_i - V_i^*$ in $cm^3/mole$.

Gibbs-Duhem's equation is applicable to individual phases and has an important consequence: To reduce by one the number of independent thermodynamic variables, for each phase that exists in equilibrium. This number, also called degree of freedom (DF) determines the number of variables required to define the system's state. Let us illustrate the restriction in DF caused by the Gibbs-Duhem's equation as follows. For a single-phase pure component, $DF = 2$ (e.g., pressure and temperature). Instead, if the pure component is distributed among P phases in equilibrium,

DF is reduced by one, for each coexisting phase. That is, $DF = 3 - P$, a formula that includes the case of a single-phase system. If now, the number of components distributed among the P phases is C, one has to add $C - 1$ independent compositions, leading to the general expression of the so-called *phase rule*, $DF = 2 - P + C$. Notice that the number of independent compositions corresponds to $C - 1$ and not C, due to the restriction imposed by the Gibbs-Duhem's relation.

From the derivation of the phase rule, it is clear that any other relations between the thermodynamic degrees of freedom that restrict the number of independent variables have an effect equivalent to phase coexistence–for instance, when one component is stoichiometrically related with others, or, which is the same, when there is an equilibrium condition between different chemical species. This will be clearer in chapter 8 when discussing chemical equilibrium and it will be seen that it consists of additional relations between the chemical potential of the involved substances.

A system consisting of $CaO(s)$, $CaCO_3(s)$, and $CO_2(g)$ has three phases and three different chemical species, so it would have two degrees of freedom. This reasoning is valid only at low temperatures when chemical equilibrium between the three species

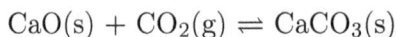

$$CaO(s) + CO_2(g) \rightleftharpoons CaCO_3(s)$$

is not able to operate because the chemical processes in both directions of reaction are very slow. However, at higher temperature (e.g., 900 K), the fact that chemical equilibrium between the three species present restricts one more degree of freedom, only a single independent variable, either temperature or pressure, remain.

Eqn. (1.18) allows analyzing very conveniently the cases where other types of work exist, besides the mechanical work. For systems having elastic energy contributions, the value of Y in the more general equation (1.18) is the force of restitution of the spring or elastomer and X its length, etc. In a following chapter, an example of elastic work will be analyzed in more detail because it allows to illustrate the capacity of thermodynamics to describe, in a general manner, the behavior of many different systems.

3.3   Phase Equilibrium in Pure Substances

In chapter 1 it was shown that when a system has two phases, α and β in equilibrium, it means that the chemical potential of the substances in both phases must be the same. For a pure substance coexisting in two phases the chemical potential of the substance in phase α, μ^α, must be the same as that in phase β, μ^β. Considering that for a mole of substance $G^\alpha \equiv \mu^\alpha$ and $G^\beta \equiv \mu^\beta$, and that under equilibrium conditions chemical potential in phases α and β are identical, matter transfer between the two phases in coexistence is described according to eqn. (3.9) by,

$$V^\alpha dp - S^\alpha dT = V^\beta dp - S^\beta dT$$

From this expression it may be concluded that

$$\frac{dp}{dT} = \frac{S^\alpha - S^\beta}{V^\alpha - V^\beta} = \frac{\Delta_{tr}S}{\Delta_{tr}V} \tag{3.11}$$

where $\Delta_{tr}X$ represents the change of property X when the phase transition takes place $\beta \to \alpha$. Since we are dealing with a phase transition process, which does not involve a change in Gibbs energy per mole, we have $T\Delta_{tr}S = \Delta_{tr}H$, and replacing $\Delta_{tr}S$ in the previous equation it yields the Clausius equation,

$$\frac{dp}{dT} = \frac{\Delta_{tr}H}{T\Delta_{tr}V} \tag{3.12}$$

This type of transitions are referred as first order transitions, because the first derivatives of G with respect to the intensive thermodynamic parameters are discontinuous, V and S in this case.

Eqn. (3.12) can be used for transitions between a gaseous phase and a condensed phase (gas-liquid or gas-solid) or between two condensed phases (solid-liquid or between two solid allotropic phases). If one of the phases is gaseous and the temperature is far away from the critical temperature of that substance, it is possible to assume that the gas molar volume is much larger than that of the condensed phases, $\Delta_{tr}V \simeq V(g)$; also, if it is possible to use the ideal gas equation to express the gaseous volume, $V(g) \simeq RT/p$, the Clapeyron-Clausius is obtained,

$$\frac{d \ln p}{dT} = \frac{\Delta_{tr}H}{RT^2} \tag{3.13}$$

which expresses how the vapor pressure of a condensed phase varies with temperature.

It should be made clear why in eqns. (3.11), up to (3.13), the change of pressure with temperature has been represented as a total derivative and not a partial one. That is because p (and T) changes following an equilibrium path between the two phases (coexistence) involve only one degree of freedom; hence, it is correct that it is given by a total derivative.

The upper graphs in figure 3.2 illustrate the phase diagram p vs. T for CO_2 and for metallic tin. It is very interesting to analyze what happens with the chemical potentials (or molar Gibbs energies) when temperature changes at constant pressure, and also when pressure changes at constant temperature; this is illustrated in the

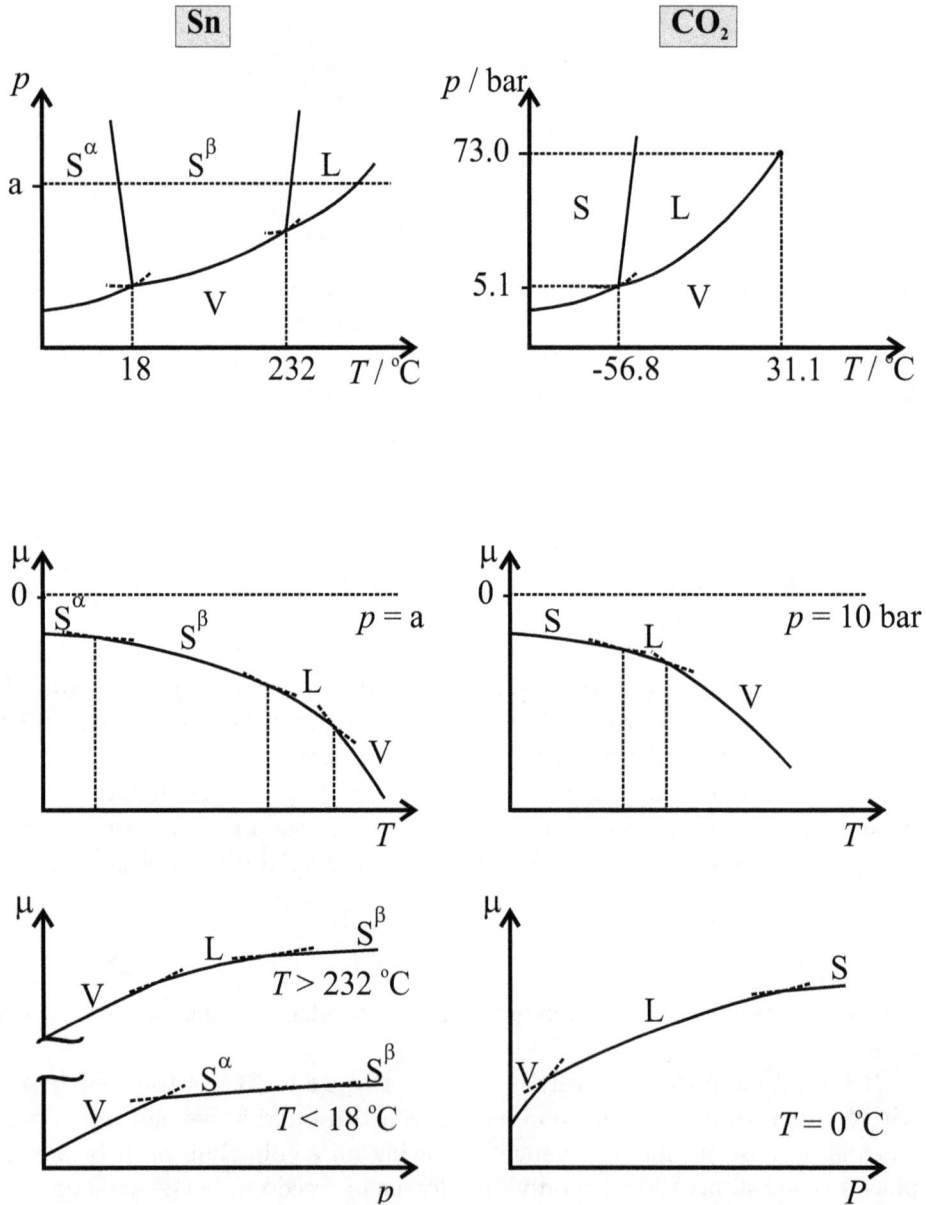

Figure 3.2: Phase diagrams and changes of the chemical potential for CO_2 and elemental Sn, as a function of T and p.

lower panels of figure 3.2. This analysis exemplifies how the general principles, already presented and discussed, are used in some particular cases. The form of the curves of chemical potential as a function of temperature are related to the fact that the slopes must be negative, because according to eqn. (3.9),

$$\left(\frac{\partial \mu^\alpha}{\partial T}\right)_p = -S^\alpha < 0$$

and that the stability condition determines the curvature of $(\mu^\alpha)_p = f(T)$, because the following expression must always be valid

$$\left(\frac{\partial^2 \mu^\alpha}{\partial T^2}\right)_p = -\frac{C_p^\alpha}{T} < 0$$

Analogously, it is necessary that $(\mu^\alpha)_T = f(p)$ has a positive slope because, according to eqn. (3.9), $(\partial \mu^\alpha / \partial p)_T = V$, and this is a positive quantity, while the curvature of this curve, equal to $-\kappa_T^\alpha$, must be negative. For the case of tin, the stable phase a lower temperature (α-Sn) has a larger molar volume than the solid phase stable at higher temperature (β-Sn); that is why the curve denoting coexistence of both solids has a negative slope–it is like ice and liquid water.

Using these graphs, including the case of Sn having two allotropic crystalline phases, the range of stability of each phase can be established and also their coexistence with other phases, or whether a metastable phase is formed.

3.4   Cubic Equations of State. The van der Waals equation

Now, we will analyze in greater detail the phase transition between a liquid and its vapor. It is very convenient to use the equation proposed by J. D. van der Waals, which involves a very precise summary of the behavior of fluids over all the range of densities. For one mole of fluid the equation he proposed is

$$\left(p + \frac{a}{V^2}\right)(V - b) = RT \tag{3.14}$$

Eqn. (3.14) was the origin of a series of equations of state which are frequently used in chemical engineering and which are denominated generically as *cubic equations of state*.

Figure 3.3 shows a curve corresponding to the van der Waals equation for one isotherm of CO_2. As observed, the curve is very asymmetric, which complicates somewhat its correct illustration. Cubic equations of state describe the liquid-vapor transition with a continuous curve, but since this is a first-order transition, a first derivative of μ has to present a discontinuity; in this case, since T is constant, the discontinuity is observed in the molar volume of each phase.

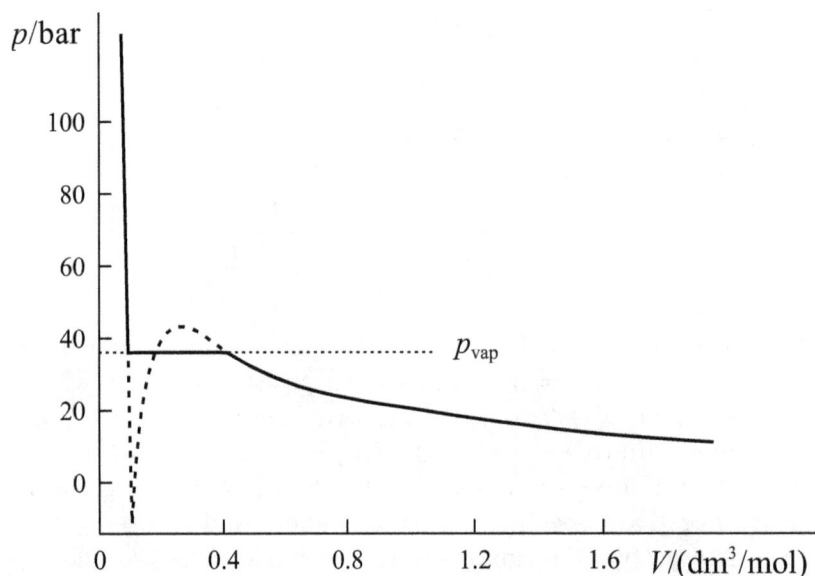

Figure 3.3: Change of pressure as a function of the molar volume of CO_2, at 258 K. The solid curve represents the experimental isotherm; the dashed curve was calculated with the van der Waals equation. The pressure p_v at that temperature was calculated using Maxwell's method of equal areas (cf. text). The difference in the behavior of the metastable liquid and gas phases should be noted.

What is the liquid volume and the gas volume in equilibrium? To clarify this point it is necessary to establish the equilibrium pressure value, or vapor pressure; this is known to have a fixed value at each temperature. The way to solve this problem is known as Maxwell's construction. Since the chemical potential of the two phases in equilibrium has to be the same for an isotherm, it is necessary that

$$\mu^\beta - \mu^\alpha = \int_\alpha^\beta V\,\mathrm{d}p = 0 \tag{3.15}$$

Also we have $\mathrm{d}(pV) = p\mathrm{d}V + V\mathrm{d}p$, hence

$$\int_\alpha^\beta V\,\mathrm{d}p = p_v(V^\beta - V^\alpha) - \int_\alpha^\beta p\,\mathrm{d}V = 0 \tag{3.16}$$

That is, the area of the shown isotherm illustrated in figure 3.3 that is lying above the horizontal line corresponding to the value of the vapor pressure at that temperature, and the area lying below it, must be equal.

As discussed in the last section of chapter 1, the compressibility of a pure fluid must be positive if there is a mechanically stable phase. For a fluid, whose behavior may be described by a cubic state function, like that of van der Waals's, it is observed that part of the isotherm is represented by a function $p = f(V)$ with a positive slope (cf. figure 3.3) (i.e., its compressibility would be negative, and hence that region does not represent equilibrium states). As illustrated in the figure, that region lies between a maximum for the gas volume and a minimum for liquid volumes. Between the maximum and the minimum there must be an inflection point where the curvature of the isotherm changes sign. The extreme points obey the necessary condition for the compressiblity,

$$\left(\frac{\partial p}{\partial V}\right)_T = 0 \tag{3.17}$$

and the inflection point implies

$$\left(\frac{\partial^2 p}{\partial V^2}\right)_T = 0 \tag{3.18}$$

As the temperature increases and the critical isotherm is approached, the phase transition between the liquid and its vapor disappears, the extreme points approach each other, and at the critical temperature T_c they coincide and, obviously, this also happens with the inflection points. That is, at V_c, p_c, and T_c, the two previous conditions are satisfied simultaneously. Using the van der Waals equation under the conditions (3.17) and (3.18), it is possible to get the following relations

$$V_c = 3b \qquad a = \frac{27RT_c b}{8} \qquad \frac{RT_c}{p_c V_c} = \frac{8}{3} = 2.67 \tag{3.19}$$

The van der Waals equation (3.14) has two parameters specific to the particular substance, in the same way as the intermolecular potential expressing the interaction among apolar molecules (dispersion, or van der Waals forces), which are represented at molecular level by the Lennard-Jones potential (eqn. (2.23)). The parameter a is related to the attractive contribution of intermolecular forces (i.e., with ϵ_{ij} in eqn. (2.23)); the parameter b is related to the repulsive contribution (i.e., with the molecular size given by σ_{ij} en la eqn. (2.23)). According to the principle of corresponding states discussed in chapter 2, if the van der Waals coefficients are expressed in terms of the critical parameters, a universal equation of state will be obtained as a function of reduced variables. If the thermodynamic variables of fluids are divided by their critical values, the following reduced variables can be defined as $\theta = (T/T_c)$ for the reduced temperature, $\pi = (p/p_c)$ for the reduced pressure, and $\phi = (V/V_c)$ for the reduced volume. The van der Waals equation of state can be written as a function of such reduced variables as follows,

$$\left(\pi + \frac{3}{\phi^2}\right)(3\phi - 1) = 8\theta \qquad (3.20)$$

where the van der Waals parameters were expressed in terms of the critical parameters, replacing eqns. (3.19) in eqn. (3.14).

It may be added that the van der Waals equation also describes correctly the thermodynamic behavior of metastable fluid phases, including the range over which they are unstable.

According to the microscopic qualitative analysis described in chapter 1, page 28, from the van der Waals equation it is possible to obtain the contributions to the intermolecular potential energy due to the kinetic energy and to the intramolecular potential energy. If the van der Waals (3.14) equation for the pressure is written as

$$p = \frac{RT}{V-b} - \frac{a}{V^2}$$

it is simple to demonstrate that

$$\left(\frac{\partial p}{\partial T}\right)_V = \frac{R}{V-b}$$

and

$$\left(\frac{\partial U}{\partial V}\right)_T = \frac{a}{V^2}$$

clearly illustrating the origin of each of the two rhs terms; these were discussed when the thermodynamic equations of state were presented in section 1.4.

State Equation of an Elastomer

In chapter 1, page 30, it was pointed out that a thermodynamic equation of state may be obtained for an elastomer, which will be analogous to eqn. (1.14), taking into account explicitly that the system is able to perform elastic work. When a tension is applied to an elastic material, this extends and it does elastic work.

$$\mathrm{d}w_{elast} = J\mathrm{d}l$$

where J is a force or tension. The total work, elastic and mechanical, is $\mathrm{d}w = J\mathrm{d}l - p\mathrm{d}V$, but in general, for polymers having an elastic behavior, the second term on the rhs is negligible. Hence, for these materials,

$$\mathrm{d}U = T\mathrm{d}S + J\mathrm{d}l$$

and

$$\mathrm{d}A = -S\mathrm{d}T + J\mathrm{d}l$$

The equation of state for tension J may be obtained from the following relations,

$$J = \left(\frac{\partial U}{\partial l}\right)_S = \left(\frac{\partial A}{\partial l}\right)_T = \left(\frac{\partial U}{\partial l}\right)_T - T\left(\frac{\partial S}{\partial l}\right)_T$$

because $A = U - TS$. The following Maxwell's relation

$$\left(\frac{\partial S}{\partial l}\right)_T = -\left(\frac{\partial J}{\partial T}\right)_l$$

is important to obtain a very useful version of the equation of state of an elastometer,

$$J = \left(\frac{\partial U}{\partial l}\right)_T + T\left(\frac{\partial J}{\partial T}\right)_l$$

Using this equation, it is possible to know the change in the thermodynamic quantities without any calorimetric measurement, through the relation of those changes with the system's mechanical properties.

It is convenient to compare the state equation for an elastomer with the thermodynamic equation for gases (1.14) and especially with the van der Waals equation. It may be seen that for the elastomer there is also a term in the state equation that is related to intermolecular interactions, while the other term accounts for the isothermal change of entropy with the system's dimension or elongation–the extensive variable volume for the case of gases taken by elongation in the case of elastic work. We discuss, in detail, results from the work by Anthony, Caston, and Guth[2],

---

[2]R. L. Anthony, R. H. Caston, and E. Guth, "Equations of state for natural and synthetic rubber-like materials. I. Unaccelerated natural soft rubber", *J. Phys. Chem.*, 46 (1942): 826-840.

and also Pellicer, Manzanares, Zúñiga, Utrillas, and J. Fernández[3], about the elasticity of vulcanized natural rubber.

Experimental results are shown in the following two graphs. The first, figure 3.4, illustrates the change of the elastic force J as a function of the percentual increase of the elastomer's elongation. The second one, figure 3.5, shows the two contributions to the state equation for the elastomer at 20 °C. It is observed that $(\partial U/\partial l)_T \simeq 0$, hence $J \simeq -T(\partial S/\partial l)_T$ increases up to elongations between 300 and 400%. The process of elastic stressing of the elastomer is essentially entropic, and experiment indicates that as the elastomer elongates, the number of possible configurations of the polymeric chain diminishes due to the fact that $(\partial S/\partial l)_T < 0$. This conclusion is in agreement with what was discussed about entropy; when no stress is applied the elastomer's chains will be partially coiled, while upon application of stress the chains will tend to become aligned, producing a decrease of entropy.

It can also be interesting to establish whether the polymer length increases or decreases with temperature at constant tension. Starting with the function $G' = U - JL - TS$, we obtain

$$\mathrm{d}G' = -l\mathrm{d}J - S\mathrm{d}T$$

This is an example illustrating how one can define thermodynamic functions that allow the resolution of a problem in a more convenient way. Remembering the relation between the cross derivatives in the previous equation, we get

$$\left(\frac{\partial l}{\partial T}\right)_J = \left(\frac{\partial S}{\partial J}\right)_T$$

As already discussed, S decreases with increasing J because there are fewer possible molecular configurations; hence,

$$\left(\frac{\partial l}{\partial T}\right)_J < 0$$

This interesting property of elastic materials frequently has practical consequences. For instance, the steel rails used by fast trains do not need to have small free spaces–as they used to be in conventional trains–due to maintenance problems and the need of a smoother displacement of trains.

[3]J. Pellicer, J. A. Manzanares, J. Zúñiga, P. Utrillas, and J. Fernández, "Thermodynamics of Rubber Elasticity", *J. Chem. Educ.*, 78 (2001): 263-267.

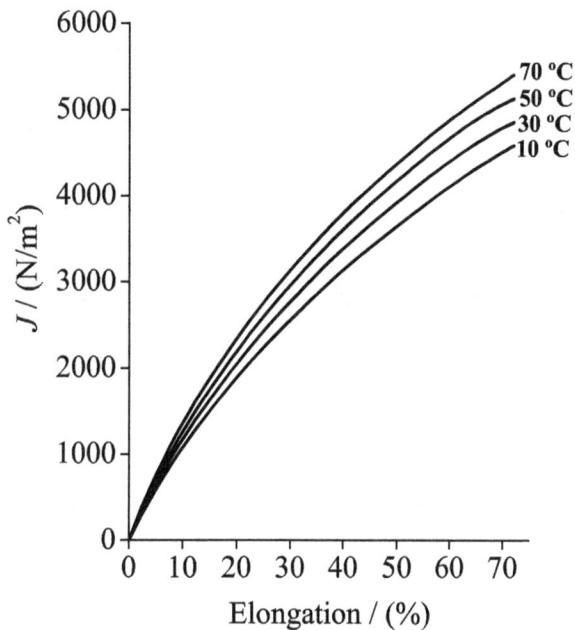

Figure 3.4: Variation of tension, J, as a function of the percentage of elongation of an elastomer, plotted for four isotherm.

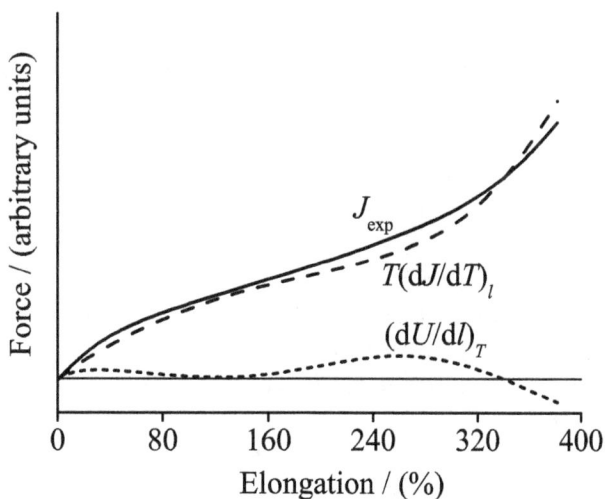

Figure 3.5: Contributions to the equation of state for an elastomer.

One solution that permits eliminations of the dilation space consists in using prestressed steel for rail fabrication. Under these conditions, what happens with the rail when the temperature rises?

3.5   Metastability and Glass Transition

Now, we will analyze systems that are in metastable states, that is, systems constituted by phases that are not the most stable ones thermodynamically, but nonetheless present relative stability. These type of systems are important because there are many materials that are used daily that are in metastable states, like the case of many metals and alloys, glasses, and polymeric systems (e.g., many plastics). Condensed matter frequently present many relative minima in their thermodynamic properties, which should be a global minimum in complete equilibrium. For instance, for processes at constant T and p, Gibbs energy G can have minima corresponding to different values of volume, only one of them will be the absolute minimum (i.e., corresponding to the most stable state). The other minima also correspond to relatively stable but not to the most stable. On the other hand, metastable systems have low mobility compared to that observed in a normal liquid and this implies that the system goes through all microstates more slowly and can get trapped in a relative minimum. In these systems, time has a very important role even when not explicitly. It is clear that to reach a metastable state, for example by cooling the system, the temperature must decrease more rapidly than the time necessary to reach the most stable equilibrium.

A microscopic view of the metastable states in terms of the systems' Gibbs energy is the following: The Gibbs energy surface of some systems, represented as a function of different degrees of freedom (a hypersurface), presents various local minima separated from the others that correspond to states in the metastability region. The system can remain for a long time *exploring* that region, and fluctuations will take the system from a local minimum to another until finally, in one of these fluctuations reaches the state that corresponds to the absolute minimum in G, that is, the equilibrium state characterizing the stable phase. Figure 3.6 is a bidimensional scheme of the hypersurface of chemical potentials or molar Gibbs energies for a single-component system.

Some simple models of fluids, like that of van der Waals's, are useful to illustrate the characteristics of metastable regions in the liquid-vapor transition. Thus, for instance, in figure 3.3, the curve p vs. V for CO_2 at 258 K, is a particular isotherm from the many that may be plotted for this system. In order to illustrate clearly

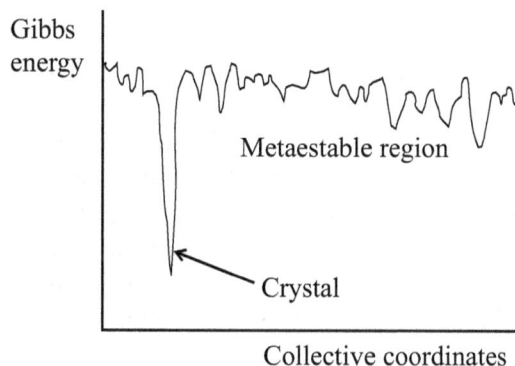

Figure 3.6: Gibbs energy as a function of the collective coordinates.

the metastable behavior, the temperature of the isotherm in figure 3.7 is close to the critical temperature of CO_2 ($T_c = 304.3$ K) and the curve is drawn only for the density interval of vapor and liquid close to the critical density. The liquid region AB and the vapor region CD for that isotherm correspond to metastable phases because, despite that $\kappa_T = -1/V(\partial V/\partial p)_T > 0$ (cf. criterion for thermodynamic stability in section 1.7), vapor is the more stable phase for conditions in region AB, while the liquid is the more stable phase in region CD.

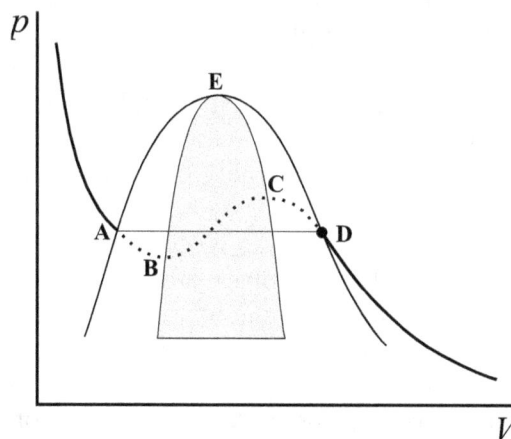

Figure 3.7: Binodal (AED) and spinodal (BEC) curves in the phase diagram of p as a function of V.

It is possible to have liquid CO_2 at pressures lower than the corresponding vapor pressure (point A), because it will be metastable liquid all along the region AB, such that no vapor bubbles form. This situation can be maintained during a certain time without the transition to vapor phase, which is the thermodynamically stable one (point D). In the same manner, it is possible to have metastable gaseous CO_2 at a pressure greater than $p_A = p_D$ (metastable vapor in region CD). On the other hand, the region BC is unstable because on this curve $\kappa_T < 0$; that is, the volume would increase by enhancing the pressure, something that is not possible.

One must observe that it is also possible, in principle, to keep liquid Sn under 232 ^oC (see figure 3.2) without observing the separation of stable solid under these conditions.

In figure 3.7, the curve AED is the result of joining the values of the liquid and the gas phases in equilibrium at each vapor pressure, from that isotherm up to the critical isotherm. AED is called binodal curve or curve for the coexistence of phases. The inner bell-shaped curve corresponds to the extreme points limiting the instability region, and this curve is defined as the spinodal curve. The region between this curve and the binodal curve defines the region of metastability of CO_2, extending until the critical temperature, T_c, at point E in figure 3.7. The spinodal and the binodal curves meet at that point.

Many systems having practical interest, which we are usually in contact with, are really metastables. An example is the common glass, which is a metastable phase (amorphous solid) corresponding to the stable crystalline solid (stable phase) of an alkaline silicate. These silicates melt at high temperatures, and when they are cooled down quickly they are not able to adopt the periodic crystalline structure, which is the stable phase, they form amorphous solids; however, they can be in a metastable state for quite a long time. It is analogous, albeit not so extreme with what happens with the metastable phase of carbon and diamond, where the conversion to the stable graphite phase is not observed over a reasonable time (fortunately for people that invest in it) due to the very slow mobility of atoms or molecules in the metastable phase.

In the example of CO_2, which was recently analyzed, the transition from the metastable to the stable phase involves processes of drops nucleation or bubble formation,[4] which requires a short time to transform the metastable phase into the stable one, thus contrasting with the case of the transformation of metastable solid phases, like the silicates.

Liquid water at ambient pressure has an interval of thermodynamic stability between 273 K and 373 K; nevertheless, it can exist as metastable liquid over a wide temperature range. Effectively, when cooling liquid water below its fusion temperature under some particular conditions, for instance inside capillaries, it can

---

[4]These processes are analyzed in section 7.5. It may be convenient to read that section for a better understanding of the present one.

be maintained as a supercooled liquid down to 232 K (-41 ^{o}C); on the other hand, water droplets in clouds may be liquid down to 238 K (-35 ^{o}C), but at some stage metastable water spontaneously crystallizes to form stable ice Ih.

On the other hand, at atmospheric pressure drops of liquid water can become superheated when dispersed in an organic solvent having a high boiling temperature. These drops vaporize violently at temperatures above 373 K, which is the normal boiling point of water, as illustrated in the following example.

Superheated Liquid Water

The experiments with metastable fluids are carried out under very careful conditions to eliminate all possibilities of heterogeneous nucleation (i.e., nucleation induced by other particles or impurities) either of ice crystals in the case of overcooled water or of vapor bubbles when water is overheated.

Apfel,[5] in 1972, did some experiments consisting of a vertical tube holding benzyl benzoate at atmospheric pressure ($T_b = 597$ K); this liquid is immiscible with water. An electric surrounding resistance generates a gradient of temperature along the tube, so that its top is hotter than its bottom. Microdrops of water, 200 to 500 μm in diameter, were injected with a syringe and ascended through the tube at speeds of some cm/s.

Under these conditions, it may be observed that the microdrops explode; that is, they become unstable at different heights in the column corresponding to temperatures between 513 and 552 K.

One can estimate the maximum temperature at which water may be overheated, using data of compresibility for the metastable region. Just like other thermodynamic properties of fluids, the compressibility is continuous between the binodal and the spinodal curves and, in this region, the experimental data for water can be represented by the equation

$$\kappa_T \propto [588 - T/K]^{-0.97}$$

According to this expression, the limit of overheating of water would be 588 K, a value quite close to that obtained experimentally.

---

[5]R. E. Apfel, "Water Superheated to 279.5 ^{o}C at Atmospheric Pressure", *Nat. Phys. Science*, 238 (1972): 63-64.

In the case of supercooled water, if the temperature decreases below 232 K, homogeneous nucleation of ice can occur; that is, the formation of ice crystals begins in the liquid phase and not in the interface with another substance (e.g., an impurity). However, if the freezing is quick enough (faster than 10^7 K/s), it is possible to overcool liquid water below 232 K. During this process, the viscosity of supercooled water increases exponentially with increasing temperature, reducing the rate of nucleation, until it reaches the temperature T_g (glass transition temperature), where the viscosity becomes virtually equivalent to that of a solid. Below T_g, the system has the microscopic structure of a liquid (its molecules are ordered in a periodic lattice), but having the mechanical properties of a solid. As quoted by P. Debenedetti,[6] a supercooled liquid is a metastable phase too hot to be a solid, and too cold to be a liquid. At such temperatures, close to 136 K, we have amorphous or glassy water.

It is not trivial to understand why pure water needs to be cooled very rapidly to reach the glassy state while some aqueous solutions with added solutes can be vitrified by cooling it at moderate rates (as we will see). Also, many pure substances can undergo a glass transition (going from supercooled liquid directly to glass) at rates of cooling much smaller than water (7×10^3 K/s for SiO_2 and only 50 K/s for phenyl salicilate). It may be argued that for the glass transition to occur, the rate of cooling must be larger than the rate of nucleation so that there is no chance that crystalline nuclei can be formed. On the other hand, the velocity of nucleation decreases as the liquid temperature becomes lower because its viscosity increases, so that if nucleation involves collective molecular displacements, it will be a very slow process at low temperatures and the system will go through the glass state instead of crystallizing. Water's nucleation seems very fast because it only requires slight modifications of angular orientation of the hydrogen bonds. That is why water needs to be cooled very fast if we want it to attain the amorphous state.

Independently of the type of substance, the transition between supercooled liquid and glass involves a discontinuity in C_p and in thermal expansivity $\alpha = V^{-1}(\partial V / \partial T)_p$ (i.e., in the second derivatives of Gibbs energy, with respect to pressure and temperature, cf. figure 3.8).

This transition from supercooled liquid to amorphous or glassy solid its not a thermodynamic transition occurring between phases in equilibrium; it can be described as a dynamic transition. T_g is a temperature that depends to a certain extent on the kinetics of the cooling process and on the time of relaxation of the property being observed to determine the glass transition. A consequence of this feature is that T_g changes when the time scale of the experiment is modified: This temperature becomes smaller in experiments that take a longer time.

---

[6]P. Debenedetti, Workshop on Supercooled Fluids, Buenos Aires (2012).

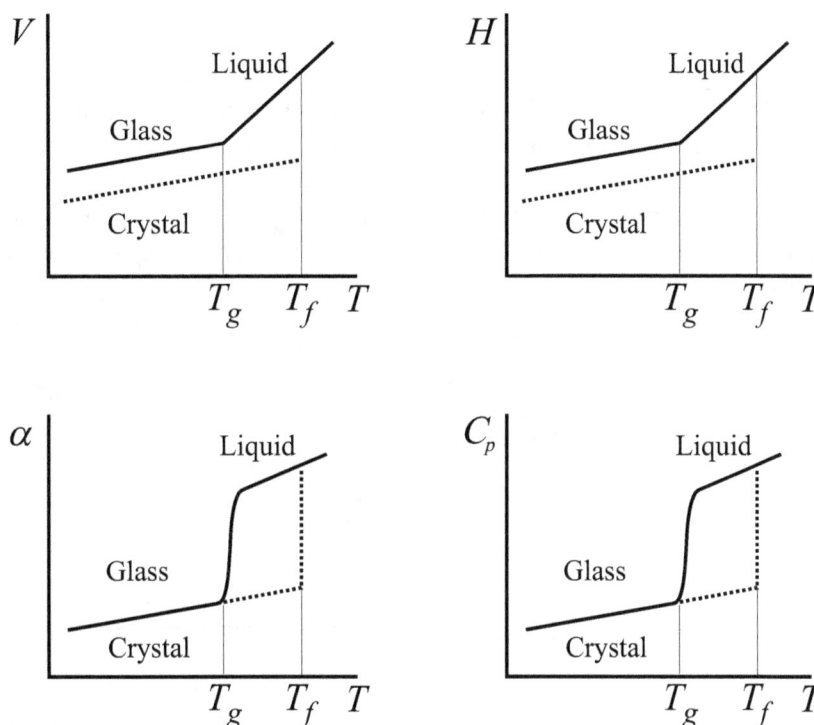

Figure 3.8: Change with temperature of volume, enthalpy, and their derivatives for a glass forming substance.

We will analyze two types of systems where the glass transition has a great practical importance. The first concerns polymers, where the transition temperature is one of the most relevant rheological and thermodynamic properties, because it establishes its possible applications. For these materials, it is necessary to study the transition from the glassy state to the plastic one performing measurements of the elastic modulus (or Young modulus), E, of the material. The glass transition in polymers corresponds to the transition from the plastic (vitreous or glassy) solid state to that often denoted as rubbery. Sometimes it may be called elastic state despite that polymers are plastics because they undergo elastic deformations (reversible) as well as viscous deformations (irreversible).

It is convenient to emphasize that for polymers the glass transition is due to the impediments for the movement of the molecular segments of the polymeric chain. This is similar to what happens close to the glass transition temperature of a super-cooled liquid, when its molecules are strongly impeded to go freely from one place to another due to the fluid's high viscosity and the low kinetic energy of the molecules.

The Young modulus relates the tension on the material to its elastic deformation $E = (\partial J/\partial l)_T$, so that E is a measure of the material's deformation susceptibility (note the similarity of E with the isothermal compressibility if J and l are replaced respectively by pressure and volume). When plotting E as a function of temperature for a polymer a typical curve is depicted in figure 3.9. Region 1 corresponds to the glassy state, where the polymer is hard and can be broken, as the disposable polyethylene cups.

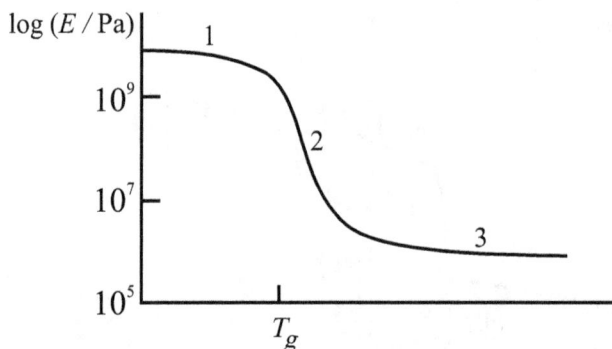

Figure 3.9: Change in the Young modulus with temperature, including the glass transition region.

The elastic modulus is ca. 10^9 Pa close to the glass-rubber transition. At that point, the only active degrees of freedom are vibrational and short-range rotations that involve small segments of polymeric chains.

Region 2 indicated in figure 3.9 is the glass-rubber transition and the Young modulus decreases abruptly in a range of 20–30 K. It is the beginning of coordinated molecular movements involving polymer chain segments of growing length in this interval. T_g depends on the nature of the polymer.

Region 3 corresponds to an elastic state where E attains values around 10^6 Pa. If the polymer is cross linked, the elastic-rubbery region extends until the polymer thermally decomposes and the value in the plateau represents the equation of state of the elastomer, $E = nRT$, n being the number of active segments in the polymeric chain. The elastic bands used popularly are in this region of the diagram; its thermodynamic properties have already been discussed. Table 3.1 shows values of T_g for different polymers. PVC has a T_g, which is above room temperature, so at this temperature region it is a plastic only because liquid plastifiers, that decrease T_g, are added to the polymer.

Another set of systems where the glass transition is of practical importance is the aqueous solutions containing vitrifying agents. These solutes, typically polyols

Table 3.1: Glass transition temperature of polymers

Polymer	$T_g(^oC)$
Polyethylene	-130
Polyisoprene	-73
Polyuretane	-60
Polyethylenglycol	-41
Polyvinylchloride (PVC)	82
Polystyrene	100
Polymethylmetacrylate	105

or carbohydrates, which are very soluble in water, have the property of increasing water viscosity and also the temperature of glass transition. This behavior is used in industry and very efficiently in nature, to avoid the crystallization of ice at low temperature, thus preserving cellular structures, pharmaceutical compounds, and biomolecules in aqueous suspensions (cryopreservation). Figure 3.10 is a phase diagram of the system sucrose-water. The glass transition of the pure disaccharide is 62 ^oC, and if water is added to the sucrose solution, T_g decreases monotonously, and the same happens when a solvent (plastifier) is added to a polymer.

According to figure 3.10, if an aqueous solution of sucrose of composition A is cooled rapidly when it reaches point B over the solubility curve of sucrose, no solid

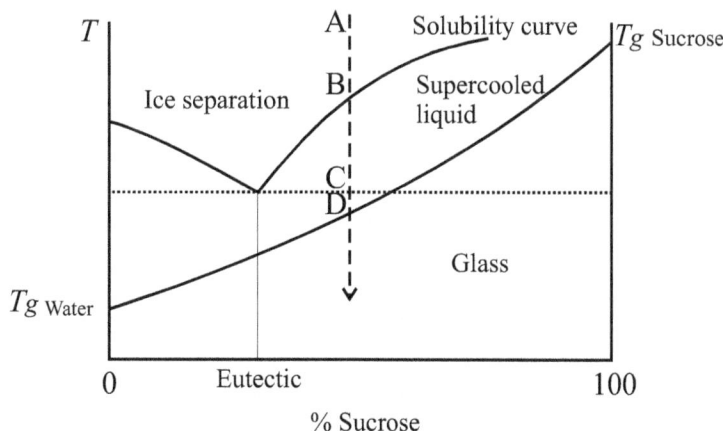

Figure 3.10: Supplemented phase diagram of the system sucrose-water as a function of the mass percentage of sucrose, including the glass transition curve.

solute separates and the solution becomes metastable. A solution of composition C can stand at such point for hours or even days before sugar crystallizes. If the system is cooled even more, the composition D is reached on the glass transition curve, and the system vitrifies. In this state, the glassy solution remains essentially forever if the temperature is not greater than T_g. It may be observed that under those conditions the system is at much lower temperature than the fusion temperature of water and even than the eutectic sucrose-water; nevertheless, ice does not separate in the system. The same happens if the system is cooled when it has a composition that corresponds to the eutectic. The solution may be vitrified without crystal separation. This is essential to preserve the cellular walls from mechanical damage produced by the formation of microcrystals of ice, and consequently it is the base cryopreservation of food, biological organs, tissues, etc.

3.6   Effect of Pressure on the Vapor Pressure of a Condensed Phase

When a solid is in equilibrium with its vapor, the vapor pressure $p^*(\mathrm{s},T)$ is only a function of temperature, as already discussed, and is evident according to the phase rule. But, what happens if the solid is mechanically pressurized at constant temperature? According to eqn. (1.12) the change of molar Gibbs energy of a pure solid (or its chemical potential) due to a variation in mechanical pressure, enhances the chemical potential of the solid; hence, in order to maintain equilibrium between the solid and its vapor, it is necessary that the vapor pressure increases. This effect is known as the Poynting effect; it is seen that the increase of pressure on the solid phase and its escaping tendency or vapor pressure will increase according to

$$\mu_{\mathrm{s}}(p,T) = \mu_{\mathrm{s}}[p_{\mathrm{s}}^*(T)] + \int_{p_{\mathrm{s}}^*}^{p} V_{\mathrm{s}}(T)\mathrm{d}p \qquad (3.21)$$

and for an incompressible solid, the integral is equal to $V_{\mathrm{s}}(p - p_{\mathrm{s}}^*)$. The vapor pressure at equilibrium with the solid at pressure p due to a mechanical force will be (cf. chapter 4)

$$\mu_{\mathrm{v}}(p,T) = \mu^{\ominus}(T) + RT\ln\frac{p_{\mathrm{v}}}{p^{\ominus}} = \qquad (3.22)$$
$$= \mu^{\ominus}(T) + RT\ln\frac{p_{\mathrm{s}}^*}{p^{\ominus}} + V_{\mathrm{s}}[p - p_{\mathrm{s}}^*]$$

then,

$$p_{\mathrm{v}} = p_{\mathrm{s}}^* \exp\left(\frac{V_{\mathrm{s}}[p - p_{\mathrm{s}}^*]}{RT}\right) \qquad (3.23)$$

The Poynting effect is also observed when the condensed phase is a liquid in equilibrium with its vapor, but the illustration of this case appears more *artificial* because it requires applying pressure only to the liquid phase, keeping the pure vapor as the only component in the gas phase. For instance, it would require that in the liquid-vapor interface there is a membrane that does not allow the liquid to pass through it due to the surface tension.

The Poynting effect becomes important for the case of condensed phases subjected to high pressure. It's the case of many equilibria between crystalline minerals in the Earth's lithosphere. We will describe in chapter 8 how this effect is used to get thermodynamic properties of those systems. It is also important in modern technologies that make use of processes that occur at high pressure (e.g., extraction of substances using supercritical fluids, as discussed in chapter 6).

The fact that the vapor pressure at equilibrium with a condensed phase changes with the external pressure according to eqn. (3.23) may be thought as a violation of the phase rule, which establishes that for a pure substance having two phases in equilibrium there is only one thermodynamic degree of freedom, either p or T. Nevertheless there is no violation; when the phase rule was derived it was explicitly established that apart from the composition variables there were two thermodynamic variables having the same value in all the phases in equilibrium. If this is not true, as happens in the Poynting effect, we must add the new variable acting on the system, and not just two (p and T).

Problems

Problem 1

What is the value of the boiling temperature of water at the Aconcagua summit (7,000 m) where the temperature is 273.2 K? The average molar mass of air is 29 g/mole and $\Delta_v H(\mathrm{H_2O}) = 40.8$ kJ/mole.

Hint: In order to derive the change of atmospheric pressure with the height, the gravitational work must be accounted for.

Answer: $T = 349.9$ K

Problem 2

Liquid NH_3 coexists with its vapor at 240 K. The densities of both phases are: $\rho(l) = 40.01$ mole/dm^3; $\rho(v) = 0.0527$ mole/dm^3. Calculate $(\partial U/(\partial V)_T$ for both phases of NH_3 using the van der Waals equation.

Data: a = 4.225 bar dm^6/mole^2; $b = 0.03713$ dm^3/mole

Answer: $(\partial U/\partial V)_T(\text{l}) = 6.763$ kbar $= 676.3$ MPa; $(\partial U/\partial V)_T(\text{v}) = 11.7$ mbar $= 1,170$ Pa

Problem 3

For a mixture of propanone (1) and trichloromethane (2) having the composition $x_2 = 0.559$ at 298.2 K, it is known that $V_1^* = 73.993$ cm^3/mole, $V_2^* = 80.665$ cm^3/mole, $V_1 = 74.072$ cm^3/mole, and $V_2 = 80.300$ cm^3/mole. Calculate:

a) The mixture's volume of that composition and having 2.00 moles of propanone

b) The change of the mixture's volume when mixing 2.00 moles of propanone with trichloromethane to obtain a composition $x_2 = 0.559$

Answer: a) $V = 352.5$ cm^3; b) $\Delta V = 0.80$ cm^3

Problem 4

A mixture of 1.2 moles of gaseous CO_2 and 95 moles of He are compressed isothermically until the pressure reaches 55 bar at 170.2 K. Considering it is a mixture of ideal gases, calculate how much CO_2 will separate as a solid.

Data: $p^*(CO_2, 170.2 \text{ K}) = 0.1284$ bar; $V^*(CO_2, 170.2 \text{ K}) = 27.2$ cm^3/mole

Answer: 0.953 moles of solid CO_2

Chapter 4

Gaseous Mixtures

4.1 The Chemical Potential

The chemical potential of one component i in a gaseous mixture depends on temperature, pressure, and composition. However, for the case of gaseous mixtures, the effect of pressure and composition may be combined and expressed using partial pressure, p_i, which is defined by $p_i = x_i p$. The change of μ_i with pressure is given by

$$\left(\frac{\partial \mu_i}{\partial p}\right)_{T,\mathbf{n}} = \left[\frac{\partial}{\partial p}\left(\frac{\partial G}{\partial n_i}\right)_{T,p,n_{j\neq i}}\right]_{T,\mathbf{n}} = \left[\frac{\partial}{\partial n_i}\left(\frac{\partial G}{\partial p}\right)_{T,\mathbf{n}}\right]_{T,p,n_{j\neq i}} = V_i \qquad (4.1)$$

where $\mathbf{n}$ indicates the set of the quantities of matter corresponding to each one of the substances forming the mixture. Since,

$$(\mathrm{d}p_i)_{\mathbf{n}} = x_i (\mathrm{d}p)_{\mathbf{n}}$$

in eqn. (4.1), we can replace the derivative with respect to p by the derivative with respect to p_i, leading to

$$V_i = \left(\frac{\partial \mu_i}{\partial p_i}\right)_{T,\mathbf{n}}\left(\frac{\partial p_i}{\partial p}\right)_{\mathbf{n}} = x_i \left(\frac{\partial \mu_i}{\partial p_i}\right)_{T,\mathbf{n}} \qquad (4.2)$$

Finally, for a mixture of ideal gases $V_i = RT/p$, and hence,

$$\left(\frac{\partial \mu_i}{\partial p_i}\right)_{T,\mathbf{n}} = \frac{RT}{x_i p} = \frac{RT}{p_i} \qquad (4.3)$$

In order to calculate the changes in the chemical potential of component i when there is a change in the mixture's composition, eqn.(4.3) must be integrated, leading to

$$\mu_i = \mu_i^{\ominus} + RT \ln \frac{p_i}{p^{\ominus}} \qquad (4.4)$$

where $\mu_i^{\ominus}$ is the integration constant, which in principle can depend on the variables which were held constant during the integration process, in this case T and the mixture's composition. The complete meaning of eqn. (4.4) will be analyzed in following sections.

For an ideal gas mixture, according to the Euler's integration method discussed in the previous chapter, G is given by

$$G = \sum_i n_i \mu_i = \sum_i n_i \left(\mu_i^{\ominus} + RT \ln \frac{p_i}{p^{\ominus}} \right) \qquad (4.5)$$

4.2   Meaning and Molecular Consequences of the Mixing Process

It is important to calculate the change in the thermodynamic properties referring to the mixing process, especially in the case of ideal gases mixtures, a process for which it is possible to eliminate all those contributions related to intermolecular interactions. Obviously, this is a paradigmatic case frequently employed when the concept of entropy is discussed in introductory courses of chemistry–the mixture of two ideal gases at constant T, V, and p is spontaneous but does not cause changes in the internal energy or enthalpy of the mixture. We want to elucidate the molecular features of the mixing process of two ideal gases; hence, it will be necessary to use mechanical statistics tools.

Using eqns. (2.5), (2.7), (2.8), and (2.9), it is possible to write the change of A that occurs when two ideal gases, occupying volumes V_1 and V_2 and having the number of molecules N_1 and N_2, respectively, are mixed at constant temperature and volume $V = V_1 + V_2$. Under these conditions, $\Delta A = \Delta G$. The partition function of the initial state corresponds to the two ideal separated gases, that is Q_{N_1} and Q_{N_2}, being

$$Q_{N_1} = \frac{[(q_{\text{int}})_1]^{N_1}}{N_1!} V_1^{N_1}$$

and an equivalent expression for Q_{N_2}, resulting from replacing the subindexes 1 by 2.

For the final state, where the two gases occupy the whole volume, the mixture's partition function $Q_{(N_1+N_2)}$ results,

$$Q_{(N_1+N_2)} = \frac{[(q_{\text{int}})_1]^{N_1}}{N_1!} \frac{[(q_{\text{int}})_2]^{N_2}}{N_2!} V^{(N_1+N_2)}$$

In order to write the partition function of the mixture, it has been taken into account that particles 1 and 2 are distinguishable. For example, one may consider two different chemical compounds that occupy the total volume $V = V_1 + V_2$. Using the relation between the Helmholtz energy and the partition functions, given by eqn. (2.5), it is possible to calculate the change in Helmholtz energy (per molecule) due to the mixing process, $\Delta_{\text{mix}}A = A_{(N_1+N_2)} - N_1 A_1 - N_2 A_2$, as

$$\Delta_{\text{mix}}A = -kT \ln \frac{Q_{(N_1+N_2)}}{Q_{N_1} Q_{N_2}} = -kT \ln \frac{V^{(N_1+N_2)}}{V_1^{N_1} V_2^{N_2}}$$

The mixing process of ideal gases, which is a spontaneous process not involving a change in the energy of the system, produces an increase in its entropy. This is due to the fact that in the mixture of the two gases, molecules of both species are able to occupy a larger geometric space. This means that the number of microstates corresponding to the thermodynamic state of the mixture is larger than when the gases were separated and, hence, it also increases the thermodynamic probability of the mixture state. This is in accordance to what was discussed when the Boltzmann equation (2.32) for entropy was presented.

Since we are dealing with ideal gases $V_i = x_i V$, we obtain

$$\Delta_{\text{mix}}A = N_1 kT \ln x_1 + N_2 kT \ln x_2 < 0 \tag{4.6}$$

and since $(\partial A/\partial T)_{V,\mathbf{x}} = -S$, we have

$$\Delta_{\text{mix}}S = -N_1 k \ln x_1 - N_2 k \ln x_2 > 0 \tag{4.7}$$

That is, $\Delta_{\text{mix}}A = -T\Delta_{\text{mix}}S$, verifying that the mixing process of ideal gases is clearly entropic. The mixing process of two ideal gases is depicted in figure 4.1.

What happens if components 1 and 2 of the mixture are the same substance? This is the Gibbs paradox, and even when it seems evident that there is no mixing when the two gases are of the same chemical substance, how can we explain the difference with the previous process? The best solution is to repeat the analysis made previously for the present example. The initial state is the same as in the previous example, N_1 particles of species 1 separated by an impermeable barrier from N_2 particles of species 2, even when 1 and 2 are the same substance. On the

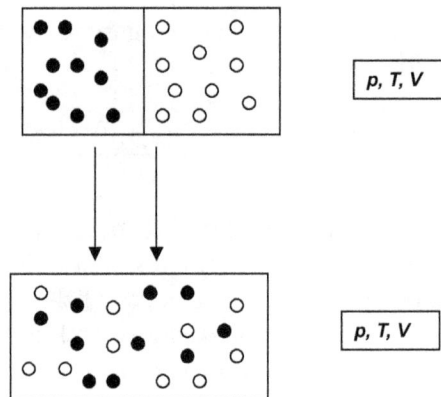

Figure 4.1: Mixing processes of two ideal gases. The space in which each particle is found is larger in the mixture, and this increases the thermodynamic probability of the mixture's corresponding state. The number of irrelevant microstates is the same when gases are apart.

other hand, it is also true that by removing the barrier, each molecule will be able to move over the total volume of the system inside the container where the mixing takes place; hence, the number of microstates corresponding to the thermodynamic state will also increase. However, now since particles 1 and 2 are indistinguishable, this makes that all the configurations exchanging states between particles do not always *lead to new microconfigurations*. The final state will have a much greater number of microstates, but these may be irrelevant because the two molecules are identical, and hence they do *not* produce an increase in the thermodynamic probability of the mixed state. Now, the partition function of the final state is

$$Q_{(N_1+N_2)} = Q_N = \frac{(q_{\text{int}})^N}{N!} \, V^N$$

Using the Stirling's approximation[1] to express $N!$, it may be shown that the increment in the number of microconfigurations due to the increase in volume is compensated exactly by the increment in the number of irrelevant microconfigurations.

Eqn. (2.5) may be generalized for mixtures of C components occupying volume

---

[1]When N is very large, $N!$ may be approximated by $N^N \mathrm{e}^{-N}$.

V, leading to

$$\beta A = -\ln \left[\prod_{i=1}^{C} \frac{[(q_{\text{int}})_i]^{N_i}}{N_i!} V^N \right] \qquad (4.8)$$

Differentiating the previous expression with respect to N_i at constant $(T, V, N_{j \neq i})$, we have for the chemical potential per molecule of the i component,

$$\beta \mu_i = -\ln \left(\frac{\text{e} \, (q_{\text{int}})_i}{N_i} \right) + 1 - \ln V \qquad (4.9)$$

where e is the base of natural logarithms. This equation may be written as

$$\beta \mu_i = -\ln(q_{\text{int}})_i - \ln \frac{V}{N_i} \qquad (4.10)$$

which for ideal gas mixtures becomes

$$\mu_i = -kT \ln(q_{\text{int}})_i + \ln p_i \qquad (4.11)$$

This equation yields the chemical potential *per molecule*, and if it is multiplied by Avogadro's number it can be compared with eqn. (4.4) that refers to one mole of the component i. Thus,

$$\mu_i^{\ominus}(T) = -RT \ln \left(\frac{(q_{\text{int}})_i}{p^{\ominus}} \right) \equiv f(T) \qquad (4.12)$$

So, we verify that $\mu_i^{\ominus}$ only depends on temperature; this result can only be obtained through statistical mechanics. Eqn. (4.12) also provides the possibility of calculating $\mu_i^{\ominus}(T)$ by using the molecular properties of species i. This is an example of the complementary role of phenomenological thermodynamics and statistical mechanics, as discussed within the third principle of thermodynamics (section 2.12).

4.3   Mixture of Real Gases: Fugacity and Standard State

The usual way to describe the behavior of a non-ideal system consists in keeping the mathematical form of the equations used to describe ideal systems and calculate the deviations of the real systems from the ideal ones. Thus, for mixtures of real gases, the chemical potential of each component is described with an expression similar to the differential equation (4.3), which defines a new variable that replaces the partial pressure,

$$\text{d}\mu_i = RT \text{d} \ln f_i \qquad (4.13)$$

where the quantity f_i is known as the fugacity of component i, which replaces the partial pressure of that component when the system is non-ideal. The integrated form of the previous equation is,

$$\mu_i(p, T, \mathbf{x}) = \mu_i^\ominus(T) + RT \ln \frac{f_i(p, T, \mathbf{x})}{p^\ominus} \tag{4.14}$$

Once again, the symbol $\mathbf{x}$ is used to designate the set of mole fractions. Since it is not possible to determine energies in an absolute way, only an energy difference can be calculated: The integration has been carried out between the standard pressure, $p^\ominus$, and the pressure p of the system. But, eqn. (4.14) has one additional condition: That at pressure $p^\ominus$ the system must behave as an ideal one. This means that the integration process, depicted in figure 4.2,[2] consists in starting with pressure p, and decreasing the pressure until the mixture behaves ideally; then the integration proceeds *by the ideal curve* until reaching the value $p_i^\ominus$. That is, the integration is carried out between the state of the system and a state at the standard pressure which belongs to ideal behavior. The standard pressure, in principle, can have any value, but the usual convention is 0.1 MPa = 1 bar. As a pressure unit, the *atmosphere* (noted as atm) is not part of the SI units, and should be abandoned; however, it is useful to remember how to convert atm in the accepted units: 1 atm = 1.01325 bar = 101.325 kPa.

In order to carry out the process detailed in the previous paragraph, it is important to take into account that it is necessary to have a law for ideality; hence, one should know the state at which the real system behaves as ideal, denoted as the reference state. For gaseous mixtures, the reference state corresponds to a very low pressure, so that the behavior of the gases is properly represented by the state equation for ideal gases. The quantity $\mu_i^\ominus(T)$ is the standard chemical potential of the component i at the pressure of $p^\ominus$; this refers to the state $\ominus$ where the gas i shows ideal behavior at pressure $p^\ominus$. At pressure $p^\ominus$, real gases do not necessarily behave as ideal ones, and hence the standard state is a hypothetical state of the system that has no real existence.

Following the same methodological scheme, gas fugacity can be written as the partial pressure of the gas multiplied by a correction factor, which takes into account the non-ideality of the gas behavior. Thus,

$$f_i = \Phi_i p_i \tag{4.15}$$

where the factor Φ_i is called the fugacity coefficient. We know that if the pressure

---

[2]In figure 4.2, a general concentration unit C_i has been used as well as a standard concentration $C_i^\ominus$; for gases C_i becomes p_i and $C_i^\ominus$ becomes $p_i^\ominus$.

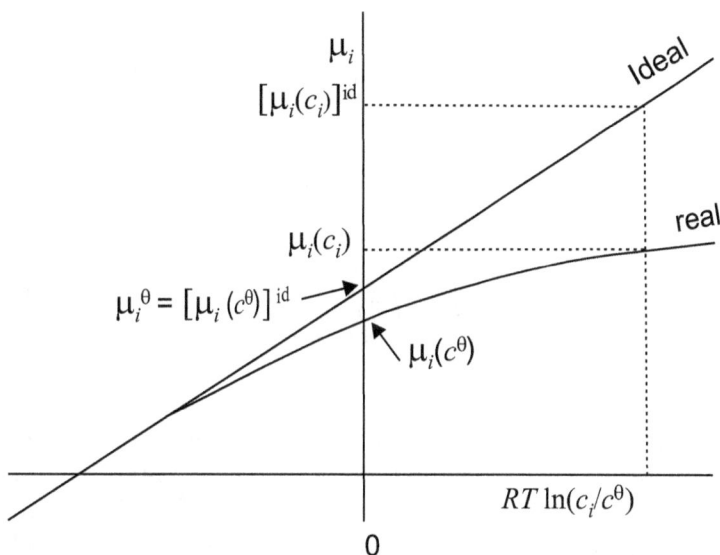

Figure 4.2: Chemical potential of a component as a function of the concentration in the real and the ideal systems. The supraindex θ denotes standard state.

is low, $\Phi_i = 1$, because in that condition the gas behaves ideally. From the previous argument, a correct thermodynamic definition for $\mu_i^{\ominus}(T)$ is

$$\mu_i^{\ominus}(T) = \lim_{p \to 0} \left[\mu_i(p, T, \mathbf{x}) - RT \ln \frac{p_i}{p^{\ominus}} \right] \tag{4.16}$$

It is important to see how the fugacity coefficient of each component of a mixture of real gases may be calculated. The calculation is carried out for constant temperature and composition of the mixture using eqn. (4.1) and (4.13); finally, we obtain

$$\mathrm{d} \ln f_i = \frac{V_i}{RT} \mathrm{d}p \tag{4.17}$$

and subtracting $\mathrm{d} \ln p$ to both members of the previous equation, it leads to

$$\mathrm{d} \ln \frac{f_i}{p} = \left(\frac{V_i}{RT} - \frac{1}{p} \right) \mathrm{d}p \tag{4.18}$$

The integration is made starting at very low pressure ($p \to 0$), such that the gas behaves ideally, and ending at the system pressure p. Since at very low pressures $f_i = p_i$, we have

$$\ln \frac{f_i}{p} - \ln \left(\frac{p_i}{p} \right)_{p \to 0} = \int_{p \to 0}^{p} \left(\frac{V_i}{RT} - \frac{1}{p} \right) \mathrm{d}p \tag{4.19}$$

Now, considering that $(p_i/p) = x_i$ remain constant in the integration,

$$\ln \Phi_i \equiv \ln \frac{f_i}{p_i} = \int_{p\to0}^{p} \left(\frac{V_i}{RT} - \frac{1}{p} \right) dp \qquad (4.20)$$

For a pure gas we will have,

$$\ln \Phi_i^* \equiv \ln \frac{f}{p} = \int_{p\to0}^{p} \left(\frac{V_i^*}{RT} - \frac{1}{p} \right) dp = \int_{p\to0}^{p} \left(\frac{Z^* - 1}{p} \right) dp \qquad (4.21)$$

where V_i^* is the molar volume of the pure component i and $Z^* = (pV_i^*/RT)$ is its compressibility factor. To calculate the fugacity coefficient of pure gases, it is possible to use the principle of corresponding states (cf. chapter 2) and use general graphs that give the values of the fugacity coefficients as a function of the reduced temperature $\theta = (T/T_c)$ and the reduced pressure $\pi = (p/p_c)$; the subindex c indicates the critical values of the corresponding variables.

Frequently, the chemical potential of a liquid is expressed in terms of the fugacity. According to eqn. (4.20), its calculation implies an integration from a very low pressure, $p \to 0$, where the substance will be in the gaseous phase, up to the pressure p, where it will be a liquid. In spite of the change of phase that takes place when the pressure increases isothermally, there are no conceptual difficulties to carry out the integration using an adequate equation of state, because the fugacity of the vapor and that of the liquid in coexistence are the same (i.e., $\mu_i(\text{v}) = \mu_i(\text{l})$ at equilibrium). Once a cubic equation is chosen for integration (e.g., the van der Waals equation), using the same arguments discussed in the case of the Maxwell construction (eqns. (3.15) and (3.16)), one can take advantage of the fact that the integral in the region of phase coexistence is equal to zero.

4.4 Second Virial Coefficient and Real Gaseous Mixtures

In chapter 2 we make use of an equation of state, that takes into account the deviations from ideality by incorporating, progressively, the interactions of two, three, etc. molecules, as the density of the fluid increases. This is the virial equation of state, which is clearly related with the intermolecular interactions, as shown for the second virial coefficient in section 2.9. The general expression for the compressibility factor according to the virial equation is,

$$\frac{pV_m}{RT} = 1 + \frac{B(T)}{V_m} + \frac{C(T)}{V_m^2} + \frac{D(T)}{V_m^3} + \dots \qquad (4.22)$$

where V_m is the molar volume of the gas. The coefficients B, C, D... are respectively the second, third, fourth, etc. virial coefficients, which only depend on temperature. This expression is very useful for gases because it allows to explain easily the deviations from ideality on the basis of intermolecular interactions. Eqn. (4.22) cannot be applied when the density of the fluid is very large; for such a situation, it is better to use cubic equations of state originated in the van der Waals model. We will use eqn. (4.22) only as a guide, indicating what is the limiting behavior of real systems. To do this, it is only necessary to analyze the first deviation from ideality (i.e., the term that contains the second virial coefficient). So then,

$$\frac{pV_m}{RT} = 1 + \frac{B(T)}{V_m} \simeq 1 + \frac{B(T)p}{RT} \tag{4.23}$$

The last term on the right is expressed as a function of the pressure, by replacing the volume with the ideal gas formula.

As already discussed in chapter 1, an equation of state must be formally identical when describing the behavior of mixtures or that of pure compounds, hence eqns. (4.22) and (4.23) are valid in both cases. For mixtures, the virial coefficients will have a dependence on the mixture composition. For instance, the second virial coefficient in a binary mixture, B_{mix}, results in

$$B_{\mathrm{mix}} = x_1^2 B_{11} + 2x_1 x_2 B_{12} + x_2^2 B_{22} \tag{4.24}$$

This indicates that in a binary mixture the interactions between molecular pairs correspond to two particles of compound 1, two particles of compound 2, and the pair formed by one molecules of particle 1 and a second one of type 2. The mole fractions that multiply each of the second virial coefficients in eqn. (4.24) represent the probability that an encounter between the two particles indicated by the subindexes, on the basis that the probability for this to happen is independent of the nature of the first chosen particle. This implies that the distribution of molecules in the gaseous system is completely random. A quantity that is interesting to define is

$$\Delta_{12} = 2B_{12} - B_{11} - B_{22} \tag{4.25}$$

With this definition, eqn. (4.24) results in

$$B_{\mathrm{mix}} = x_1 B_{11} + x_2 B_{22} + x_1 x_2 \Delta_{12} \tag{4.26}$$

For a multicomponent system, the variation of Gibbs energy with pressure, at fixed temperature and composition, is expressed by $dG = V dp$, which is the same expression as for a pure compound. If the gaseous mixture obeys the equation of state (4.23) we get

$$\mathrm{d}G(p, T, \mathbf{n}) = \sum_i n_i \left(\frac{RT}{p} + B_{\mathrm{mix}} \right) \mathrm{d}p \qquad (4.27)$$

Integrating between a low pressure $p \to 0$, where the system behaves ideally, and the pressure p, we obtain

$$G(p, T, \mathbf{n}) - G(p \to 0, T, \mathbf{n}) = \sum_i n_i \left[RT \ln \left(\frac{p}{p \to 0} \right) + pB_{\mathrm{mix}} \right] \qquad (4.28)$$

In this expression it has been considered that $p \to 0$ may be taken as $p \simeq 0$. The term $G(p \to 0, T, \mathbf{n})$, which refers to ideal gases, is calculated with eqn. (4.5) and, in the particular case of a binary system, we arrive to

$$\begin{aligned} G(p, T, \mathbf{n}) &= n_1 \mu_1^{\ominus} + n_2 \mu_2^{\ominus} + \qquad (4.29) \\ &\quad + (n_1 + n_2) RT \ln \frac{p}{p^{\ominus}} + n_1 RT \ln x_1 + \\ &\quad + n_2 RT \ln x_2 + p \left(n_1 B_{11} + n_2 B_{22} + \frac{n_1 n_2}{n_1 + n_2} \Delta_{12} \right) \end{aligned}$$

where B_{mix} has been replaced using eqn. (4.26).

Using this expression, it is possible to calculate the chemical potentials of both components. For component 1,

$$\mu_1 \equiv \left(\frac{\partial G}{\partial n_1} \right)_{p, T, n_2} = \mu_1^{\ominus} + RT \ln \frac{p_1}{p^{\ominus}} + p(B_{11} + x_2^2 \Delta_{12}) \qquad (4.30)$$

and similarly for component 2, changing, in the previous equation, the subindexes where it corresponds. If the gaseous system only contains component 1 (pure gas) at the same p and T as the mixture,

$$\frac{G}{n_1} = \mu_1^{\ominus} + RT \ln \frac{p}{p^{\ominus}} + pB_{11} \qquad (4.31)$$

We observe that if $\Delta_{12} = 0$ in eqn. (4.30), gas 1 in the mixture deviates from ideality as if it were pure, at the same total pressure and temperature than the mixture. In the following section we will come back to these expressions.

4.5 Mixture Quantities and Excess Functions

The strategy normally used to express the properties of mixtures consists in thinking about them as terms corresponding to pure components, including some correction term to express the difference in interactions that occur in the mixing process and are absent in the pure component systems (i.e., the interactions between the various components of the mixture). If the properties of a mixture can be fully expressed on the basis of the properties of the pure compounds, we say *the mixture is ideal*. Thus, for every extensive quantity X, it is possible to define $\Delta_{\mathrm{mix}}X$ as the respective quantity for the mixture,

$$\Delta_{\mathrm{mix}}X = X(p,T,\mathbf{n}) - \sum_i n_i X_i^* \tag{4.32}$$

which is denoted as $\Delta_{\mathrm{mix}}X^{\mathrm{id}}$ for an ideal mixture. The corresponding expressions of the properties A and S for ideal mixtures have already been given by eqns. (4.6) and (4.7). When mixing is not ideal, we define an excess quantity X^{ex} as the difference between the property value in the real mixture and that in the ideal mixture;

$$\Delta_{\mathrm{mix}}X = \Delta_{\mathrm{mix}}X^{\mathrm{id}} + X^{\mathrm{ex}} \tag{4.33}$$

For instance, the free energy of mixture $\Delta_{\mathrm{mix}}G$, for a binary mixture of real gases, at constant p and T, is given by

$$\Delta_{\mathrm{mix}}G = n_1\left[\mu_1^\ominus(T) + RT\ln\frac{f_1}{p^\ominus}\right] + n_2\left[\mu_2^\ominus(T) + RT\ln\frac{f_2}{p^\ominus}\right] - \tag{4.34}$$

$$-n_1\left[\mu_1^\ominus(T) + RT\ln\frac{f_1^*}{p^\ominus}\right] - n_2\left[\mu_2^\ominus(T) + RT\ln\frac{f_2^*}{p^\ominus}\right]$$

where the supraindex* indicates a pure substance. Then we have

$$\Delta_{\mathrm{mix}}G = n_1 RT\ln\frac{f_1}{f_1^*} + n_2 RT\ln\frac{f_2}{f_2^*} \tag{4.35}$$

and, using the fugacity coefficients that express the differences between real and ideal behaviors, and the definition of partial pressure, we have

$$\Delta_{\mathrm{mix}}G = n_1 RT\ln\frac{\Phi_1}{\Phi_1^*} + n_2 RT\ln\frac{\Phi_2}{\Phi_2^*} + n_1 RT\ln x_1 + n_2 RT\ln x_2 \tag{4.36}$$

As in eqn. (4.6)

$$\Delta_{\mathrm{mix}}G^{\mathrm{id}} = n_1 RT\ln x_1 + n_2 RT\ln x_2 \tag{4.37}$$

and hence, the excess property becomes

$$G^{\text{ex}} = n_1 RT \ln \frac{\Phi_1}{\Phi_1^*} + n_2 RT \ln \frac{\Phi_2}{\Phi_2^*} \tag{4.38}$$

where all the deviations from ideality are contained in the fugacity coefficients.

The thermodynamic quantities that do not contain the ideal mixing term (i.e., those that have an entropic origin), like U, H, and V, exhibit $\Delta_{\text{mix}} X^{\text{id}} = 0$ and $\Delta_{\text{mix}} X = X^{\text{ex}}$.

Enthalpies of Gaseous Mixtures and Determination of Second Virial Coefficients

It is possible to use calorimetry to determine the second virial coefficents of gaseous mixtures and, consequently, obtain values for the cross coefficient B_{12}, which yields information about the interaction between the two components.

It requires a differential flow calorimeter. The two gases that form the mixture enter into the first calorimeter and they mix inside, so it is possible to measure the enthalpy of mixture from the amount of heat given (or taken out) to keep the temperature constant inside the calorimeter. The measurement method is a differential one because it makes use of a second calorimeter to eliminate all effects due to small differences of pressure between the two gases, which would produce thermal perturbations (Joule-Thomson effect).

In this way, it is possible to obtain $\Delta_{\text{mix}} H(p, T, x)$ at different pressures, and have access to $H^{\text{ex}}(p, T, x)$. These values are used to calculate the excess enthalpy of the mixture of the two gases, given by

$$H^{\text{ex}}(p, T, x) = H(p, T, x) - (1 - x)H_1^*(p, T) - xH_2^*(p, T)$$

remembering that $\Delta_{\text{mix}} H^{\text{id}} = 0$.

The virial equation up to the second coefficients may be used as an equation of state if pressure is not too high; then, we'll have

$$\frac{pV}{RT} = 1 + \frac{B}{V} \approx 1 + \frac{Bp}{RT} - \left(\frac{Bp}{RT} \right)^2$$

Using eqn. (1.17), it is possible to relate $H(p, T, x)$ with the equation of state

$$H(p, T, x) = \int_{p \to 0}^{p} \left[V - T \left(\frac{\partial V}{\partial T} \right)_{p,y} \right] \mathrm{d}p$$

since gases at $p \to 0$ behave ideally and H is pressure independent. With this integrated expression, and using the state equation up to the second virial coefficient and its temperature variation, the following equation is obtained,

$$
\begin{aligned}
H^{\mathrm{ex}}(p, T, x) &= x(1-x)\, p\,(2\phi_{12} - \phi_{11} - \phi_{22}) - \\
&\quad - \frac{p^2}{RT} \left[B_{\mathrm{mix}}\phi_{\mathrm{mix}} - (1-x)B_{11}\phi_{11} - xB_{22}\phi_{22} \right]
\end{aligned}
$$

where $\phi = B - T(\mathrm{d}B/\mathrm{d}T)$ is the isothermal Joule-Thomson coefficient, B_{mix} is given by eqn. (4.24) and, in a similar manner, it is possible to express ϕ_{mix}.

This experimental procedure to evaluate the cross interactions in gaseous mixtures of HCl and H_2O, at $p = 0.1$ MPa and a composition $x = 1 - x = 0.50$, was used by Wormald[3]. This example has been selected to show how to treat the second virial coefficients in gases where interactions exist beyond those of Lennard-Jones.

In this case, the pure components, as well as the mixtures, exhibit dipolar interaction and can form hydrogen bonds besides dispersion forces. The intermolecular interactions due to their molecular permanent dipolar moments can be taken into account of by means of the Stockmeyer equation for the potential energy of two dipolar molecules,

$$u_{ij}(r_{ij}, \omega_{ij}) = u_{ij}^{\mathrm{LJ}} + \frac{\mu_i \mu_j}{r_{ij}^3} f(\omega_{ij})$$

where u_{ij}^{LJ} is the contribution of the dispersion forces given by the Lennard-Jones equation (eqn. (2.23)), and $f(\omega_{ij})$ is a function of the relative angular orientation of both molecular dipoles μ_i and μ_j.

In order to consider the existence of hydrogen bonds between the two different molecules, a quasichemical model can be used to study the

---

[3]C. J. Wormald, "Water-Hydrogen Chloride Association. Second Virial Cross Coefficients for Water-Hydrogen Chloride from Gas Phase Excess Enthalpy Measurements", *J. Chem. Thermodynamics*, 35 (2003): 417-431.

mixture's behavior. This model presupposes the existence of a chemical equilibrium of the type

$$X - H + Y - H \rightleftharpoons X - H \cdots Y - H$$

involving the two atomic groups acting as hydrogen donors (X) and hydrogen acceptors (Y), with a constant K_{ij}. The second virial coefficient is then given by

$$B_{ij} = B_{ij}^{\text{nc}} - \frac{RTK_{ij}}{a}$$

and ϕ_{ij} can now be written as

$$\phi_{ij} = \phi_{ij}^{\text{nc}} + \frac{K_{ij}\,\Delta_{\text{hb}}H}{a}$$

where B_{ij}^{nc} and ϕ_{ij}^{nc} are the contributions not involving hydrogen bonds, as given by the Stockmayer intermolecular potential, and $\Delta_{\text{hb}}H_{ij}$ is the formation enthalpy of hydrogen bonds between the molecules i and j. When the interactions i-j are between molecules of the same component $a = 1$; instead, if it is a cross interaction, $a = 2$.

This experimental determination lets us obtain the molecular parameters characterizing the non-chemical cross interaction, the equilibrium constant K_{ij} and $\Delta_{\text{hb}}H_{ij}$, as illustrated in the following table:

Table 4.1: Properties of equimolar gaseous mixtures of HCl(1)–H$_2$O(2), at 383 K

i–j	B_{ij} (cm^3/mol)	Φ_{ij} (cm^3/mol)	K_{ij} (MPa^{-1})	$\Delta_{\text{hb}}H_{ij}$ (kJ/mol)
1–1	-83	-252	0.026	-8.36
2–2	-409	-1938	0.36	-16.2
1–2	-252	-1043	0.42	-13.3

4.6   Ideal and Real Gaseous Mixtures

According to eqn. (4.38), the mixing process will be ideal if $\Phi_i(p, T, x_i) = \Phi_i^*(p, T)$. This condition implies that

$$f_i(p, T, x_i) = x_i f_i^*(p, T) \tag{4.39}$$

This relation was proposed in 1923 by G. N. Lewis and M. Randall as basis for the calculation of fugacities in gaseous mixtures using the properties of the pure components. It is known as the Lewis and Randall rule and it relates the deviation from ideality of a gas in the mixture with the deviation of the pure component, at the same temperature and total pressure, under the assumption that the mixture is ideal. As already mentioned on page 114, generalized plots of fugacity coefficients are frequently used, employing the reduced pressure and reduced temperature to estimate the deviations from ideality of the pure gases. This is based on the principle of corresponding states discussed in chapter 2. According to the Lewis and Randall rule, the same generalized diagram could be used to calculate the fugacity coefficients of the components of a gaseous mixture. In order to use the generalized diagram of fugacities, it is necessary to restrict its use to those gases that satisfy the corresponding states principle, for instance non-polar gases.

Since eqns. (4.20) and (4.21) allow a rigorous determination of the fugacity coefficients in a gaseous mixture and for the pure components, respectively, the rule of Lewis and Randall implies that $V_i(p, T, x_i) = V_i^*(p, T)$ (i.e., the partial molar volume of the gas in the mixture is equal to the molar volume of the pure gas at the same p and T). This is Amagat's law, which is obviously valid for ideal gases and also for gases that comply with the second virial coefficient if $\Delta_{12} = 0$. An examples of the use of Lewis and Randall rule referring to a mixture of n-butane with nitrogen is illustrated in figure 4.3, which plots the compressibility factor Z of the binary mixture as a function of its composition, at different pressures and constant temperature. It may be observed that at high pressures Z varies linearly with composition, and this implies that the relation $V_i(p, T, x_i) = V_i^*(p, T)$ applies in these p and T ranges. The fact that Amagat's law applies despite the gaseous mixture is far from ideality and may be explained by taking into account that at very high pressure molecules are very close and, at those distances, the intermolecular potentials are essentially repulsive. Hence, the deviation from ideality is mainly due to the volume excluded by the molecules, which are very close one to the other. The concentration change may therefore be thought as the change of molecules of a given size by molecules having a different size. As a result, the volume will change linearly with the composition, and Amagat's law will apply. Albeit, it *does not* imply that the Lewis and Randall rule will also apply because, according to eqns. (4.20) and (4.21), for the validity of the equality $\Phi_i = \Phi_i^*$ it is necessary that $V_i(p, T, x_i) = V_i^*(p, T, x_i)$ *over all the range of pressures*–from p to $p \to 0$. This is not observed in figure 4.3, where at intermediate and low pressures the linearity of Z with the composition does not hold.

One concludes that Lewis and Randall's rule can only be applied to systems with small deviations of ideality, where it is possible to apply the second virial coefficient

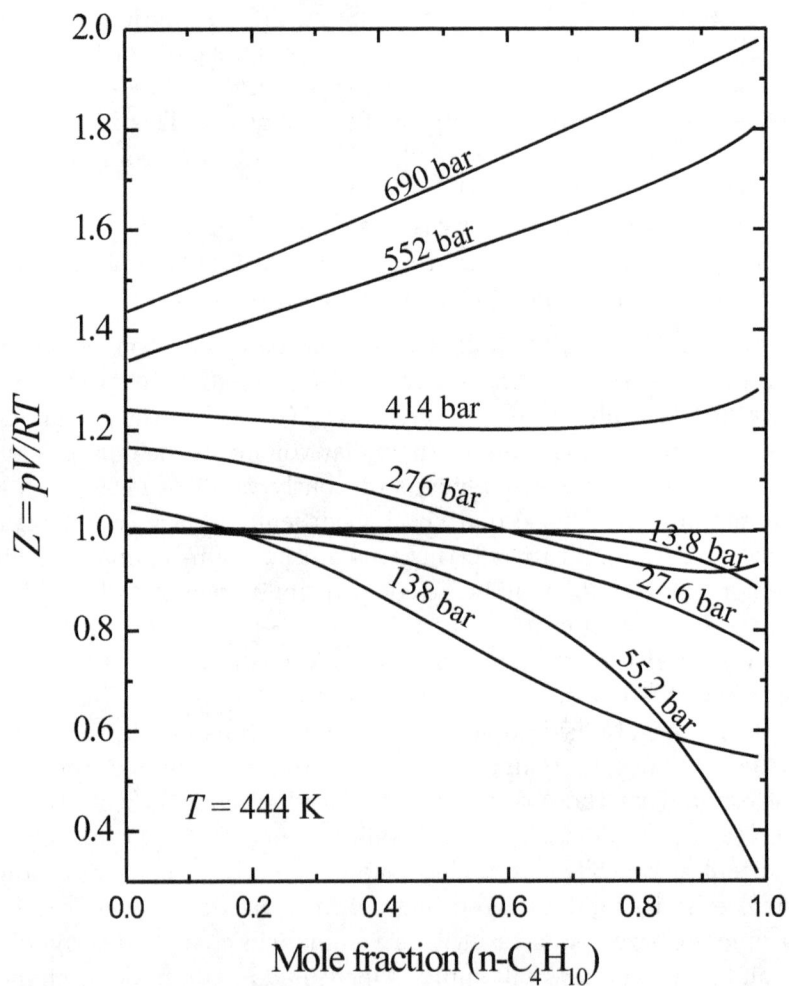

Figure 4.3: Compressibility factor Z of the mixture N_2-nC_4H_{10} as a function of composition, at different pressures and $T = 444$ K.

equation and $\Delta_{12} = 0$. This condition can be interpreted, according to eqn. (4.25), as an indication that the excess of 1-2 interactions, given by Mayer's function (chapter 2), is the arithmetic mean of the excess interactions between 1-1 and between 2-2. Albeit, there are only a few systems with negligible Δ_{12} values (see table 4.2).

Table 4.2: Second virial coefficients and the Lewis and Randall rule

Mixture	T (K)	B_{12}	B_{11}	B_{22}	Δ_{12}
		(cm^3/mol)			
Ar-Xe	298	-43	-15.8	-130.2	60
CH_4-C_3H_8	298	-139	-42	-399	163
N_2-CO_2	303	-41.4	-4.0	-119.2	40.4
N_2-C_4H_{10}	323	-69.0	-0.30	-630	492
C_2H_6-C_8H_{18}	403	-354	-85	-1641	1018
C_3H_8-C_8H_{18}	403	-450	-205	-1641	946
C_4H_{10}-C_8H_{18}	403	-735	-360	-1641	531
C_5H_{12}-C_8H_{18}	403	-953	-570	-1641	305
C_6H_{14}-C_8H_{18}	403	-1173	-838	-1641	133

As already discussed, Lewis and Randall's rule has a limited range of applicability and its general use is quite questionable. The fugacity calculation is mainly performed on the basis of eqn. (4.20), by use of a proper equation of state, generally one of the cubic equations. In order to calculate the fugacity coefficients, it is convenient to have an alternative equation to (4.20), where the variable of integration is V, rather than p. This can be obtained in a similar way to eqn. (4.1), but starting with the Helmholtz energy instead of using Gibbs's. The final result is

$$RT \ln \Phi_i = \int_V^\infty \left[\left(\frac{\partial p}{\partial n_i} \right)_{T,V,n_{j \neq i}} - \frac{RT}{V} \right] dV - RT \ln Z \qquad (4.40)$$

Cubic equations of state have two parameters, depending on the molecular nature of the gas being considered, like the van der Waals equation, which are related to the parameters that describe the intermolecular potential. Nonetheless, only in the simple van der Waals equation, the relation linking them is rather trivial. This equation is useful to illustrate some aspects of the use of equations of state to calculate fugacities. The van der Waals equation of state, see eqn. (3.14), has two parameters, a and b. The first one is related to the attractive part of the intermolecular potential, in particular the depth of the minimum intermolecular energy curve vs. the intermolecular distance of two molecules, given by eqn. (2.23). The parameter b, frequently denoted as co-volume, is directly related to the molecular diameter. However, it is clear that for polar molecules, or those participating in hydrogen bonding, the relation has to be more complex. When a cubic equation is used for gaseous mixtures, it is necessary to have combination rules allowing us the decomposition of the parameters according to the contributions of those corresponding to the pure components i and j. Thus, the following expressions can be written for the parameters a and b of the van der Waals equation of state in gaseous mixtures:

$$b = \sum_i x_i b_{ii}$$

$$a = \sum_{i,j} x_i x_j a_{ij}$$

as well as the combination rules for a_{ij} and b_{ij}

$$a_{ij} = (1 - \xi_{ij})(a_{ii}a_{jj})^{1/2}$$

$$b_{ij} = (1 - \eta_{ij})\frac{(b_{ii}b_{jj})}{2}$$

The simplest criterion consists in assuming that ξ_{ij} and η_{ij} are zero. In this case, there is no need to have any other information about the mixture, and it is enough with the knowledge of the values of a and b for the pure components. A weakness of this procedure is that small changes in the values of the mixture's parameters ξ_{ij} and η_{ij}, strongly affect the values obtained for the calculated properties.

Problems

Problem 1

Employing the state equation with the second virial coefficients, and using the values reported in table 4.2, calculate the compressibility factor $Z = pV/RT$ for a mixture of ethane with n-octane, at $T = 403$ K and $p = 1$ bar, when the mole fraction of ethane is $x = 0.30$. Describe the mixture considering that a) the rule of Lewis and Randall is valid and b) the mixture does not comply with the rule of Lewis and Randall.

Answer: a) $Z = 1.037$; b) $Z = 1.031$

Problem 2

Calculate the fugacity coefficients of the two compounds in a gaseous mixture of HCl and H_2O, having $x_{HCl} = 0.35$ at 1 bar of total pressure and $T = 383$ K. Use the second virial coefficients equation of state and the data reported in table 4.1.

Answer: $\Phi(HCl) = 0.997$; $\Phi(H_2O) = 0.987$

Problem 3

It is known that for gaseous ethane at $T = 500$ K the compressibility factor (for pressures in MPa) can be expressed by

$$Z = -0.0154p + 5.312 \times 10^{-4}\, p^2 + 2.10 \times 10^{-7}\, p^3$$

Calculate the fugacity coefficient of C_2H_6 at 10, 25, and 40 MPa.

Answer: 0.8797; 0.7946; 0.7899

Problem 4

a) Calculate the fugacity coefficients of pure H_2O and CH_3NO_2 at $T = 374$ K and $p = 1$ bar, using the state equation having second virial coefficients.

b) Discuss if the gases are in their standard states at that temperature.

Data: $B(H_2O, 374 \text{ K}) = -459.8 \text{ cm}^3/\text{mol}$; $B(CH_3NO_2, 374 \text{ K}) = -1500.3 \text{ cm}^3/\text{mol}$

Answer: a) $\Phi(H_2O, 374 \text{ K}) = 0.9853$; $\Phi(CH_3NO_2, 374 \text{ K}) = 0.9529$

Chapter 5

Mixtures in Condensed Phases and Their Equilibrium With the Vapor

5.1 Ideal Behavior

The effect of ideal mixing in gaseous systems, which was analyzed in chapter 4, also contributes to the mixing process between the components in condensed phases. When the condensed phases are liquids, the extension of our discussion of gaseous mixtures is straight forward. The behavior of solid mixtures will depend on the composition in a similar way to the dependence observed in the liquid mixtures. In solids, the molecules have a very low mobility and the interconversion between microconfigurations at temperatures far from the melting temperature is a much slower process than it is in fluids. Hence, it is possible to consider that in a solid the thermodynamic state is generated as a result of the *freezing* of microconfigurations that form a set of those describing the thermodynamic state in which the solid mixture is found. The relation between the temporal average of a property and its macroscopic equilibrium value (eqn. (2.1)) must be interpreted in this case as if throughout the macroscopic solid, molecular microconfigurations are observed that, on average represent those that are observed in the total system if the particles exchange their positions as they do in a fluid. On the other hand, this is the reason why in the solid state it is frequent to find systems in a metastable condition; once they have crystallized, the system's evolution toward the lowest Gibbs energy state is very slow.

In the case of liquid mixtures it is very frequent that they are found in equilibrium with the vapor; that is why it is convenient for the study of liquid mixtures under the conditions where the liquid and the vapor phases coexist. When mixing two substances, their chemical potentials will decrease when compared with the same pure substances at the same p and T (eqn. (1.24)); hence, the partial pressure of each component in the vapor mixture will be less than the vapor pressure of each pure compound, at the same temperature. The vapor pressure can be considered as the escaping tendency of the substance from a condensed phase to the vapor phase. If the substance is *diluted* by the presence of a second compound, one should expect that its escaping tendency in the mixture will be smaller. This will be valid whenever the average interactions are very similar in the mixture and in the pure component. This discussion leads us to formulate Raoult's law as the relation corresponding to the ideality law for liquid mixtures.

$$p_i = x_i p_i^* \tag{5.1}$$

Evidently, this ideality law is connected with the behavior of the pure liquid components, and the ideal behavior is attained when the mole fraction of one component tends to unity; at this point, ideal behavior occurs independently with the type of interactions existing in the mixture.

It is possible to define another ideality law. Let us assume that a small amount of methanol is added to a large volume of water; the interactions between the molecules of water and methanol are rather strong because the small amount of methanol molecules will form hydrogen bonds with the H_2O molecules and their dipoles will interact also with those of water molecules surrounding a CH_3OH molecule–frequently this process is referred as the hydration of methanol. Under these conditions the escaping tendency of methanol will be reduced compared with that existing in the pure alcohol, because solvation stabilizes the CH_3OH molecule in the aqueous solution. As the number of methanol molecules increases, it will only lead to an increase in the amount of hydrated alcohol in the aqueous solution, and its escaping tendency will increase proportionally. This picture applies until two CH_3OH molecules are sufficiently close to each other so their mutual interaction cannot be neglected. Therefore, at infinite dilution, a simple relation between p_i and the mole fraction of component i can be expected, and it is possible to write

$$p_i = x_i k_i^{\mathrm{H}} \tag{5.2}$$

where the constant k_i^{H} is known as Henry's constant, which depends on the nature of components in the mixture and on the temperature. Eqn. (5.2) is known as Henry's ideality law, which is valid in very diluted solutions.

In what follows we will refer to binary systems, and, in these cases, the mole fraction of component 2 in the mixture will be denoted by y in the vapor phase, and by x in the liquid phase.

For the case of completely miscible liquid mixtures, it is common to use Raoult's law as the ideality equation because both reference states may be approached easily to the two pure components. Using eqn. (5.1), we get for the total pressure over the binary mixture,

$$p = p_1^* + x(p_2^* - p_1^*) \tag{5.3}$$

as a function of the composition of the liquid phase, and

$$p = \frac{p_1^* p_2^*}{p_2^* - y(p_2^* - p_1^*)} \tag{5.4}$$

as a function of the mole fraction of component 2 in the vapor phase. Eqn. (5.3) yields a straight line while eqn. (5.4) is a hyperbolic curve.

Real systems present deviations from ideality, which can produce a total pressure greater (positive deviations) or lower (negative deviations) than those calculated with Raoult's law. Figure 5.1 illustrates a tridimensional diagram in the coordinates $(p, T, x/y)$ for a real mixture having positive deviations. Bidimensional planes are shown that illustrate isotherms–in clear grey–and isobars–in dark grey.

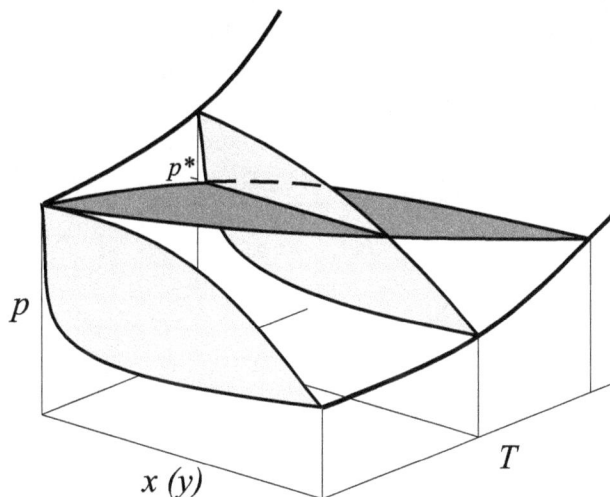

Figure 5.1: Liquid-vapor phase diagram. Scheme in $(p, T, x/y)$ for a typical binary system.

5.2   Chemical Potential in an Ideal Liquid Mixture

Whenever there is equilibrium between a liquid and its vapor phase, the chemical potential of the components must be the same in both phases. This thermodynamic condition establishes a set of useful relations by means of equations (5.1) and (5.2). For Raoult's ideality, we have.

$$\mu_i(\text{v}) = \mu_i^{\ominus}(T) + RT \ln \frac{p_i}{p^{\ominus}} = \mu_i^{\ominus}(T) + RT \ln \frac{x_i p_i^*}{p^{\ominus}} = \mu_i(\text{l}) \tag{5.5}$$

As a consequence, we obtain

$$\mu_i = \mu_i^*(T) + RT \ln x_i \tag{5.6}$$

and the standard chemical potential of component i is μ_i^*, when the ideal behavior is given by Raoult's law and then μ_i^* corresponds to the chemical potential of the pure liquid i because eqn. (5.6) indicates that $\mu_i = \mu_i^*$ when $x_i = 1$. The relation between this chemical potential and that of the gaseous component is

$$\mu_i^*(T) = \mu_i^{\ominus}(T) + RT \ln \frac{p_i^*}{p^{\ominus}} \tag{5.7}$$

If a pressure larger than the vapor pressure is applied to the condensed phase, the chemical potentials will increase due to the Poynting effect (cf. section 3.6). In this case, it is convenient to imagine the process by dividing it in two stages. First, the pure liquid component is taken from its vapor pressure p_i^* to the final pressure p, and then the two components are mixed together, that is,

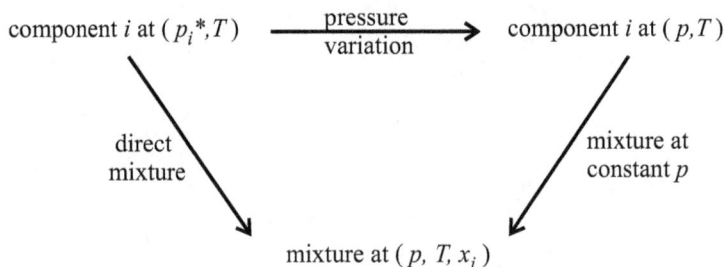

Hence,

$$\mu_i = \mu_i^*(p_i^*, T) + \int_{p_i^*}^{p} V_i^* \mathrm{d}p + RT \ln x_i \tag{5.8}$$

An analogous derivation is made when the ideality law is that of Henry, it yields,

$$\mu_i(\text{v}) = \mu_i^{\ominus}(T) + RT \ln \frac{p_i}{p^{\ominus}} = \mu_i^{\ominus}(T) + RT \ln \frac{x_i k_i^{\mathrm{H}}}{p^{\ominus}} = \mu_i(\text{l}) \tag{5.9}$$

$$\mu_i = \mu_i^{\infty}(T) + RT \ln x_i \tag{5.10}$$

and the standard chemical potential of component i is μ_i^∞ when using Henry's law as the ideal one; μ_i^∞ corresponds to the chemical potential of component i at infinite dilution in the other component. That is,

$$\mu_i^\infty(T) = \mu_i^\ominus(T) + RT \ln \frac{k_i^{\mathrm{H}}}{p^\ominus} \tag{5.11}$$

and the general equation that also accounts for the Poynting effect results

$$\mu_i = \mu_i^\infty(p_j^*, T) + \int_{p_j^*}^{p} V_i^\infty \mathrm{d}p + RT \ln x_i \tag{5.12}$$

In this case, the integral is taken from the vapor pressure of component j, because in the reference state the component i is infinitely diluted in the other component and, in the absence of an external pressure, p will equal the vapor pressure of j.

5.3   Mixtures of Real Liquids and Standard States

In order to describe the behavior of real liquid mixtures in equilibrium with their vapors, one employs also the strategy of using equations having the same form as those describing ideal mixtures and replacing the concentrations by activities, defining the activity coefficient. Since here we deal with systems having two phases in equilibrium, it is convenient to separate the deviations from ideality due to deviations in the gaseous mixture, from those due to non-ideal behavior of the condensed phase. Whenever it is presumed that the vapor mixture is not described correctly by the ideal gases law, it is convenient to replace pressures by fugacities, as described in chapter 4. Then, eqn. (5.1) becomes

$$f_i(p, T, y_i) = x_i f_i^*(p_i^*, T) \tag{5.13}$$

and Henry's equation(5.2) is now

$$f_i(p, T, y_i) = x_i k_i^{\mathrm{H}}(p_j^*, T) \tag{5.14}$$

Eqn. (5.13) should not be confused with the rule of Lewis and Randall (eqn. (4.39)) that is given by the following equation

$$f_i(p, T, y_i) = y_i f_i^*(p, T)$$

because in eqn. (5.13) the mole fraction corresponds to the liquid phase and $f_i^*(p_i^*, T)$ is the pure vapor's fugacity of i at its equilibrium vapor pressure. From now on, in order to describe the behavior of liquid mixtures we assume that the vapor in the mixture behaves ideally, and only in those cases where the vapor is clearly non-ideal, fugacities will be used instead of pressures.

In the next section, the molecular features of the liquid mixtures presenting deviations from ideality will be discussed. Nevertheless, we may say that deviations from ideal behavior will be observed in the liquid mixture whenever the molecular volumes of the components are not similar, or when the interaction energy between molecules of the two components is different to the average of the two interaction energies of the pure constituents. For non-ideal mixtures, equations with the same appearance to the ideal ones will be used, replacing the mole fractions by activities; i.e., by the concentrations multiplied by the activity coefficients

$$a_i = x_i \gamma_i$$

thus generating an activity scale. Since for liquid mixtures there are two different laws of ideality, we need to define two scales of activities, each having its reference state–pure component for Raoult's law and component at infinite dilution for Henry's law.

For real binary mixtures eqn. (5.1) transforms into

$$p_i = x_i \gamma_i^{\mathrm{R}} p_i^* \tag{5.15}$$

and the corresponding activity is

$$a_i^{\mathrm{R}} = x_i \gamma_i^{\mathrm{R}} \tag{5.16}$$

Hence, the general equation for the chemical potential goes from eqn. (5.8) to

$$\mu_i(p, T, x_i) = \mu_i^*(p_i^*, T) + \int_{p_i^*}^{p} V_i^* \mathrm{d}p + RT \ln a_i^{\mathrm{R}} \tag{5.17}$$

The standard state is defined by

$$\mu_i^*(p_i^*, T) = \lim_{x_i \to 1} \left[\mu_i(p, T, x_i) - RT \ln x_i - \int_{p_i^*}^{p} V_i^* \mathrm{d}p \right] \tag{5.18}$$

Analogously, for the case where the ideal law is that of Henry, using eqn. (5.2) one gets,

$$p_i = x_i \gamma_i^{\mathrm{H}} k_i^{\mathrm{H}} \tag{5.19}$$

$$a_i^{\mathrm{H}} = x_i \gamma_i^{\mathrm{H}} \tag{5.20}$$

$$\mu_i(p, T, x_i) = \mu_i^{\infty}(p_j^*, T) + \int_{p_j^*}^{p} V_i^{\infty} \mathrm{d}p + RT \ln a_i^{\mathrm{H}} \tag{5.21}$$

$$\mu_i^\infty(p_j^*, T) = \lim_{x_i \to 0} \left[\mu_i(p, T, x_i) - RT \ln x_i - \int_{p_j^*}^{p} V_i^\infty \mathrm{d}p \right] \tag{5.22}$$

In chapter 4, the standard chemical potential was introduced for the components of gaseous mixtures and it was emphasized that, in general, standard states are only *hypothetical states*. This is due to the fact that they imply ideal behavior under conditions that usually cannot be guaranteed. An exception to this is the standard state of the Raoult activity, because the pure component, either liquid or solid, is in a real state of the system. On the other hand, the standard state for the activities based on Henry's law is always hypothetical.

Generally, for mixtures in condensed phases,

$$\mu_i(p, T, \mathcal{C}_i) = \mu_i^\circ(p^*, T) + RT \ln \frac{a_i^\circ}{\mathcal{C}^\circ} + \int_{p^*}^{p} V_i^\circ \mathrm{d}p \tag{5.23}$$

$$\mu_i^\circ(p^*, T) = \lim_{\mathcal{C}_i \to \text{ideal}} \left[\mu_i(p, T, \mathcal{C}_i) - RT \ln \frac{\mathcal{C}_i}{\mathcal{C}^\circ} - \int_{p^*}^{p} V_i^\circ \mathrm{d}p \right] \tag{5.24}$$

In these expressions, $\mathcal{C}_i$ indicates a scale of concentrations for component i, the supraindex $\circ$ implies a standard state, and the reference state corresponds to $\mathcal{C}_i \to \mathcal{C}^\circ$ (the concentration where the system behaves ideally according to the law adopted for ideality). These expressions are completely general for mixtures in any of the three states of aggregation; however, in the case of gaseous mixtures, the last term (Poynting effect) must not be added.

For liquid mixtures, using Raoult's law for the ideal behavior, we have $\Delta_{\mathrm{mix}} X^{\mathrm{id}} = 0$ when $X = U$, H or V. If $X = S$, $\Delta_{\mathrm{mix}} X^{\mathrm{id}}$ has the value corresponding to the ideal mixture given by eqn. (4.7). When Henry's law describes ideal behavior, $\Delta_{\mathrm{mix}} X$ is constant for $X = U$, H or V, being equal to the difference of X between the reference state of the mixture–infinite dilution–and its value for the pure component.

It is important to be able to go from one scale of activity to the other. The calculation is simple and based on the fact that μ_i and p_i are independent physical quantities of the particular scale being used to express the activities. Using eqns. (5.15) and (5.19), we obtain,

$$p_i = x_i \gamma_i^{\mathrm{R}} p_i^* = x_i \gamma_i^{\mathrm{H}} k_i^{\mathrm{H}} \tag{5.25}$$

Hence,

$$\frac{\gamma_i^{\mathrm{R}}}{\gamma_i^{\mathrm{H}}} = \frac{k_i^{\mathrm{H}}}{p_i^*} \tag{5.26}$$

and for the difference of standard chemical potentials,

$$\mu_i^* - \mu_i^\infty = RT \ln \frac{p_i^*}{k_i^{\mathrm{H}}} \tag{5.27}$$

a result that can also be obtained with eqns. (5.7) and (5.11).

5.4   Lattice Model. Simple and Regular Solutions

In order to proceed with the habitual strategy, it is necessary to propose a model that allows relating the molecular features of the components in the liquid mixture with its macroscopic properties. The tool normally used for this purpose again arises from statistical thermodynamics. The model requires that the two components of the mixture do not deviate very much from ideal behavior, that is, that the cross interaction energy should not be very different from the average of the two interactions of the pure compounds and that the molar volume of the component be similar. This model is a simplification in that it requires that each molecule spends a longer time surrounded by its neighboring molecules, and although the molecules are able to diffuse from one position to another, its properties are determined by these restrictions. In this way, the lattice model (see figure 5.2) is formulated assuming the liquid mixture may be divided in cells or sites, one for each molecule, which is at all times surrounded by z neighbors interacting with it–this limits interactions to first neighbors only. In order to be able to calculate the mixture's properties, it is necessary to write its partition function $Q_{(N_1+N_2)}$,

$$Q_{(N_1+N_2)} = (q_{\text{int}})_1^{N_1} (q_{\text{int}})_2^{N_2} F \exp[-\beta(N_{11}w_{11} + N_{22}w_{22} + N_{12}w_{12})]$$

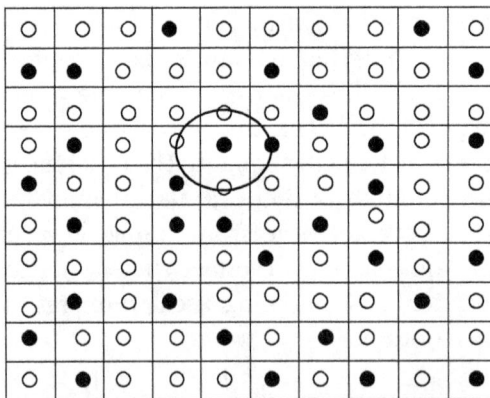

Figure 5.2: Lattice model for liquid mixtures. Each molecule occupies a lattice site, and is surrounded by z first neighbor molecules with which it interacts.

where $(q_{\text{int}})_i^{N_i}$ is the internal partition function of molecule i (cf. page 49), which only depends on temperature, and N_{ij} represents the number first-neighbor interactions between molecules i and j. Since each molecule occupies a single site, it is necessary that $V_1^* \approx V_2^*$. The F factor denotes the degeneration of the system's state (i.e., the number of distinguishable microconfigurations that correspond to the system's equilibrium state). F is given by

$$F = \frac{(N_1 + N_2)!}{N_1! N_2!}$$

In principle, the permutation of molecules leads to new configurations, but only those that permute molecules of different compounds are relevant. Finally, the argument of the exponential series expansion will depend on the number of first-neighbor interactions, N_{ij}, multiplied by the corresponding energy, w_{ij}. If each molecule has z (first) neighbors, then the total interactions with molecules of type 1 will be

$$zN_1 = 2N_{11} + N_{12}$$

and with molecules of type 2

$$zN_2 = 2N_{22} + N_{12}$$

It is convenient to define the quantity ω, which represents the difference between the cross interaction energy and the average of the interactions between the pure components

$$\omega = 2w_{12} - w_{11} - w_{22}$$

This is a measure of how large is the difference between the cross interaction in the mixture and the arithmetic mean of the interactions of the pure components–as already seen, ideality implies $\omega = 0$.

With these relations and definitions, the partition function of the binary mixture is

$$
\begin{aligned}
Q_{N_1 + N_2} &= (q_{\text{int}})_1^{N_1} (q_{\text{int}})_2^{N_2} \frac{(N_1 + N_2)!}{N_1! N_2!} \times \exp\left(-\frac{\beta z}{2} N_1 w_{11}\right) \times \\
&\quad \times \exp\left(-\frac{\beta z}{2} N_2 w_{22}\right) \times \exp\left(-\frac{\beta}{2} N_{12}\omega\right)
\end{aligned}
$$

Using eqn. (2.5) and Stirling's approximation for the factorial number of particles (cf. page 110), we get

$$
\begin{aligned}
A &= -kT N_1 \ln(q_{\text{int}})_1 + \frac{1}{2} z N_1 w_{11} - kT N_2 \ln(q_{\text{int}})_2 + \qquad (5.28) \\
&\quad + \frac{1}{2} z N_2 w_{22} + \frac{1}{2} N_{12}\omega - kT \left[N \ln N - N_1 \ln N_1 - N_2 \ln N_2\right]
\end{aligned}
$$

where $N = N_1 + N_2$. If the particle distribution in the mixture is *totally random*, x_i will be the probability that one molecule chosen randomly be of type i; hence, the fraction of contacts between molecules of type 1 with those of type 2 will be

$$\frac{N_{12}}{N} = 2x_1 x_2 z$$

Now, differentiating A with respect to N_1 at constant V, T, and N_2 (eqn. (5.28)), μ_1 per molecule can be calculated as

$$\mu_1 = -kT \ln \left[(q_{\text{int}})_1 \exp \left(-\frac{w_{11} z}{2kT} \right) \right] + z x_2^2 \omega + kT \ln x_1 \tag{5.29}$$

If the molar chemical potential is desired, the previous expression becomes,

$$\mu_1 = -RT \ln \left[(q_{\text{int}})_1 \exp \left(-\frac{N_A w_{11} z}{2RT} \right) \right] + z x_2^2 \omega N_A + RT \ln x_1 \tag{5.30}$$

where N_A is the Avogadro number.

If the previous equation is compared with eqn. (5.17), omitting the Poynting term, we have

$$\mu_1^* = -RT \ln \left[(q_{\text{int}})_1 \exp \left(-\frac{N_A w_{11} z}{2RT} \right) \right]$$

and

$$RT \ln \gamma_1^{\text{R}} = z x_2^2 \omega N_A \tag{5.31}$$

According to eqn. (5.22), it is possible to calculate the chemical potential at infinite dilution, μ_i^∞, from eqn. (5.30),

$$\mu_1^\infty = \lim_{x_1 \to 0} (\mu_1 - RT \ln x_1) = -RT \ln \left[(q_{\text{int}})_1 \exp \left(-\frac{N_A w_{11} z}{2RT} \right) \right] + z \omega N_A \tag{5.32}$$

Then we get

$$RT \ln \gamma_1^{\text{H}} = z (x_2^2 - 1) \omega N_A \tag{5.33}$$

According to eqn. (5.31), it is possible to write G^{ex} for a binary mixture on the basis of the lattice model; thus,

$$G^{\text{ex}} = RT (x_1 \ln \gamma_1^{\text{R}} + x_2 \ln \gamma_2^{\text{R}}) = B x_1 x_2 \tag{5.34}$$

For systems that obey eqn. (5.34), where B is equal to $z \omega N_A$, G^{ex} results a symmetric mathematical function in composition, having an extremum in $x_1 = x_2 = 0.5$. There are a number of binary mixtures that exhibit this simple behavior, which is due to the first order deviation from ideality according to Raoult, and in that sense, is equivalent to the equation of state for gaseous mixtures using the second virial coefficient correction. These types of liquid mixtures are denominated simple mixtures. If we consider that B strictly would not be temperature dependent,

$$S^{\mathrm{ex}} = - \left(\frac{\partial G^{\mathrm{ex}}}{\partial T} \right)_{p,\mathbf{x}} = 0$$

and then $H^{\mathrm{ex}} = G^{\mathrm{ex}}$; the latter being the characteristic of the denoted regular mixtures. It has been shown that even systems having excess properties so symmetric, as given by eqn. (5.34), they mainly have $S^{\mathrm{ex}} \neq 0$, albeit small; this means that B depends on temperature and hence there are only a small number of rigorously regular mixtures. This is not unexpected, taking into account that for the model so simple, ω really represents an intermolecular *effective* energy. In this case,

$$S^{\mathrm{ex}} = - \left(\frac{\mathrm{d}B}{\mathrm{d}T} \right) x_1 x_2 \tag{5.35}$$

and

$$H^{\mathrm{ex}} = x_1 x_2 \left[B - T \left(\frac{\mathrm{d}B}{\mathrm{d}T} \right) \right] \tag{5.36}$$

All the thermodynamic quantities of simple and regular mixtures conserve the symmetry in composition. It should be remembered that it is convenient to denote mixtures having $S^{\mathrm{ex}} = 0$ as regular mixtures, only using the denomination of simple mixtures to those that can be represented by eqn. (5.34), independently of the value of S^{ex}.

5.5   Behavior of Mixtures: Thermodynamic Requirements

The Gibbs-Duhem equation (3.10) requires that the chemical potential of both components of a binary mixture be related to each other at constant p and T by[1]

$$x_1 \frac{\partial \mu_1}{\partial x_1} = -x_2 \frac{\partial \mu_2}{\partial x_1} \tag{5.37}$$

Let us assume that component 1 follows Raoult's law (eqn. (5.8)) over all the composition range, in this case

$$\frac{\partial \mu_1}{\partial x_1} = \frac{RT}{x_1}$$

---

[1] In order to simplify the notation, the indexes indicating constant p and T are omitted in the derivation. This criterion will be followed from now on whenever it is clear which variables are kept constant.

Replacing this relation in eqn. (5.37) we get

$$-\frac{\partial \mu_2}{\partial x_1} = \frac{\partial \mu_2}{\partial x_2} = \frac{RT}{x_2}$$

Integrating the last equation one finally gets

$$\mu_2 = \mu_2^* + RT \ln x_2$$

which applies for all values of x_2. Hence, if component 1 follows Raoult's law for every composition, component 2 also obeys it over the whole range of composition.

If component 1 follows Raoult's law over a reduced composition range (x_1 close to unity), then the integration of the Gibbs-Duhem relation can only be accomplished in that restricted range of composition, thus obtaining for component 2

$$\mu_2 = \text{constant} + RT \ln x_2$$

a relation that is applicable for x_1 close to unity (i.e., $x_2 \to 0$), and the integration constant will be μ_2^∞ because it corresponds to the value of $(\mu_2 - RT \ln x_2)$ when $x_1 \to 1$ and $x_2 \to 0$, according to eqn. (5.22). Thus, in the interval of compositions where component 1 follows Raoult's law, component 2 follows Henry's law.

These thermodynamic conditions can be extended for other real mixtures, assuming that the activity coefficient in the Raoult scale can be represented as a series in composition given by

$$\ln \gamma_1^R = B_0 + B_1 x_2 + B_2 x_2^2 + B_3 x_2^3 + ... \tag{5.38}$$

Since we know that every component behaves ideally in the limit of pure substance; i.e.,

$$\lim_{x_1 \to 1} \ln \gamma_1^R = 0$$

and then $B_0 = 0$.

The Gibbs-Duhem cannot only be applied to the total chemical potential of the mixture's components, but also, separately, to the ideal and excess terms that rule the behavior of a mixture. The application of this relation to the ideal part of the chemical potentials results evident if one considers that the ideal mixture is thermodynamically consistent with the general treatment, and this implies that the excess chemical potentials will also obey the Gibbs-Duhem equation. Consequently, using eqn. (5.37) we obtain

$$\frac{\partial \mu_1^{ex}}{\partial x_2} = -\frac{x_2}{x_1}\frac{\partial \mu_2^{ex}}{\partial x_2}$$

That is,

$$\frac{\partial \ln \gamma_1}{\partial x_2} = -\frac{x_2}{x_1} \frac{\partial \ln \gamma_2}{\partial x_2} \tag{5.39}$$

or also

$$\left(\frac{\partial \ln \gamma_1}{\partial x_2} \bigg/ \frac{\partial \ln \gamma_2}{\partial x_2} \right) = -\frac{x_2}{x_1}$$

For the activity coefficients in Raoult's scale in the limit $[x_1 \to 1, x_2 \to 0]$, component 1 will behave ideally; that means that in the region where component 2 is dilute, we have the the expression $(\partial \ln \gamma_1 / \partial x_2) \to 0$, or that $(\partial \ln \gamma_2 / \partial x_2) \to -\infty$. For the case of mixtures of components that are not ionic, the first condition prevails; that is, for $x_2 \simeq 0$ it gives

$$\lim_{x_1 \to 1} \frac{\partial \ln \gamma_1}{\partial x_2} = 0$$

Since eqn. (5.38) expresses $\ln \gamma_1^{\mathrm{R}}$ in powers of the composition, it is possible to write

$$\frac{\partial \ln \gamma_1^{\mathrm{R}}}{\partial x_2} = B_1 + 2B_2 x_2 + 3B_3 x_2^2 + \cdots$$

and necessarily $B_1 = 0$ when $x_2 \to 0$.

Using the Gibbs-Duhem relation we arrive at

$$\ln \gamma_2^{\mathrm{R}} = B_2 x_1^2 + 3B_3 \left(\frac{x_1^2}{2} - \frac{x_1^3}{3} \right) + \cdots$$

Also, the activity coefficient in Henry's scale can be written as

$$\ln \gamma_2^{\mathrm{H}} = -2B_2 x_2 - \left(\frac{3B_3}{2} - B_2 \right) x_2^2 + \cdots$$

This clearly shows that the first correction to ideality is given by the coefficient B_2 in both components in the case of simple and regular mixtures, described by the lattice model, and, in such cases, $B_2 = B/RT$. Also it is useful to emphasize that the two previous relations indicate that whenever a component deviates positively from Raoult's law, it will deviate negatively from Henry's law.

The Gibbs-Duhem relation, which is central to the description of the behavior of mixtures, giving access to the value of one component's activity coefficient when those of the other components are known (over a given composition range), can also be used to verify the thermodynamic consistence of experimental results; that is, when independent experimental data is available for each one of the components of the mixture.

It is convenient to analyze the consequences of the thermodynamic consistency and, for doing so, we will restrict the discussion to the case of binary mixtures. Consider that the activity coefficients of the two compounds forming the mixture have been determined independently, at constant p and T. If the data were thermodynamically consistent, eqn. (5.39) would necessary be satisfied; however, the practical use of this relation requires differentiation of the experimental values of $\ln \gamma_i$, and this is not very easy to do and, anyway, not very precise. That is why, frequently, to establish the thermodynamic consistency of the activity coefficients data, a relation derived from the previous one is used; this is a simpler expression, although it is a *necessary* but not *sufficient* condition, for the results to be consistent. To obtain it from eqn. (5.39), it is necessary to integrate by parts the expression

$$x_1 \mathrm{d}\ln\gamma_1 + x_2 \mathrm{d}\ln\gamma_2 = 0$$

to exchange the variables used for the integration. The resulting expression is,

$$\int_0^1 \ln\frac{\gamma_2^{\mathrm{R}}}{\gamma_1^{\mathrm{R}}}\mathrm{d}x_2 = 0 \tag{5.40}$$

This expression shows that when one component has positive (negative) deviations over the complete composition range, one can only affirm that the other component will not be able to have negative (positive) deviations over the complete composition range.[2]

Another consequence of the Gibbs-Duhem equation is that it gives an interesting relation to establish the consistency of experimental data that only requires the values of the equilibrium concentrations in both phases. G^{ex} of a binary mixture can be expressed on the basis of eqn. (5.15) as

$$\frac{G^{\mathrm{ex}}}{RT} = (1-x)\ln\frac{p_1}{(1-x)p_1^*} + x\ln\frac{p_2}{xp_2^*} \tag{5.41}$$

where $x = x_2$. Differentiating this expression and eliminating some terms using the Gibbs-Duhem relation, an equation is obtained in which the partial pressures can be replaced by the composition in the vapor and the total pressure. Finally, we get,

$$\frac{1}{RT}\left(\frac{\partial G^{\mathrm{ex}}}{\partial x}\right)_{p,T} = \ln\left(\frac{p_2(1-x)p_1^*}{p_1 x p_2^*}\right) = \ln\left[\frac{y/x}{(1-y)/(1-x)}\frac{p_1^*}{p_2^*}\right]$$

[2]A classical article illustrating the behavior of many binary mixtures was published by M. L. McGlashan, "Deviations from Raoult's law", *J. Chem. Ed.*, 40, (1963): 516-518.

Now, integrating over the whole range of composition one gets,

$$\frac{1}{RT} \int_0^1 \frac{\partial G^{\mathrm{ex}}}{\partial x} \mathrm{d}x = \int_0^1 \ln\left(\alpha\, \frac{p_1^*}{p_2^*}\right) \mathrm{d}x$$

where $\alpha = [y(1-x)/(1-y)x]$ is the relative volatility given in terms of the compositions in the liquid phase (x) and in the vapor phase (y). The integration of the left-hand side term yields

$$\int_0^1 \ln\left(\alpha \frac{p_1^*}{p_2^*}\right) \mathrm{d}x = 0$$

It is possible to obtain the same result by replacing in eqn. (5.40) the activity coefficients according to eqn. (5.15). The final expression shows that, without knowledge of the partial pressures and only knowing the compositions in both phases, it is possible to establish the thermodynamic consistency of the data.

It is important to remember that the application of the Gibbs-Duhem relation, in the form given by eqn. (5.37), requires that when the composition changes, p as well as T remain constant. This, in general, is not strictly valid because the system's total pressure changes with composition. However, when dealing with systems having low vapor pressures (smaller or near the ambient pressure), the variation of p does not impact, appreciably, on eqn. (5.37).

5.6   Systems With Azeotropes

When the binary mixtures have identical equilibrium compositions in both, the liquid and the vapor, phases they form an azeotropic mixture. A consequence of this is that its components cannot be separated by distillation because the vapor, as well as the liquid, do not change their composition. Figure 5.3 illustrates that, at constant temperature, a binary mixture that presents an azeotrope in the liquid-vapor equilibrium, exhibits a maximum in the pressure-composition curve–when the deviation from ideality is positive–and a minimum–when the deviation is negative.

Observing the figure it is clear that the possibility that a binary mixture presents azeotropes depends on the magnitude of the deviation from the ideal behavior (Raoult's law), but also on the difference of vapor pressures of the pure components $(p_1^* - p_2^*)$.[3] A case where the last effect is observed is the mixture of benzene and perfluorobenzene at 343 K, where the vapor pressure of the pure components are 71.851 kPa and 73.408 kPa, respectively, at that temperature; this system presents

---

[3]In case the two components have the same vapor pressure at a certain temperature, any deviation of ideality (even if it is very small) would lead to the formation of an azeotropic mixture.

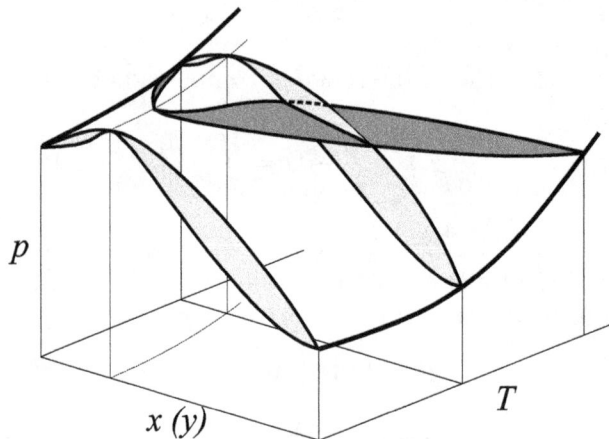

Figure 5.3: Liquid-vapor $(p, T, x/y)$ phase diagram of a binary system having an azeotrope.

two azeotropes: One having a maximum and the other with a minimum total pressure. This example illustrates the fact that one must be careful to infer the intensity of intermolecular interactions on the basis of the formation of azeotropes, that is, not always the presence of azeotropes implies that the molecular interactions are very strong.

Azeotropes present some advantages to determine the behavior of the mixtures. For a component of a real mixture having ideal vapors, it is possible to write

$$p_2 = x\gamma_2^{\mathrm{R}} p_2^* = yp$$

That is,

$$\gamma_2^{\mathrm{R}} = \frac{yp}{xp_2^*} \tag{5.42}$$

Then, at the azeotropic composition it results that

$$(\gamma_2^{\mathrm{R}})_{y=x} = \frac{p}{p_2^*}$$

This means that it is possible to know the value of the activity coefficient at the azeotropic composition, knowing only the value of the total pressure. For a simple mixture, that is enough to obtain the value of the B coefficient, which characterizes the behavior of the mixture at all compositions (cf. eqn. (5.34)).

It is important to analyze the azeotropic behavior on the basis of the Gibbs-Duhem relation to obtain an expression, known as the Gibbs-Konowalow relation.

The equilibrium between the liquid and the vapor phases ensures the equality of the chemical potentials of each component in both phases; also, their variations will be equal, *whenever the condition of coexistence of the two phases is maintained.* Then, for a binary mixture whose composition variable is the mole fraction of component 2, x, in the liquid phase and y in the vapor phase, one has $\mathrm{d}\mu_i(\mathrm{l}) = \mathrm{d}\mu_i(\mathrm{v})$. Applying the eqn. (3.10) to both phases which coexist in equilibrium,

$$\text{liquid}: \quad S(\mathrm{l})\mathrm{d}T - V(\mathrm{l})\,\mathrm{d}p + [(1-x)\mathrm{d}\mu_1 + x\mathrm{d}\mu_2] = 0 \tag{5.43}$$
$$\text{vapor}: \quad S(\mathrm{v})\mathrm{d}T - V(\mathrm{v})\mathrm{d}p + [(1-y)\mathrm{d}\mu_1 + y\mathrm{d}\mu_2] = 0$$

Subtracting the two previous expressions, we have for the binary mixture

$$[S(\mathrm{l}) - S(\mathrm{v})]\mathrm{d}T - [V(\mathrm{l}) - V(\mathrm{v})]\mathrm{d}p + (x-y)(\mathrm{d}\mu_2 - \mathrm{d}\mu_1) = 0$$

Writing $\mathrm{d}\mu_i$ for a binary mixture, according to eqn. (3.9) applied to the liquid phase,

$$[S(\mathrm{l}) - S(\mathrm{v})]\mathrm{d}T - [V(\mathrm{l}) - V(\mathrm{v})]\mathrm{d}p + (x-y)\times \tag{5.44}$$

$$\times \left[-S_2(\mathrm{l})\mathrm{d}T + V_2(\mathrm{l})\mathrm{d}p + \frac{\partial\mu_2}{\partial x}\mathrm{d}x + S_1(\mathrm{l})\mathrm{d}T - V_1(\mathrm{l})\mathrm{d}p - \frac{\partial\mu_1}{\partial x}\mathrm{d}x \right] = 0$$

For a binary mixture at constant temperature, one has

$$(V(\mathrm{l}) - V(\mathrm{v}) + (x-y)[V_1(\mathrm{l}) - V_2(\mathrm{l})]) \left(\frac{\partial p}{\partial x}\right)_{T,\sigma} = (x-y)\left[\frac{\partial\mu_2}{\partial x} - \frac{\partial\mu_1}{\partial x}\right] \tag{5.45}$$

and at constant pressure,

$$(S(\mathrm{v}) - S(\mathrm{l}) + (x-y)[S_2(\mathrm{l}) - S_1(\mathrm{l})]) \left(\frac{\partial T}{\partial x}\right)_{p,\sigma} = (x-y)\left[\frac{\partial\mu_2}{\partial x} - \frac{\partial\mu_1}{\partial x}\right] \tag{5.46}$$

In both equations, the subindex σ indicates that the differentiation occurs along the curve of liquid-vapor coexistence. These equations show that, for the azeotropic composition, where $x = y$, $(\partial T/\partial x)_{p,\sigma}$ and $(\partial p/\partial x)_{T,\sigma}$ are zero, that is, the boiling temperature (at constant pressure) or the total vapor pressure (at constant temperature) go through extreme points. This behavior is equivalent to that of a pure liquid in equilibrium with its vapor. This analogy with the one-component system can be extended because the change of the azeotrope's vapor pressure with temperature results analogous to that observed for the $(\mathrm{d}p/\mathrm{d}T)_\sigma$ derivative in the Clapeyron equation for a pure liquid.

5.7 Partial Immiscibility

The liquid mixtures can show a different behavior to those described so far. It may occur that at certain composition the liquid mixture separates in two dense liquid phases. Analyzing the curves of vapor pressure for the mixture of water with the series of aliphatic primary alcohols, it is observed (figure 5.4) that the positive deviations from ideality are enhanced as the alcohol's molecular weight increases, that is, as the molecules of both components become more different, not only in size, but also in the magnitude of their interactions with water. When n-butanol is mixed with water at room temperature, it is observed that in the diagram of $a_i = (p_i/p_i^*)$ *vs* x_i, at a certain composition, the activities of the components become constant until the system reaches a quite different composition from the previous one, after which p_i and a_i start increasing again with the concentration of n-butanol (i.e., with x_2). This observation suggests immediately that, in that composition interval, a phase separation occurs and the chemical potential of each component remains constant. This phenomenon is quite common and requires a careful analysis.

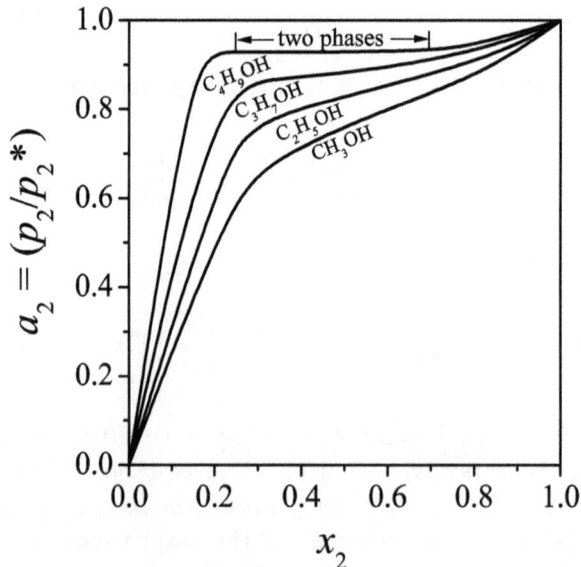

Figure 5.4: Activity of aliphatic alcohols in water, for the four first members of of the family. For the n-butanol-H_2O mixture, partial immiscibility with water is observed.

If a change in composition makes a single-phase system separate in two phases, it means that the single-phase system is less stable than that constituted by two liquid phases, and hence it will spontaneously evolve toward the more stable state–having two liquid phases. Using the expression for the molar free energy, at constant temperature and pressure, we get

$$G_m = (1 - x)\mu_1 + x\mu_2$$

The derivatives of the Gibbs function give the condition for phase separation,

$$\left(\frac{\partial G_m}{\partial x}\right)_{p,T} = \mu_2 - \mu_1 \tag{5.47}$$

and

$$\left(\frac{\partial^2 G_m}{\partial x^2}\right)_{p,T} = \left(\frac{\partial \mu_2}{\partial x}\right)_{p,T} - \left(\frac{\partial \mu_1}{\partial x}\right)_{p,T} = \left(\frac{\partial \mu_2}{\partial x_2}\right)_{p,T} + \left(\frac{\partial \mu_1}{\partial x_1}\right)_{p,T} \tag{5.48}$$

While two liquid phases having each one a constant composition coexist, the chemical potentials of both components will be constant and, therefore, the derivative of G_m in eqn. (5.47) will be a constant; hence, G_m is a straight line in that interval of concentration, as illustrated in figure 5.5.

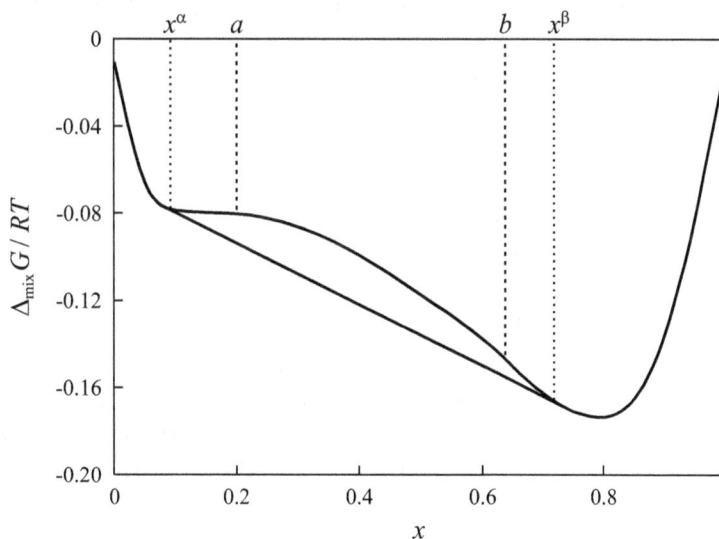

Figure 5.5: Gibbs energy of mixing as a function of composition. x^α, composition of the phase α; x^β, composition of the phase β. Between a and b the system is thermodynamically unstable.

In the last section of chapter 1 it was shown that the second derivatives of the thermodynamic functions establish the stability condition of material systems. It is obvious that $G_m(p, T, x)$ presenting a minimum at the equilibrium state must have a positive curvature, so its second derivative, given by eqn. (5.48), must be greater than zero when the liquid phase is stable, and will be equal to zero when it splits into two liquid phases. Figure 5.5 shows how $\Delta_{\mathrm{mix}}G = G_m - G_m^*$ varies when the separation of the two liquid phases occurs. It is important to realize the similarity between this phenomenon and the liquid-vapor transition described as an example for the van der Waals equation of state. Figure 5.5 illustrates that the real behavior is characterized by a region represented by a straight line between the two compositions (x^α and x^β) where there is liquid phase separation. If G_m for the mixture is described by an equation of state, one obtains a curve like the upper one in figure 5.5. We observe that over the range between a and b the curvature is negative; this composition interval presents material instability and the system necessarily undergoes phase separation. There are two other intervals where the curvature of G_m is still positive [$(x^\alpha$-$a)$ and $(b$-$x^\beta)$], but the binary system having a single liquid phase has a higher molar Gibbs energy than it would if it separates into two liquid phases. Thus, the two-phase region is more stable; nevertheless, the one-phase state can exist, albeit under a metastable condition. Between both compositions where the curvature goes from positive values to negative ones, there must exist a point where the third derivative of G_m with respect to composition is zero, $(\partial^3 G_m / \partial x^3)_{p,T} = 0$.

Figure 5.6 illustrates how the phase separation evolves when temperature increases, at constant pressure. It could be seen that at a given temperature the system returns to a state having a single liquid phase; this occurs when the system's temperature goes above T_c, called the consolute temperature, that, in the previous example, is an upper consolute temperature.

At T_c, the system goes through a critical point analogous to that observed in the vapor-liquid system of a pure compound. At temperatures above an upper T_c, the liquid-liquid equilibrium does not exist. It should be pointed out that there exist solutions with lower consolute temperature (i.e., they separate in two liquid phases above T_c and below such temperature there is a single liquid phase). According to what has been discussed, at the consolute temperature $(\partial^2 G_m / \partial x^2)_{p,T} = (\partial^3 G_m / \partial x^3)_{p,T} = 0$ –these are the conditions that allow the determination of T_c and the composition at the consolute point.

Once again, we will use simple solutions as an example of the conditions in which the liquid phase separation in binary systems takes place. It is possible to write for simple solutions that

$$G_m = G_m^{\mathrm{id}} + G_m^{\mathrm{ex}}$$

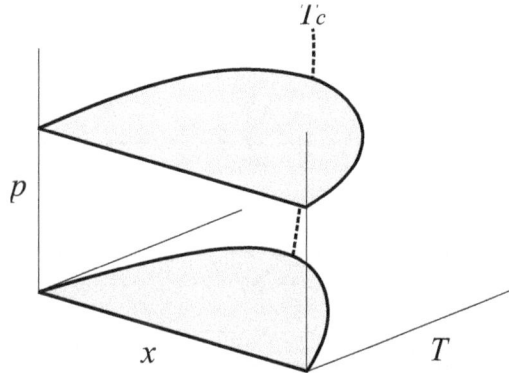

Figure 5.6: Liquid-vapor phase diagram for a binary system with partial miscibility.

Using for G_m^{ex} eqn. (5.34),

$$G_m = x_1\mu_1^* + x_2\mu_2^* + x_1RT\ln x_1 + x_2RT\ln x_2 + Bx_1x_2$$

It is possible to establish the condition that defines the critical temperature T_c, by fixing to zero the second and third derivatives of G_m, with respect to composition. The expressions for simple solutions result in

$$\frac{\partial G_m}{\partial x} = \mu_2^* - \mu_1^* + RT\ln\frac{x}{1-x} + B(1-2x)$$

$$\frac{\partial^2 G_m}{\partial x^2} = \frac{RT}{x(1-x)} - 2B = 0$$

$$\frac{\partial^3 G_m}{\partial x^3} = \frac{RT(2x-1)}{x^2(1-x)^2} = 0$$

The two latter equations yield values of the critical composition, $x_c = 0.5$, and the critical temperature, $T_c = B/2R$. Simple solutions present properties that are so symmetrical in x that the critical composition is exactly 0.5.

The expression obtained for T_c is normally interpreted as an indication that a simple solution with negative deviations cannot have phase separation, because in this case $B < 0$ and T_c would have to be negative. J. M. Prausnitz (cf. suggested list of books for consultation) has discussed the consequences of a change with temperature of the B coefficient for simple solutions. He showed that in principle all the types of phase separations that have been observed experimentally can be reproduced just by modeling the temperature dependence of B. Figure 5.7 illustrates the phase diagrams to be obtained for various dependencies of B on T. It may be seen

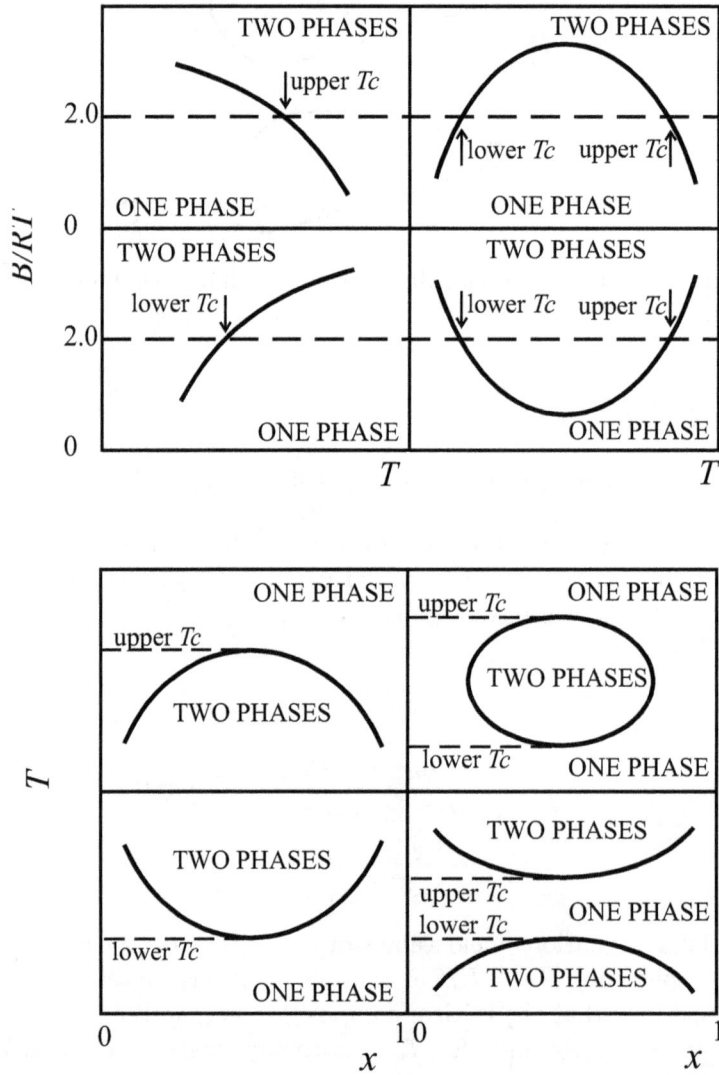

Figure 5.7: Different phase diagrams showing partial miscibility for simple solutions, when B exhibits a temperature dependence.

that simple mixtures can present upper or lower T_c, and also the known cases of binary systems with two consolute temperatures (upper and lower), which constrains the appearance of two liquid phases in equilibrium to a temperature interval. This exercise is interesting because it shows the consequences of interpreting with excessive rigor, the fact that, for the lattice model, B is independent of the temperature. On the other hand, this analysis does not give any clue about the molecular causes that produce the different behaviors.

J. S. Rowlinson and F. L. Swindon[4] have done the same analysis for simple solutions, but on the basis of the magnitudes of excess enthalpy and entropy. They concluded that simple solutions that present a lower consolute temperature also have a negative excess enthalpy, less than $-(R/2)$.

In the case of the phase transition that converts a single liquid phase into two liquid phases, there is an analogous problem than in the vapor-liquid phase transition for a pure substance. The state equations are continuous and, hence, they neither mark where the transition occurs, nor establish the volumes of the phases or the composition of the two phases in the mixture. For this case, we must use a method similar to the Maxwell construction, as described in section 3.4.

5.8   van Laar's Equation

An equation to calculate the activity coefficients of the components of a liquid mixture that has been successful, but is limited by its simplicity, is the van Laar equation. The expressions for the activity coefficients given by J. J. van Laar[5] is based on a simple application of the van der Waals equation of state for pure and mixed fluids. The derivation is very instructive about how a rather simple molecular model like the van der Waals, in this case, allows the deduction of a practical expression to calculate the deviations from ideality of the solutions.

The following scheme illustrates the model proposed for the process of mixing of two pure liquids to obtain the desired mixture, assuming that each liquid is vaporized by means of an isothermal expansion until they behave like an ideal gas.

---

[4]See list of books suggested for consultation in page 345.

[5]cf. J. M. Prausnitz, R. N. Lichtenthaler and E. Gomes de Azevedo, "Molecular Thermodynamics of Fluid-Phase Equilibria", Prentice-Hall, 1986.

```
pure liquid 1          I: expansion          ideal gas 1
    +              ─────────────────▶            +
pure liquid 2          I: expansion          ideal gas 2
                   ─────────────────▶

    │                                          │
    │                                          │
  II: mixture                                II: mixture
    │                                          │
    ▼                                          ▼

  liquid           III: compression      binary mixture of
binary mixture    ◀─────────────────        ideal gases
                  mixture at ( p, T, x_i )
```

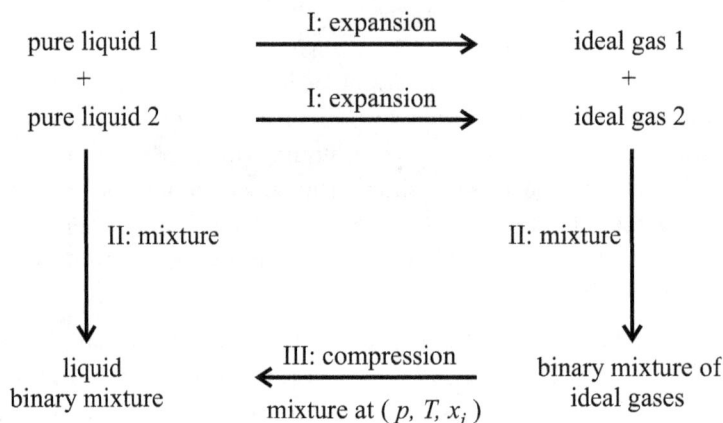

The ideal gases are mixed–we already know the thermodynamic behavior of those mixtures–and finally the gaseous mixture is compressed until it returns to the liquid state. To evaluate the change of the thermodynamic properties that occur in the processes of compression and expansion, the van der Waals equation was used.

This model is used to calculate the excess magnitudes based on

$$G_m^{ex} = U_m^{ex} + pV_m^{ex} - TS_m^{ex}$$

It is assumed that V_m^{ex} and S_m^{ex} are zero, and hence in this case the purpose is to describe the primary deviations from ideality. Then, according to the previous relation, the problem resides in the calculation of $U_m^{ex} = \Delta_I U + \Delta_{II} U + \Delta_{III} U$. Applying to the van der Waals equation (3.14) the thermodynamic relation that gives $(\partial U/\partial V)_T$ (eqn. (1.14)), one gets

$$\left(\frac{\partial U}{\partial V}\right)_T = T\left(\frac{\partial p}{\partial T}\right)_V - p = \frac{a}{V^2}$$

Upon integration of $(\partial U/\partial V)_T$ for each component, terms a_i/V_i^* are obtained; since this quantity refers to the liquid state, it is reasonable to assume that $V_i^* = b_i$, the co-volume in the van deer Waals equation. Thus,

$$\Delta_I U = x_1\frac{a_1}{b_1} + x_2\frac{a_2}{b_2}$$

As the second stage of the mixing process refers to mixing of two ideal gases, $\Delta_{II} U = 0$. The third stage is the inverse of stage I, now applied to the mixture of the two substances.

$$\Delta_{III} U = \frac{a_{mix}}{b_{mix}}$$

van Laar used the following combination rules to calculate the van der Waals parameters of the mixture from those of the pure components, obtaining

$$a_{\text{mix}} = \left(x_1\sqrt{a_1} + x_2\sqrt{a_2}\right)^2 \quad \text{and} \quad b_{\text{mix}} = x_1 b_1 + x_2 b_2$$

Thus, the following relation is obtained,

$$G_m^{\text{ex}} = RT(x_1 \ln \gamma_1^{\text{R}} + x_2 \ln \gamma_2^{\text{R}}) = RT\,\frac{x_1 A x_2 B}{x_1 A + x_2 B}$$

with

$$A = \frac{b_1}{RT}Q\,, \qquad B = \frac{b_2}{RT}Q\,, \qquad \text{and} \qquad Q = \left(\frac{\sqrt{a_1}}{b_1} - \frac{\sqrt{a_2}}{b_2}\right)^2$$

In this way, deviations to ideality for the van Laar model are

$$\ln \gamma_1^{\text{R}} = \frac{AB^2 x_2^2}{[A + (B - A)x_2]^2}$$

and

$$\ln \gamma_2^{\text{R}} = \frac{A^2 B x_1^2}{[A + (B - A)x_2]^2}$$

The expressions of van Laar for the activity coefficient use two specific parameters to describe each mixture; they express the deviation from ideality of the binary mixtures, but are different from those for simple mixtures, because this equation takes into account the differences in the molecular size of the pure components. When $b_1 = b_2$, it results $A = B$, and the mixture is a simple one.

5.9   Solutions of Polymers and Macromolecules

The lattice model for mixtures discussed in section 5.4 is a first and adequate approximation for mixtures having similar molecular volumes. On the other hand, the van Laar equation (section 5.8) takes explicitly into account the effect of different molecular sizes through their two parameters, but it is applied within the conceptual scheme of the van der Waals state equation, and assumes that the mixing entropy of both components is the ideal one. However, when the ratio of molar volumes of the two components in the mixture is very different from unity, the van Laar equation is unsuccessful.

In a binary mixture of propane and polyethylene, for instance, the interaction between C_3H_8 and the sites in the polymeric chain, represented by $-(CH_2-CH_2)_n-$, will comply acceptably with the approximation of ideality for intermolecular energies. Thus, ω will be small, and the corrections to the excess properties will be only moderate and, from this point of view, the lattice model should be applicable (section 5.4). However, the other hypothesis of the lattice model, that each molecule of both components will occupy a single site, will not apply to a polymer solution. If it is considered that the small component, which generally is the solvent (propane in our example), occupies a single site of the lattice, then the polyethylene must occupy several sites. Moreover, the number of sites the polymer will occupy will depend on its molecular mass (i.e., its degree of polymerization). It is a known fact that the thermodynamic properties of polymer solutions depend on the degree of polymerization of this macromolecule. Let's consider that the polymeric chain is formed by q segments; each of them may be formed by one or more monomers, and each segment will occupy a site of the lattice identical to those occupied by the solvent molecules. Figure 5.8 depicts a bidimensional representation of a lattice-containing solvent and a linear polymeric solute, like polyethylene, and let's imagine that a segment of polyethylene is occupying a site s in the lattice, and the two adjacent segments will be chemically joined to segment s; consequently, they *must* occupy two of the z neighboring sites to s. This means that the segments' distribution is not totally random. Hence, the degeneration of each microstate, that in section 5.4 we denominated as F, must be different in the case of polymeric chains. Effectively, now the occupation of each site is not completely random; when a site in the lattice is occupied by a polymer's segment, the occupation of the other site is not random anymore, and the other segments of the polymer chain are not free to occupy any site of the lattice; they must preserve the chemical integrity of the polymeric chain.

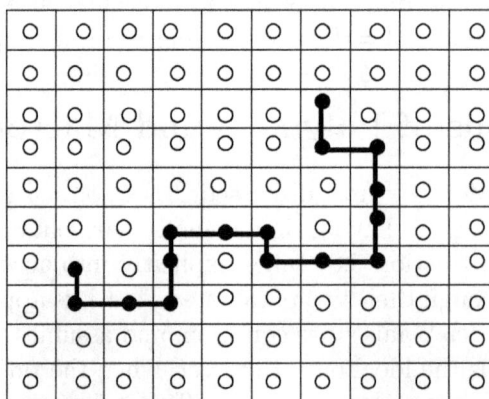

Figure 5.8: Lattice model applied to macromolecular solutes $(- \bullet - \bullet -)_n$ dissolved in molecular liquids (○).

In order to correct the value of the degeneracy of each microstate, it is necessary to analyze the probability of occupation of the sites, which is more complex than that used for the model applied to simple solutions; this problem was solved by P. J. Flory and C. Huggins[6] fifty years ago. Now it is possible to appreciate the advantage of using lattice models because an adequate consideration of the probable distribution of units is amenable to describe the behavior of linear, branched, or globular (for instance some proteins) macromolecules. The theoretical analysis of Flory and Huggins shows that for solutions where the components have very different molar volume, it is natural to choose the volume factions ϕ_i in place of the mole fractions. The volume fractions are defined by

$$\phi_i = \frac{V_i^*}{V}$$

For our example, each segment of polyethylene will occupy a site in the lattice, and considering that each monomer is a segment of the chain[7] and that each segment occupies a single site, the fraction of volume ϕ_2 is

$$\phi_2 = \frac{qn_2}{n_1 + qn_2} = \frac{qx}{1 - x + qx}$$

and the volume fraction of the solvent is

$$\phi_1 = \frac{n_1}{n_1 + qn_2} = \frac{1 - x}{1 - x + qx}$$

expressions where q denotes the degree of polymerization, n_1 moles of solvent, and n_2 moles of the polymeric solute. Flory and Huggins demonstrated that in these cases the entropy of mixing is

$$\Delta_{\text{mix}}S = -R(n_1 \ln \phi_1 + n_2 \ln \phi_2) \tag{5.49}$$

Figure 5.9 shows the curves of $\Delta_{\text{mix}}S$ per mole of mixture as a function of the mole fraction of polymer, for different values of q. Two aspects are notable and characteristic of the thermodynamic properties of polymers: The curves as a function of mole fraction are more skewed and asymmetric the more positive the mixing entropy is, which is attained as the degree of polymerization, q, increases. For solutions of polymers chemically similar, like the system we used as an example, the mixing process results in a $\Delta_{\text{mix}}H$ close to zero, and these mixtures are called athermal. In this case, the mixture's Gibbs energy is determined by the mixing entropy and the deviations from Raoult's behavior will be negative.

---

[6]P. J. Flory, *Principles of Polymer Chemistry*, Cornell University Press, Ithaca, NY, 1953.

[7]This assumption is not totally correct, since more than one monomer is required to form a segment with a totally random orientation.

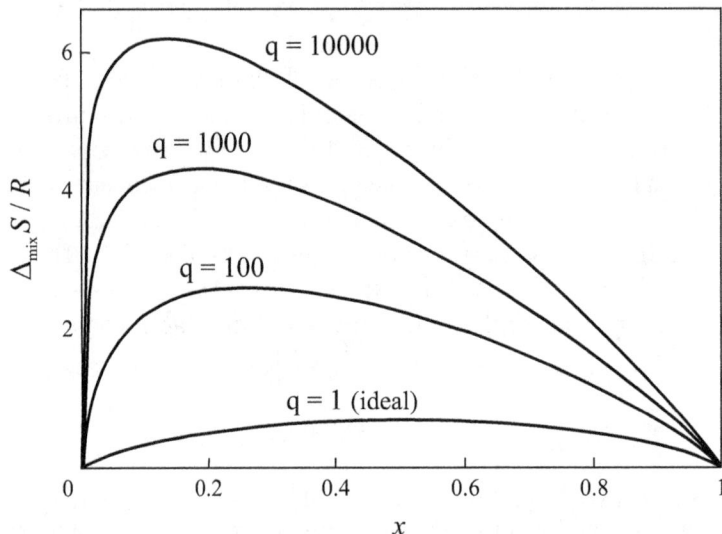

Figure 5.9: Mixing entropy of a polymer, with a degree of polimerization q, with another component.

Before finishing the section of macromolecules, it is useful to consider the following consequence of Flory and Huggins theory: The mixing entropy of the polymers, as illustrated in figure 5.9, is more positive than $\Delta_{\mathrm{mix}}S^{\mathrm{id}}$; how is it possible that the mixing entropy is greater than the ideal one that corresponds to a process that is completely random? The apparent paradox is due to the fact that in lattice models site statistics is used, and when one of the components has a much larger size than the other, the total number of disposable sites per mole of mixture is much larger than the total number of molecules present in the solution, because each polymer molecule requires q sites. Calculating the number of microconfigurations generated in a random distribution of $(n_1 + qn_2)$ moles of sites for the $(n_1 + n_2)$ moles of molecules of both types, it is clear that $\Delta_{\mathrm{mix}}S$ is bigger than the value obtained for the case when $(n_1 + n_2)$ moles of molecules occupy the same number of lattice sites. Thus, the molar entropy of mixing must be compared with the ideal entropy of mixing for the total number-of-moles of sites, $(n_1 + qn_2)$, and not for the total number-of-moles of molecules, $n_1 + n_2$, which is much less.

This last consideration suggests to go back to the expression of the ideal entropy of mixing, given by eqn. (4.7), for polymeric solutes, in terms of mole fractions of sites, instead of mole fractions of the components forming the solution. Within the approximation that assumes that monomers and solvent molecules occupy the same type of site, the mixing entropy *per mole of sites* is written

$$\Delta_{\text{mix}}S = -R \left[\frac{x_s}{q} \ln x_s + (1 - x_s) \ln(1 - x_s) \right] \tag{5.50}$$

where $x_s = \phi_2$ is the mole fraction of monomers. In this expression, it can be observed that $\Delta_{mix}S$ is less than $\Delta_{mix}S^{id}$ for a mole of independent sites ($q = 1$).

In this way, it is possible to establish a relation between the lattice model and the Flory-Huggins equation for polymer solutions. Since p and V do not change, we can write $\Delta_{\text{mix}}A$ per mole as

$$\Delta_{\text{mix}}A = RT \left[\frac{x_s}{q} \ln x_s + (1 - x_s) \ln(1 - x_s) \right] + \chi(T) \, x_s(1 - x_s) \tag{5.51}$$

where $\chi(T)$ is the Flory-Huggins parameter, which is generally obtained by fitting the experimental results to this equation. Formally it may be considered that $\chi(T) = z\omega N_A \equiv B$; $\chi(T)$ also stands for an interaction energy correction to the lattice model. As indicated by eqn. (5.50), the real entropy of mixing between sites of polymer and solvent will be larger or equal to $\Delta_{\text{mix}}S$.

The expression eqn. (5.51) may be also used to describe the thermodynamic behavior and the phase separation in liquid polymeric mixtures (frequently referred as *blends*), which are of great interest for their applications. For polymers A and B, with degrees of polymerization q_A and q_B, respectively, and monomers of similar size, eqn. (5.51) becomes

$$\Delta_{\text{mix}}A = RT \left[\frac{x_A}{q_A} \ln x_A + \frac{x_B}{q_B} \ln x_B \right] + \chi(T) \, x_A x_B \tag{5.52}$$

5.10   Solubility of Non-Ionic Solids

It is interesting to consider the case of non-ionic solids formed only by non-polar molecules, having not too high melting points, like iodine, sulfur, naphthalene, etc. If the total pressure is near the vapor pressure of the pure solvent, it is not necessary to include the term with the Poynting effect in eqn. (5.23) and, since we are interested in using the ideal laws of Raoult or Henry, we employ the mole fraction to express the composition of the mixture. Applying that expression to the case of a saturated mixture, that is, a solid solute in equilibrium with dissolved solute is

$$\mu_2^*(\text{s}, p_1^*, T) = \mu_2(\text{sln}, p_1^*, T, x) = \mu_2^\circ(p_1^*, T) + RT \ln a_2 \tag{5.53}$$

In the previous expression, the standard state is denoted by a supraindex ∘ because it will be applied to the Raoult and also to the Henry standard state. The change of $\ln a_2$ with temperature will be given by

$$\left(\frac{\partial \ln a_2}{\partial T}\right)_{p\sigma} = \frac{H_2^\circ - H_2^*(\mathrm{s})}{RT^2} \tag{5.54}$$

where the relation between enthalpy and the change of (μ/T) with temperature have been taken into account. In the previous expression, H_2° is the partial molar enthalpy of the solute in its standard state. If the saturated solution is ideal with respect to Henry's law,

$$\left(\frac{\partial \ln x}{\partial T}\right)_{p\sigma} = \frac{H_2^\infty - H_2^*(\mathrm{s})}{RT^2} \tag{5.55}$$

The change of solubility with temperature is independent of the solution's composition and only depends on the particular solvent, which will determine the magnitude of the enthalpy of dissolution at infinite dilution, H_2^∞.

If the mixture is considered to be ideal according to Raoult's law, eqn. (5.54) gives

$$\left(\frac{\partial \ln x}{\partial T}\right)_{p\sigma} = \frac{H_2^*(\mathrm{l}) - H_2^*(\mathrm{s})}{RT^2} \tag{5.56}$$

and the change of solubility with temperature is independent of the solvent. This value of solubility is called the ideal solubility. The enthalpies difference in eqn. (5.56) is equal to the melting enthalpy of the pure solid. This result is easily understandable if the dissolution is thought separated in two stages: In the first one the solid is melted; in the second one the liquid solute is mixed with the solvent and, since the resulting solution behaves ideally, the mixing stage has zero enthalpy. Integrating the previous equation and assuming that the enthalpy of melting does not change appreciably from the value of its melting temperature, $T_{\mathrm{m},2}$, and T, we get

$$\ln x = \frac{\Delta_{\mathrm{m}} H_2}{R}\left[\frac{1}{T_{\mathrm{m},2}} - \frac{1}{T}\right]$$

It is possible to imagine this process in another way. The solid's vapor pressure, which is in equilibrium with the solution, is the partial pressure of the pure solute in the vapor phase $[p_2^*(\mathrm{s}) = p_2(\mathrm{sln})]$. If the solution is ideal, p_2 can be related with the solute mole fraction, its solubility, multiplied by the pure solute vapor pressure in the liquid state, that is, under metastable condition.

Let us describe a few examples of the latter behavior. At room temperature, the solubility of anthracene dissolved in benzene is $x = 0.087$ and that of phenanthrene in the same solvent is $x = 0.207$. The melting temperature of these two solids are 490 and 373 K, respectively, hence it is expected that the melting enthalpy of anthracene will be greater than for phenanthrene; moreover, for anthracene the value $(T - T_{m,2})$ is more negative and, consequently, its solubility will be less, in agreement with the experimental values.

A very illustrative example is the solubility of naphthalene at 298 K in different solvents that are summarized in table 5.1. It is evident that naphthalene solubility in different solvents is close to the ideal one in those solvents, where one would expect ideality in the resulting liquid mixtures. Also, it is seen that ideality is determined by the molecular similarity of the two components of the mixture.

Table 5.1: Solubility of naphthalene in different liquids at 298 K

Solvent	x	Solvent	x
ideal	0.300		
benzene	0.291	toluene	0.294
chlorobenzene	0.315	dibromomethane	0.302
chloroform	0.332	dichloromethane	0.318
CCl$_4$	0.257	CS$_2$	0.196
hexane	0.120	methanol	0.025

Effect of Long-Lasting Contaminants in the Environment

How is it possible to evaluate the distribution of long-lasting contaminants, like DDT or dioxins, in the environment? This is a complex problem, especially when facing so many different situations–contaminants are discharged in the soil, the water, or the air and, since they have a dangerous long persistence, they are transferred from one medium to another. This problematic situation looks particularly adequate for a thermodynamic treatment, using the hypothesis that the contaminant concentrations in the various media are those that correspond to states of equilibrium. That is, it will be assumed that different media correspond to the equilibrium condition, implying that the various media where they may be distributed constitute compartments or phases, and that all of them–air, water, soil, sediments, biota, etc. are in equilibrium[8].

---

[8] A good exercise for the reader is to evaluate whether this assumption is valid in this case.

A simple analysis of the situation will be carried out, but many times this is the only level of analysis possible.[9] When the problems we are dealing with are analyzed at this simple and approximate level, it is necessary to have the capacity to accept the validity of the hypothesis being used for the different real situations that may occur. A more immediate objective of this analysis is to know which physicochemical parameters of the contaminating substances must be known to evaluate their impact in the environment.

The basic assumption of this analysis is that equilibrium exists among the natural phases (media): Air, water, soil, sediments, and biota. Then, the condition of phase equilibrium will be employed for substance i,

$$\mu_i^\alpha = \mu_i^\beta$$

or in in terms of fugacities,

$$f_i^\alpha = f_i^\beta$$

In order to know the concentration of the i contaminant that exists in air, assumed to behave as an ideal gas, one has,

$$f_i(\text{air}) = x_i(\text{air})p = c_i(\text{air})RT$$

where c_i is the contaminant molar concentration in air.

The concentration of i in the water will be,

$$f_i(\text{water}) = x_i(\text{water})\gamma_i^{\text{R}}(\text{water})f_i^* \simeq x_i(\text{water})\gamma_i^{\text{R}}(\text{water})p_i^*$$

Considering that aqueous solutions of organic contaminants are very dilute, it is possible to assume $\gamma_i^{\text{R}}(\text{water}) = \gamma_i^{\text{R},\infty}(\text{water})$ (it is interesting to discuss in this case the use of the Raoult scale and its consequences). Then,

$$f_i(\text{water}) = x_i(\text{water})\gamma_i^{\text{R},\infty}(\text{water})p_i^* = x_i(\text{water})k_{\text{H},i}$$

The importance of knowing the value of the vapor pressure of the contaminant is clear, and the same can be said of $\gamma_i^{\text{R},\infty}(\text{water})$ or, in the corresponding expression, the Henry's constant.

---

[9] This example is based on the work by S. I. Sandler, "Unusual Chemical Thermodynamics", *J. Chem. Thermodyn.*, 31, 1999: 3-25

If dealing with a liquid contaminant, since its solubility in water is generally small, then $\gamma_i^{R,\infty}(\text{water})$ can be obtained from its solubility, because in this case

$$x_i(\text{water})\gamma_i^{R,\infty}(\text{water}) = 1$$

This is due to the existent equilibrium between the pure contaminant in liquid state (supposing that the liquid contaminant, in equilibrium with the aqueous solutions, dissolves very little water) and the saturated solutions. If the contaminant is solid, the previous fusion process should be taken into account, as in the case of ideal solubility,

$$\ln \gamma_i^{R,\infty}(\text{water}) = -\ln x_i(\text{water}) - \frac{\Delta_m H_i^*}{RT}\left(1 - \frac{T}{T_{m,i}}\right)$$

In this last expression, the change of $\Delta_m H_i^*$ with temperature does not exist[10].

What will be the contaminant's concentration in the biota present in water–for instance, in fish. It is considered that the organic contaminants will dissolve preferentially in the fatty tissues of the organisms forming in the biota. In order to know the concentration of these contaminants in the lipids, n-octanol is usually a good model substance for medium rich in lipids. Thus, it is important to know the equilibrium distribution constant of contaminants between water and n-octanol, which are partially miscible at room temperature. One phase is practically pure water and the other is a liquid phase in equilibrium that has 74 mole percent of the alcohol. Thus, the partition coefficient of the contaminants between water and n-octanol is a physicochemical parameter of importance to evaluate the effect of organic contaminants in aqueous media. Since solutions are very dilute,

$$K_i(\text{oc/water}) = \frac{c(\text{oc})\gamma_i^{R,\infty}(\text{water})}{c(\text{water})\gamma_i^{R,\infty}(\text{oc})} \approx \frac{c_i(\text{oc})}{c_i(\text{water})}$$

where $c(\text{oc})$ and $c(\text{water})$ are the molar concentrations of octanol and water, related with the concentrations of contaminants in each phase, that is $c_i(\text{oc}) = x_i(\text{oc})c(\text{oc})$, analogously for the aqueous phase. It is convenient to discuss the validity of this step. The values of the activity coefficients at infinite dilution can be obtained by chromatographic determinations.

---

[10]A good exercise for the reader is to discuss why this does not introduce a big error.

Also, relations between $K_i(\text{oc/water})$ and $\gamma_i^{R,\infty}(\text{water})$ are used, assuming that the dissolved contaminants in octanol have an unity activity coefficient (due to their low concentration). An empirical relation frequently applied, based on the previous equation, is

$$\log K_i(\text{oc/water}) = -0.486 + 0.8061 \log \gamma_i^{R,\infty}(\text{water}) \qquad (5.57)$$

In order to use this expression to the case of the concentration of component i in the biota, it is convenient to remember the assumption that i only dissolves in the lipids of the organism, then

$$K_i(\text{biota/water}) = \Phi(\text{biota}) K_i(\text{oc/water})$$

where $\Phi(\text{biota})$ represents the fraction of lipids in the biota.

Regarding the quantity of contaminant i found in the sediments of rivers and lakes, the matter in suspension in aqueous media and in soil, we assume that they are absorbed preferentially in the organic matter of the sample. So, it has been observed that the partition between some of these materials and water, $K_i(\text{mat/water})$, is given by,

$$K_i(\text{mat/water}) = \frac{c_i(\text{mat})}{c_i(\text{water})} = 0.41\rho(\text{mat})\Phi(\text{mat})K_i(\text{oc/water})$$

where 0.41 is an empirical factor related to the distribution of the contaminant between water and the corresponding material, with respect to its distribution between water and n-octanol. $\Phi(\text{mat})$ is the fraction of organic matter in the material, and $\rho(\text{mat})$ is its density (generally expressed in kg/dm^3).

Discussion of the next two examples is useful; these are taken from Sandler's article, and they guide the reader in the application of the equations given, testing their validity in some of the cases.

1. The weight fraction of PCB (polychlorated biphenyls) in the water of San Lorenzo's river (limiting Canada with the USA, a region strongly industrialized) is 0.3×10^{-9}. The distribution constant will depend on the particular PCB, but it can be considered equal to the average value $\log K_{\text{PCB}}(\text{oc/water}) = 5.5$. With this information and, considering that the fraction of lipids in aqueous biota is 0.05, then

$$c_{\text{PCB}}(\text{biota}) = 0.05 \cdot 0.3 \cdot 10^{-9} \times K_{\text{PCB}}(\text{oc/water}) = 4.7 \times 10^{-6}$$

that is, the concentration in the biota is 2×10^4 larger than that existing in the water. The mean value found for $c_{\text{PCB}}(\text{biota})$ in that estuary was

7.9×10^{-6}.

2. The concentration benzo[α]pyrene (BP) in water in the Ontario (Canada) estuaries resulted 2.82×10^4 mg/m^3. The vapor pressure of the compound is 2.13×10^{-5} Pa at 298 K, the partition between alcohol and water is $\log K_{BP}(oc/water) = 6.04$, and the value of $\gamma_{BP}^{R,\infty}(water) = 3.8 \times 10^8$. Assuming that the organic fraction and the soil's density are 0.02 and 1.5, respectively, and for the sediment these values are 0.05 and 1.42, respectively, the following results are obtained:

Table 5.2: Concentration in mg/m^3 of benzo[α]pyrene in the environment

System	A	B	C	Found
water				2.82×10^4
air	1.648	1.648	0	(1.3-7.1)
soil	3.71×10^8	9.06×10^8	111	1.1×10^8
sediment	8.78×10^8	2.14×10^9	262	$(0.8\text{-}3) \times 10^8$
biota	1.55×10^9	3.77×10^9	461	1.4×10^8

A: Using $\log K(oc/water) = 6.04$; B: $K(oc/water)$ from eqn. (5.57) with $\gamma_{BP}^{R,\infty}(water) = 3.8 \times 10^8$; C: Ideal solubility and $K(oc/water)$ from eqn. (5.57) with $\gamma_{BP}^{R,\infty}(water) = 1.0$.

It is shown that this analysis, even being very simple, is able to yield a good account of the concentration values of BP in the different media; it is also clear that if ideal behavior is assumed, the calculated values would be far from those observed.

Regarding the problem of the physicochemical parameters, which are more important to establish the impact of contaminants, they are the vapor pressure, the activity coefficient at infinite dilution in water (or Henry's constant), and the distribution constant between water and octanol.

When the biota has more that one species in the studied region, there is bioaccumulation of the contaminant. This is due to the fact that different species in the trophic chain feed from other species that already have incorporated the contaminant. Thus, bioaccumulation of the contaminant effect occurs and it must be taken into account when evaluating more complex situations. The results of different studies show that biomagnification is negligible when the distribution constant between octanol and water is less than 1×10^4 and that when it exceeds 10^6, biomagnification is very important. More details can be obtained from the original article by Sandler.

Problems

Problem 1

The solubility of solid sulfur in CCl_4 and in toluene was measured at two different temperatures. Determine whether the saturated solutions in that solvent are ideal, regular, or simple.

Data: $\Delta_m(S_8) = 987$ J/mole; $T_m = 392$ K

	$x_2(273$ K$)$	$x_2(327$ K$)$
sulphur-CCl_4	0.00203	0.01212
sulphur-toluene	0.00324	0.01797

Answer: The two solutions are practically regular

Problem 2

For the mixture of the liquids A and B at 273 K, it is known that $k_{A(B)}^H = 4.6$ bar, $p_A^* = 1.32$ bar, $k_{B(A)}^H = 3.4$ bar, and $p_B^* = 0.97$ bar. Answer and justify the following questions, and identify the approximations used (consider the vapors as ideal gases).

a) A and B do not form an ideal mixture, why?

b) Can the mixture be considered regular?

c) Estimate the total pressure over the mixture with $x_A = 0.75$.

d) Determine if (A+B) form azeotropes and, if so, calculate the azeotropic composition.

e) Estimate the value of $k_{A(B)}^H$ at 300 K from the value of p_A^* at that temperature.

Answers: a) No, b) Yes; c) 1.56 bar; d) Yes; e) $k_{A(B)}^H = 1.140$ bar

Problem 3

For mixtures of I_2 (2) and propanone (1), it is known that at 298 K, $k^H_{2(1)} = 1.77$ bar and $p^*_1 = 0.308$ bar, and that they behave as regular mixtures.

a) Calculate the decrease of the vapor pressure of propanone when dissolving 7.5 g of I_2 in 10 g of propanone

b) Calculate the solubility of I_2 in propanone at 298 K with precision better than 10 percent.

Data: $M_2 = 256$ g/mole; $M_1 = 58$ g/mole; $T_{fus}(I_2) = 383.9$ K; $\Delta_m H^*_2 = 15.52$ kJ/mole

Answers: a) $\Delta p_1 = 0.0345$ bar; b) $x_2 = 0.190$

Problem 4

Consider a mixture formed by carbon disulfide (CS_2, 1), and dimethoxymethane $(CH_3O)_2CH_2$, 2), at 305.65 K. For this mixture, it is known that the behavior of CS_2 may be modeled using the Margules equation, given by

$$RT \ln \gamma^R_1 = (2B - A)x^2_2 + 2(A - B)x^3_2$$

where A and B are constants independent of temperature.

a) Find an expression for $\ln \gamma^R_2$ in the mixture.

b) On the basis of the data, determine the values of A and B in the mixture.

c) Find if the mixture at 305.65 K presents an azeotrope. If this is so, determine its composition, calculate its vapor pressure, and determine if this is an azeotrope with a maximum or a minimum pressure.

Data (at 305.65 K): $p^*_1 = 467.02$ Torr, $p^*_2 = 587.73$ Torr, $k^H_{1(2)} = 1156$ Torr, $k^H_{2(1)} = 1867$ Torr

Answers: a) $RT \ln \gamma^R_2 = (2A-B)x^2_1 + 2(B-A)x^3_1$; b) $A = 2303.30$ J/mol, $B = 293..28$ J/mol; c) $x^{az}_1 = 0.416$, $p^{az} = 682.8$ Torr, the azeotrope has a maximum pressure

Problem 5

Water is immiscible with CS_2. In this binary system, solid I_2 was dissolved at 298 K. The values of the mole fractions of iodine listed in table 5.3 were found in each of the two phases, once the system reached equilibrium.

Table 5.3: Equilibrium mole fraction at 298 K

$10^6 \cdot x(I_2, \text{aq})$	$10^2 \cdot x(I_2, CS_2)$
4.924	0.978
9.146	1.853
11.162	2.324
15.130	3.242

Knowing that I_2 dissolved in CS_2 forms a regular solution with $(B/RT) = 1.448$, calculate:

a) The equilibrium constant for the partition of I_2 between both liquid phases (by extrapolation)

b) The activity coefficients of I_2 in water at the four concentrations. Comment the results obtained

Answers: a) $K_{x \to 0} = 1920$; b) the activity coefficients are constant around the value $\gamma(I_2,\text{aq}) = 4.31$; hence, this indicates that the value corresponds to the activity coefficient of I_2 in water at infinite dilution

Problem 6

The solutions of phenol in water are markedly non-ideal. In order to obtain the activity coefficient of phenol in water, the following equilibrium concentrations were measured for the partition of phenol between water and hexane at 293.0 K.

c(phenol-hexane) / (mole/dm^3)	5.0×10^{-5}	2.45×10^{-3}	0.0121	0.0187	0.0204
c(phenol-water) / (mole/dm^3)	0.001	0.050	0.268	0.450	0.496

By plotting the data, obtain the partition constant of phenol in hexane/water at 293.0 K, defined as $K_d = \exp[(\mu^\infty_{\text{phenol,water}} - \mu^\infty_{\text{phenol,hexane}})/RT]$. From that result, calculate the activity coefficients of phenol in water at all concentrations.

Answers: $\gamma^H_{\text{phenol,water}} = 1.000, 0.980, 0.903, 0.830, 0.823$, in order of increasing concentration

Problem 7

a) Using eqn. (5.51), obtain the expression for the mole fraction of sites at the consolute temperature of a polymer in a solution that has q monomers.

b) Using eqn. (5.52), calculate the critical composition for a mixture of polymers A and B having degrees of polymerization q_A and q_B respectively.

c) How do you explain that a polymer mixture of high molecular weight rarely forms a single phase (that is, it is totally miscible)?

Answers:

$$\text{a)} \quad x_s)_s = (1 + \sqrt{q})^{-1} \simeq (\sqrt{q})^{-1} \qquad T_s = 2\frac{z\omega N_A}{R}\left(1 + 1/\sqrt{q}\right)^{-2}$$

$$\text{b)} \quad x_s)_s = (1 + \sqrt{q_A/q_B})^{-1} \qquad T_s = \frac{2z\omega N_A}{R}\left(1/\sqrt{q_A} + 1/\sqrt{q_B}\right)^{-2}$$

Chapter 6

Solutions

6.1 Introduction

Dilute solutions are of great practical importance because they exist often in many natural systems, in the application of analytical techniques, as well as in chromatography, and in some industrial processes. Within the scope of this book, dilute solutions also are important to establish the basis of chemical potentials at infinite dilution, which corresponds to one of the standard states used for the mixtures–that related to Henry's law. The infinite dilution state is one of the more important reference states. From the molecular point of view, this state implies that the solute interacts only with the solvent, a condition in which solvation phenomena occur, and their knowledge also contribute to simplify the understanding of experimental information in molecular terms.

In this chapter, we will deal with different solutes, which show a limit of solubility. Usually, these are divided in ionic and non-ionic solutes. Ionic solutes are frequently denoted as electrolytes. The properties of both types of solutes are described often using the standard state of infinite dilution (or Henry state). For ionic solutes, we will also discuss, in a section of this chapter, the behavior of electrolyte solutions, which are relatively concentrated. For doing this, we will employ a modern view of the classical theories that also make use of infinite dilution as their standard state.

We will begin with the analysis of the solubility of non-ionic solid solutes dissolved in supercritical fluid solvents and that of gases in liquids; in both cases the solutions of interest are very dilute. Then, the features of the solutions containing ions will be discussed by emphasizing the difference of their behavior and that of non-electrolyte solutions. The characteristics of colligative properties for solutes of both types will also be considered because they provide a number of experimental methods for the determination of their activity coefficients. Finally, the Debye-Hückel model for electroytes will be described, as well as the expressions that can be derived from it to explain the solutes' excess chemical potentials. In the last section, the more modern extension of the model carried out by Pitzer will be discussed.

6.2   Supercritical Extraction

An increasing number of industrial chemical processes use great amounts of liquid solvents. These substances, after being used, are recuperated partially and imperfectly. The non-recuperated fraction is incorporated into the environment, which consequently produces contamination both at the local as well as at the global scale. This is worrisome, and that is why alternative procedures to the use of liquid solvents are a source of interest.

A process increasingly used in industry is the extraction of substances employing supercritical fluids, especially with CO_2, or other fluids that are environmentally benign. It is well-known that the dissolving capacity of a liquid substance depends strongly on its density, expressed in mole per unit of volume. But, in order to vary the density of a fluid over a wide range, it is necessary that the fluid be above its critical temperature; otherwise, when pressure is varied, a phase transition can occur. That introduces a gap between the liquid density and that of the vapor in equilibrium so that intermediate densities are not accessible. In particular, when the fluid is near to its critical temperature, where its compressibility is large, an appreciable change in density is produced for a given change in pressure. This factor is economically important because the cost of the installations grows steeply with the maximum pressure that will be used in the process. As a result, it is important to analyze solid's solubility in a supercritical fluid.

Equilibrium between the solid and the saturated solution implies that solute's fugacity must be the same in both phases, that is, $f_2^*(s, p, T) = f_2(sln, p, T, y)$. Hence, taking into account that solid's fugacity is determined by the fugacity of the vapor in equilibrium with it, one can write

$$\Phi_2^*(p, T)\, p_2^*(s, p_2^*, T) \exp \int_{p_2^*}^{p} \frac{V_2^*(s)}{RT} dp = y\, \Phi_2(p, T, y)\, p \qquad (6.1)$$

Considering that the solution is very dilute, and therefore $\Phi_2^* \simeq 1$, one obtains the following expression for the solubility y

$$y = \frac{1}{\Phi_2^\infty} \frac{p_2^*}{p} \exp \int_{p_2^*}^{p} (V_2^*/RT)\, dp = I\left(\frac{p_2^*}{p} \exp \int_{p_2^*}^{p} (V_2^*/RT)\, dp\right) \qquad (6.2)$$

The factor inside parentheses in the rhs member of the previous expression is equal to the ideal solubility, y^{id}, due to the vapor pressure p that the pure solute would have. Note that the ideal solubility is larger than the solubility given by the quotient of pressures, (p_2^*/p), because there is a contribution of the Poynting effect over the solute (cf. chapter 3) due to the mechanical pressure enhancement of the chemical potential of the solid, which increases its vapor pressure.

The factor I in eqn. (6.2), denoted as incremental factor, is the major contribution to the solid's solubility in the supercritical fluid. From the previous expression, we get that the incremental factor is given by

$$I = \frac{1}{\Phi_2^\infty} \tag{6.3}$$

The solute-solvent interactions are responsible for the value of Φ_2^∞. In the majority of processes of practical interest, Φ_2^∞ is appreciably smaller than unity; writing this quantity in terms of the state equation with second virial coefficients yields

$$\ln \Phi_2^\infty \simeq \frac{2}{V} B_{12} - \ln\left(\frac{pV}{RT}\right) = \frac{2B_{12} - B_{11}}{V}$$

This expression shows that, at least for low densities or pressures, Φ_2^∞ is less than unity for attractive solutes (those that interact with the solvent more strongly than two solvent molecules with each other). Hence, $\Phi_2^\infty < 1$ implies that $B_{12} < B_{11}$. This observation is generally valid up to pressures of 300 bar, where Φ_2^∞ is the most important factor to consider when an appreciable dissolution of the solid in the supercritical fluid is required.

The supercritical extraction is increasingly used in chemical processes requiring purification of substances, and its appeal is the fact that after dissolving the substances in the supercritical fluid, a decrease in the pressure produces an appreciable decrease of the solubility and the solute precipitates. Supercritical extraction is conceptually a very simple procedure to extract the desired substance and is sufficient to vary the supercritical fluid's density. This effect is many times accentuated by adding small amounts of another substance to the system (the cosolvent) which allows a better molecular tuning. Figure 6.1 illustrates how the solubility of caffeine varies in CO_2, at different pressures and at two temperatures. The figure also shows the behavior of the incremental factor I as a function of pressure.

6.3 Solubility of Gases in Liquids

Gases at temperatures above their critical temperatures have low solubility in liquids when no chemical reaction occurs with the solvent, for example, the more abundant atmospheric gases (with the exception of CO_2) and also CH_4 which is an important

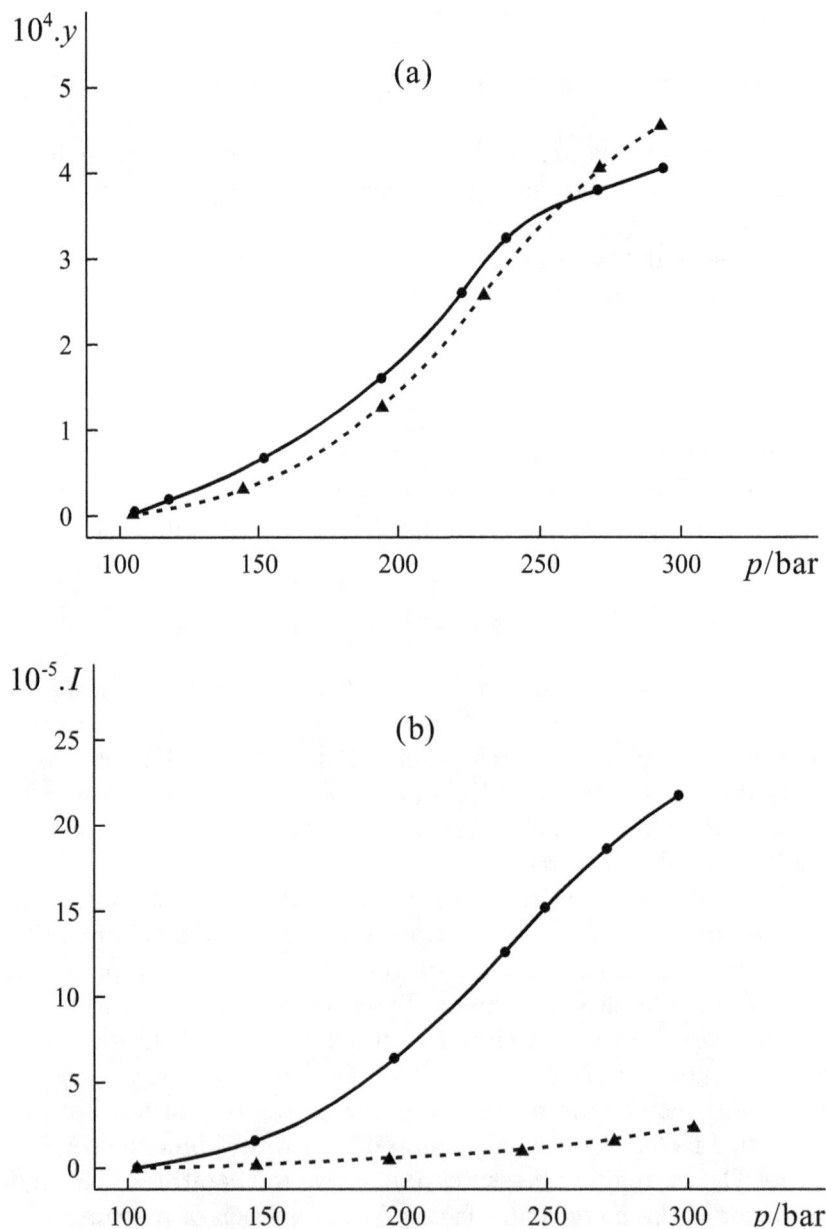

Figure 6.1: a) Solubility of caffeine in supercritical CO_2 as a function of pressure. b) Incremental factor of caffeine in supercritical CO_2. Solid curves: 60 ^oC; Dashed curves: 80 ^oC.[1]

[1] Extracted from S. Li, G. S. Varadarajan, and S. Hartland, "Solubilities of Theobromine and Caffeine in Supercritical Carbon Dioxide: Correlation with Density-Based Models", *Fluid Phase Equilibria*, 68 (1991): 263-280.

contributor to green house effect. The study of the solubility of gases in liquids was the origin of Henry's law that we used as a reference law to define one of the standard states for mixtures. If at the temperature of an experiment the solvent has very low vapor pressure, so that it can be neglected, the equilibrium between the pure gas in the gaseous phase and that dissolved allows us to write, on the basis of the equality of chemical potentials of a substance in different phases in equilibrium [see eqn. (5.12)],

$$\mu_2^\infty + RT \ln x + \int_0^p V_2^\infty \mathrm{d}p = \mu_2^\ominus + RT \ln \frac{p}{p^\ominus} \tag{6.4}$$

and taken into account the relation between both standard chemical potentials, given by eqn. (5.11), it is possible to obtain the solubility of the gas

$$\ln x = \ln \frac{p}{k_i^{\mathrm{H}}} - \int_0^p \left(\frac{V_2^\infty}{RT} \right) \mathrm{d}p \tag{6.5}$$

The Poynting effect that contributes with the last term in eqn. (6.5) can be important in some cases, for instance to calculate the gas solubility in blood under high pressure, which is important for scuba diving at great depths. It is interesting that this activity has an important physicochemical role in adopting the use of helium as a diluting gas for oxygen when the diver has to breath at very high pressures (high water depth); this is due to the low solubility of helium in an aqueous medium, including blood.

6.4   Ionic Solutes

In previous chapters, and in other sections of this chapter, we have analyzed the deviations from ideality of gases and liquid mixtures, as well as those observed in solutions of non-ionic solid and gaseous solutes. Now, we will describe the deviations from ideality of systems formed by solutions that contain ionic solutes. Ionic systems were intensely studied between the end of the nineteenth century and the beginning of the twentieth century, and the interest they received was due, to a large extent, to the fact that their solutions present macroscopic properties, which are different to those found in non-ionic solutions. Before analyzing the molecular differences, it is helpful to remember why electrolytes can be distinguished from other non-ionic solutes. Svante Arrhenius was who resolved this problem qualitatively through the hypothesis of total dissociation of strong electrolytes, when these were dissolved in polar liquids; the better known examples at that time (ca. 1912) were mainly aqueous solutions. The increasing importance of electrochemical processes that combine a chemical change with the electric work of redox reactions, and imply the circulation of an electric current through the solutions, strongly stimulated the study of ionic media. In such liquids, the electric charges are transported by the movement of ions when an electric field is applied.

One hypothesis normally incorporated to the stoichiometric calculations used in chemistry is the electroneutrality of the solutions; that means, the charge due to cations present in the solution must be equal to that of the present anions. Now, it is convenient to further analyze this situation because it has consequences in the thermodynamic treatment of ionic solutions. Let's suppose that a sphere of 1 cm radius contains NaCl dissolved in water, and imagine that the solution is not electrically neutral (i.e., a small difference exists between the cationic and the anionic charges). If we consider that the sphere has 10^{-12} mole of singly charged cations in excess to the anions, a simple calculation shows that the electrostatic potential in the surface of that sphere will be approximately 86,000 V. Obviously, a system with such an electrostatic potential will have a strong interaction with other material systems. Thus, the evidence clearly shows how a small difference between the positive and negative charges present in the solution, which would not affect appreciably the chemical potential of the salt; however, it will have a very strong effect over other material systems. Normally, in the labs, the systems having electrolytes do not show such large electrostatic interactions, and hence it becomes natural that they are considered electrically neutral.

On the other hand, if one analyzes a charged phase, its equilibrium state at constant T and p will depend on the existence of an external electrostatic potential and on the charge that each phase has; in this case, it will be possible to use electric energy and do electric work. Obviously, eqn. (1.18) must contain a term for the electric work to account for the contribution of this type of energy to the thermodynamic properties of the system. As indicated in the general equation (1.18), whenever electric work exists in a phase, the intensive quantity Y will be the electric potential of the phase, and X will be its charge.

When the system containing ions is electrically neutral (no net charge), it will not interact so strongly with external electric fields, and then it will be possible to use the same expressions that have already been used for non-ionic systems. For example, mass transport between two phases, as those described in chapter 1, can only be used in ionic systems without including explicitly the term for electric work when the transport process does not imply a change in the state of charge of the phases. In figure 6.2, we schematically illustrate the allowed processes, which are the ones that maintain, in every case, the neutrality of the phases, and the processes that are not allowed.

Mass transfer processes between phases where their charge state changes during

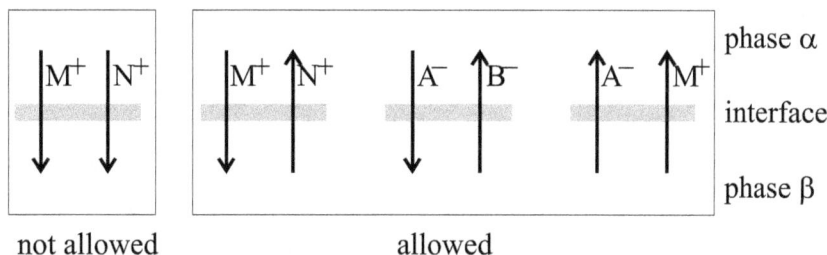

Figure 6.2: Different processes that do or do not keep electroneutrality in the phases during mass transport.

the transfer cannot be analyzed with the expressions derived before for non-electrolytes; the extension of these expressions that allow their application to ionic systems will be presented in chapter 9.

6.5 Some Differences Evidenced by Ionic Systems

Among the first properties that showed appreciable differences, depending if they are ionic or non-ionic solutions, are the colligative properties of the solutions. The depression of the melting temperature, or the osmotic pressure of the electrolyte solutions of a given concentration, are bigger than those corresponding to non-electrolyte solutions having the same molar concentration. Moreover, the ratio of both values is practically equal to the mole number of ions produced by dissociation of a mole of salt. This factor is known as the van't Hoff factor, i, and it is a demonstration of how real is the Arrhenius hypothesis of total dissociation of strong electrolytes. For example, upon dissolution of one mole of a salt of the type C^+A^- in water, there will be $i = 2$ mole of ions in the solution.

Another consequence of the total dissociation of strong electrolytes is observed in the concentration dependence shown by the partial pressure of hydrogen halides HX dissolved in water at room temperature; it presents a behavior distinct from that observed for other solutes that do not dissociate in solution. Figure 6.3 illustrates this difference: The partial pressure of HCl changes quadratically on the acid concentration.

When Henry's law describes the ideality behavior, thus fixing the activity scale, it is also agreed that the solute's partial pressure must be proportional to its concentration. If the salt $C_{\nu_+}A_{\nu_-}$ dissociates in solution according to

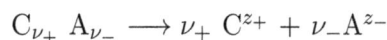

$$C_{\nu_+}\, A_{\nu_-} \longrightarrow \nu_+\, C^{z+} + \nu_-A^{z-}$$

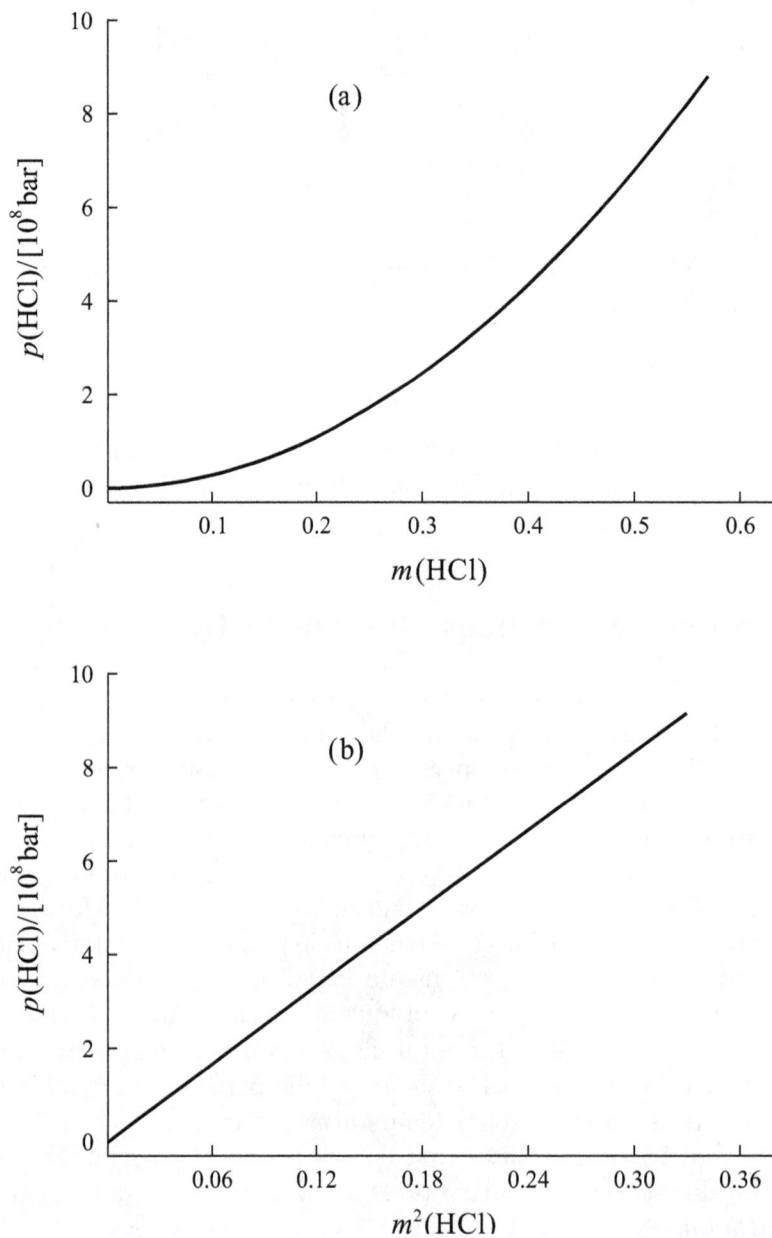

Figure 6.3: Pressure of aqueous HCl as a function of the molality m (a) and the square of molality m^2 (b).

its chemical potential can be described in terms of the activity of one mole of salt or of ν_+ mole of cation and ν_- mole of anion. Thus, we have

$$
\begin{aligned}
\mu_{CA} &= \mu_{CA}^{\infty} + RT \ln \frac{a_{CA}}{C^{\circ}} = \\
&= \nu_+ (\mu_+^{\infty} + RT \ln \frac{a_+}{C^{\circ}}) + \nu_- (\mu_-^{\infty} + RT \ln \frac{a_-}{C^{\circ}})
\end{aligned}
$$

Hence, $\mu_{CA}^{\infty} = \nu_+ \mu_+^{\infty} + \nu_- \mu_-^{\infty}$ and $a_{CA} = a_+^{\nu_+} a_-^{\nu_-}$. Since we want to use Henry's law as ideal behavior for electrolyte solutions, in order to keep linearity between activity and concentration, it is convenient to define the mean activity, $a_{\pm}$, such that

$$
a_{\pm}^{\nu} = a_+^{\nu_+} a_-^{\nu_-} \tag{6.6}
$$

where $\nu = \nu_+ + \nu_-$. Similarly, we define the mean activity coefficient, $\gamma_{\pm}^{\nu}$, as

$$
\gamma_{\pm}^{\nu} = \gamma_+^{\nu_+} \gamma_-^{\nu_-} \tag{6.7}
$$

and the mean molality

$$
m_{\pm}^{\nu} = m_+^{\nu_+} m_-^{\nu_-} = \nu_+^{\nu_+} \nu_-^{\nu_-} m^{\nu} \tag{6.8}
$$

In the last expression, use has been made of the stoichiometric relation between the salt molality and those of the ions that form the salt: $m_+ = \nu_+ m$ and $m_- = \nu_- m$.

Another notable difference between solutions that contain ions and those whose solutes have no charges, appears as a consequence of the Gibbs-Duhem relation. Eqn. (5.39) can be written as

$$
\frac{(\partial \ln \gamma_1 / \partial x_2)_{p,T}}{(\partial \ln \gamma_2 / \partial x_2)_{p,T}} = -\frac{x_2}{1 - x_2} \tag{6.9}
$$

When $x_2 \to 0$, the rhs member of the previous expression will be zero and, in these conditions, for a non-ionic solute (cf. chapter 5) $(\partial \ln \gamma_1 / \partial x_2)_{p,T} = 0$ and $(\partial \ln \gamma_2 / \partial x_2)_{p,T}$ will be a finite number. On the other hand, for electrolyte solutions we get $\lim_{x_2 \to 0} (\partial \ln \gamma_2 / \partial x_2)_{p,T} = -\infty$. This behavior is illustrated in figure 6.4, where the values of $\ln \gamma_2$ of aqueous sucrose are plotted together with the same quantity for aqueous $CaCl_2$. The difference in the slopes $(\partial \ln \gamma_2 / \partial m)_{p,T,m \to 0}$ for both solutes is remarkable.

The difference is the result of the range where the different intermolecular forces are active. For ions, the primary interactions between them are typically electrostatic, and are given by Coulomb's law, which makes the potential energy vary with the distance r separating two molecules as r^{-1}. It is said that Coulombic forces are of long range because, for particles having a molecular diameter σ, the Coulombic energy between charges of opposite signs separated a distance of 10σ is still one tenth

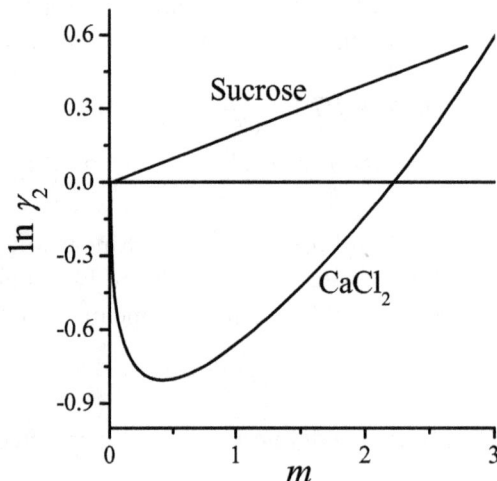

Figure 6.4: Logarithm of the activity coefficient as a function of the molality for aqueous solutions of sucrose and CaCl$_2$, at 298 K.

of the attractive energy of the two spheres in contact. On the other hand, forces of the Lennard-Jones type are of short range, and the potential energy is reduced to one tenth when the two particles are separated from contact to a distance of approximately $1,5\sigma$. This calculation shows clearly that when the attractive energy varies as r^{-6}, the interaction intensity (cf. figure 2.6) is weaker compared with Coulomb interactions.

6.6   Debye-Hückel Model

The characteristic properties of salts dissolved in polar solvents, especially in water, led Arrhenius to conceive the hypothesis of total dissociation of the salts when they are dissolved. A short time afterwards, it was established empirically that strong electrolytes, that is, those that dissociate completely in solution, show deviations from ideality even at very low salt concentrations. At the same time the deviations were observed to be linear in the square root of the ionic force, which is defined as

$$I = \frac{1}{2}\sum_i z_i^2 m_i \qquad (6.10)$$

where m_i is the molalilty of ion i.[2] For example, the logarithm of the solubility of a strong and scarcely soluble salt, MX, varies linearly with $I^{1/2}$ when adding to the solution increasing amounts of a strong electrolyte, NY, having no ions in common with MX (provided that the total strength is small).

In 1923, Peter Debye and Enrich Hückel proposed a model for ionic solutions that proved to be extremely successful and thus the deviations from ideality of the electrolytes could be calculated. The Debye-Hückel model important step was to look at the system from the perspective of a given ion in the solution, which is called the central ion. The derivation of the equations that describe the behavior of strong electrolytes, according to the Debye-Hückel model, will be given in detail in chapter 9. Here, only a qualitative description will be made, showing how the electrolyte solution can be viewed in the perspective of this model. Let us consider a central ion to have positive charge; this central ion will on average be surrounded by a distribution of charge having an opposite sign and, as a whole, the same amount of charge; in this way, the solution's electroneutrality will be preserved. It must be clear that the opposite charge is not concentrated in one particular particle but corresponds to a charge distribution surrounding the central ion. This is called *ionic atmosphere*. Figure 6.5 depicts the ionic atmosphere around a central positive ion.

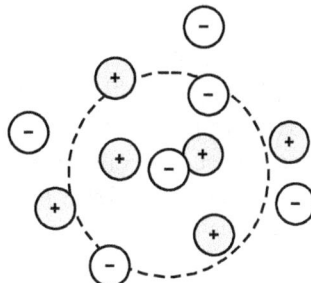

Figure 6.5: Scheme of the ionic atmosphere around a central positive cation.

The ionic atmosphere will be nearer to the central ion the higher the electrolyte concentration in the solution is. This means that the charge of the central ion will be more efficiently neutralized when the salt concentration increases (i.e., increasing the ionic strength). Consequently, the deviations from ideality reflected in the value of $\ln \gamma_{\pm}$ will be more negative as the ionic strength increases, at least for saline solutions that are not excessively concentrated.

The Debye-Hückel equation for the mean activity coefficient of an electrolyte of

---

[2]Note that for 1:1 electrolytes, like NaCl, $I = m$, the ionic strength can be also defined in terms of molarity.

cation's charge z_+ and anion's charge z_- indicates that its logarithm depends on the square root of the ionic strength, according to the equation

$$\ln \gamma_\pm = -\frac{A\, z_+ \mid z_- \mid I^{1/2}}{1 + Ba I^{1/2}} \tag{6.11}$$

where A and B are coefficients that depend only on the properties of the solvent and the temperature, the detailed expressions of which can be obtained starting with the equation that can be found on page 275. The only parameter specific of the salt is a, the distance of closest approach between the salt's cation and anion.

When the ionic force is very low, the previous expression becomes the limiting law of Debye-Hückel and reduces to

$$\ln \gamma_\pm = -A\, z_+ \mid z_- \mid I^{1/2} \tag{6.12}$$

This limiting law is applicable only in a very reduced range of electrolyte concentrations; e.g., $I \leq 10^{-3}$ molal for aqueous electrolytes having ions with single charges and at ambient temperature. It is incorrect to think that eqn. (6.12) is obtained from eqn. (6.11) for $a \to 0$, because if the centers of cation and anion could be superposed ($a = 0$), the electrolyte would be totally associated, forming neutral ion pairs.

There is a wider range of concentrations, but still a restricted range, where one can apply eqn. (6.11). Simply as a guide, at concentrations above 0.01 molal, the Debye-Hückel model cannot be applied for aqueous solutions of (1-1) strong salts at room temperature. In cases where concentrations are larger, it is necessary to add to eqn. (6.11) a linear term in the ionic force having a positive sign. The extended equation that can be used is

$$\ln \gamma_\pm = -\frac{A\, z_+ \mid z_- \mid I^{1/2}}{1 + I^{1/2}} + CI \tag{6.13}$$

The value of the Ba term in eqn. (6.11) approaches unity for water as the solvent, at room temperature, because $B = 3.286$ nm^{-1} and the size of simple ions is close to 0.3 nm. On the other hand, for electrolyte mixtures, that is, systems with more than two ionic species, eqn. (6.11) cannot be easily extended, keeping its thermodynamic consistency; the latter condition requires that

$$\frac{\partial \ln \gamma_\pm (\text{MX})}{\partial m(\text{NY})} = \frac{\partial \ln \gamma_\pm (\text{NY})}{\partial m(\text{MX})}$$

for mixtures of electrolytes MX and NY. This Maxwell relation is applicable only if both salts have the same value of Ba.

Figure 6.6 illustrates the change in $\ln \gamma_{\pm}$ with the square of molality for NaCl and $ZnSO_4$ aqueous solutions, where the limiting slope has been indicated in each case with dashed lines.

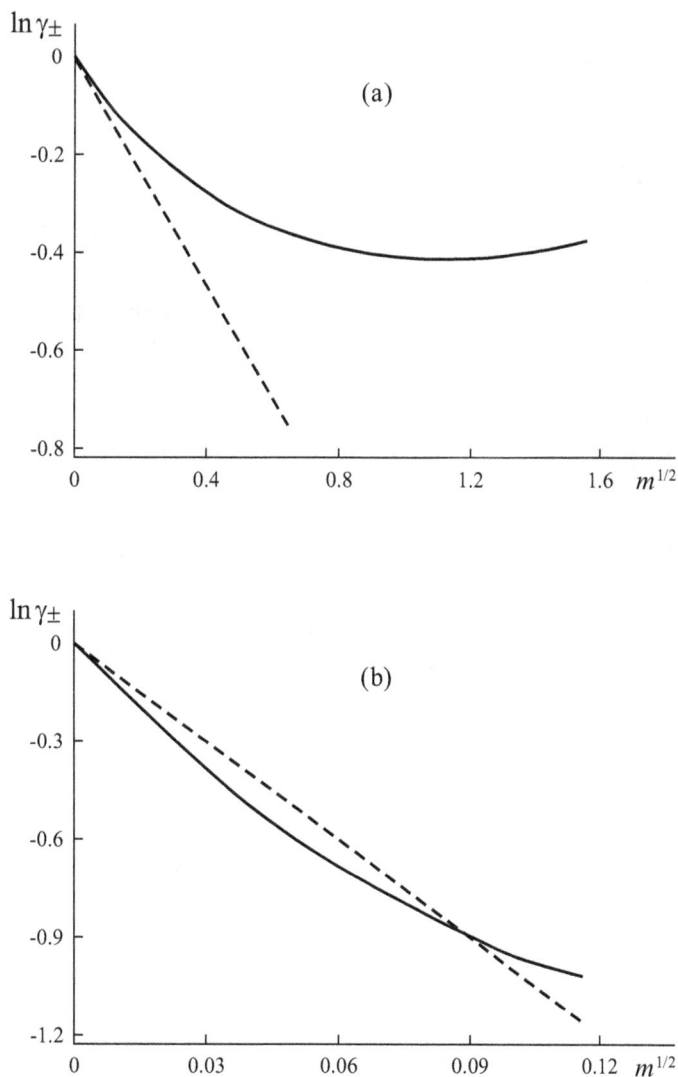

Figure 6.6: Logarithm of the mean activity coefficient, $\ln \gamma_{\pm}$, as a function of $m^{1/2}$ for aqueous solutions of NaCl (a) and $ZnSO_4$ (b). Dashed lines account for the Debye-Hückel limiting law. Note the different x-axis scale for NaCl and $ZnSO_4$.

In panel (a) of this figure, the behavior of a typical strong electrolyte, like aqueous NaCl, is described. The increase of its mean activity coefficient observed at high electrolyte concentration may be explained, for example by means of a model of hydration due to Robinson and Stokes, which assumes that a certain amount of water is fixed to the ions as hydration water; these molecules are assumed not to contribute to the total quantity of solvent molecules where the ions are dissolved. Also, it can be explained with the Pitzer treatment, to be presented in the next section.

In panel (b) of figure 6.6, the case of $ZnSO_4$ in water at low concentrations and room temperature is illustrated; it may be observed that the value of $\ln \gamma_\pm$ in dilute solutions is smaller than that predicted by the limiting law. This is due to the fact that the present ions have two charges and their electrostatic interaction is sufficiently large to obey the central approximation of the simple Debye-Hückel model (cf. chapter 9), which leads to eqn. (6.11). This approximation implies that the thermal energy must be on average bigger than the electrostatic energy between the two oppositely charged ions. N. Bjerrum in 1926 introduced the concept of ionic association in order to explain this phenomenon, assuming that the deviation from ideality is due to the fact that some ions are forming ion pairs, which have no net charge. That is, since the ionic strength in this solution is only due to free ions, it's value will be smaller than that calculated using the stoichiometric concentration of all ions. This hypothesis can be applied to explain the behavior of $ZnSO_4$ illustrated in figure 6.6b.

The two corrective models that have been presented are convenient to predict the behavior of electrolyte solutions, but the hydration model is not rigorously valid.

6.7   Concentrated Solutions of Electrolytes

In the previous section, we introduced the idea that due to the narrow range of concentrations where the Debye-Hückel equation (6.11) is applicable, it is necessary to use semiempirical equations, such as eqn. (6.13), to extend the validity of predictive equations up to moderately concentrated solutions. In 1973, fifty years after Debye and Hückel formulated their model, K. Pitzer developed the ionic interaction model which represented an important progress in the theory of ionic solutions and is specially applicable to aqueous solutions, due to the more abundant availability of experimental data.

Pitzer's model proposes that G^{ex} for a simple ionic solution of the electrolyte $M_{\nu_M}X_{\nu_X}$, containing n_1 mole of solvent, may be expressed through contributions of the long range electrostatic interaction, which depends on I (as in Debye-Hückel model), plus terms of specific interactions among the ions dissolved in the solution.

Thus, G^{ex} can be expressed by

$$G^{\text{ex}} = RT \left[w_o f(I) + \frac{2}{w_o} n_M n_X \left(B_{\text{MX}} + \frac{1}{w_o} n_M z_M C_{\text{MX}} \right) \right] \tag{6.14}$$

where w_o is the solvent's mass in kg that contain n_M and n_X mole of ions. The function $f(I)$ represents the long range electrostatic interaction and is a function of the ionic force, the temperature, and of the solvent's properties. To obtain $f(I)$, Pitzer used the same model as Debye and Hückel; that is, he started with the Poisson-Boltzmann equation (eqn. (9.4)), but he also considered the quadratic term in the series expansion of the electrostatic potential ψ_j. In this way, the following expression was obtained for the Coulombic part of the G^{ex}

$$f(I) = -\frac{4A'I}{3b} \ln \left(1 + bI^{1/2} \right) \tag{6.15}$$

where $b = 1.2 \text{ kg}^{1/2} \text{ mol}^{-1/2}$ and $A' = A \, z_+ \mid z_- \mid$ is the Debye-Hückel limiting slope [see eqn. (6.11)]. As expected, at infinite dilution eqn. (6.15) coincides with the limiting law of Debye-Hückel given by eqn. (9.12), so that there is no difference between the two models when $I \to 0$.

The coefficients B_{MX} and C_{MX} give an idea of the short range interactions among two or three ions, respectively; that is, they are similar to the virial coefficients for an ionic mixture (cation-anion). Strictly speaking, B_{MX} and C_{MX} are combinations of the true interaction coefficients between two (λ_{ij}) and three (μ_{ijk}) ions.

$$B_{\text{MX}} = \lambda_{+-} + \frac{\nu_M}{2\nu_X} \lambda_{++} + \frac{\nu_X}{2\nu_{M+}} \lambda_{--} \tag{6.16}$$

$$C_{\text{MX}} = \frac{3}{2} \left(\frac{\mu_{\text{MMX}}}{z_M} + \frac{\mu_{\text{MXX}}}{\mid z_X \mid} \right) \tag{6.17}$$

In principle, the λ_{ij} and μ_{ijk} coefficients depend on the ionic strength, but the dependence of the ternary coefficients on I is weak and can be neglected. Hence, only the coefficient B_{MX} is function of I. Pitzer proposed, based on empirical knowledge, the following dependence of B_{MX} on I, leading to

$$B_{\text{MX}} = \beta_{\text{MX}}^{(0)} + \frac{p\beta_{\text{MX}}^{(1)}}{\alpha^2 I} \left[1 - (1 + \alpha I^{1/2}) \exp(-\alpha I^{1/2}) \right] \tag{6.18}$$

where $p = 2$, $\alpha = 2 \text{ kg}^{1/2} \text{ mole}^{-1/2}$, and both $\beta_{\text{MX}}^{(i)}$ coefficients are electrolyte specific.

Summarizing, it may be said that the expression for G^{ex} in the Pitzer model has three adjustable parameters: $\beta^{(0)}$, $\beta^{(1)}$ and C. Since Pitzer's treatment starts from

G^{ex}, the expressions derived from it, such as $\ln\gamma_\pm$ and the osmotic coefficient (see chapter 9), must necessarily comply with the Gibbs-Duhem relation expressed in the form given by eqn. (5.39).

Substituting eqn. (6.15) and eqn. (6.18) in eqn. (6.14), and differentiating with respect to the mole number of the ionic species i, we get the excess chemical potentials and the corresponding expressions for the activity coefficients γ_i. The following expression for the electrolyte's mean activity coefficient is finally obtained:

$$\ln\gamma_\pm = -3\,|z_M z_X|\,A\left[\frac{I^{1/2}}{1+bI^{1/2}}+\frac{2}{b}\ln(1+bI^{1/2})\right]+m\left(\frac{4\nu_M\nu_X}{\nu}\right)B_{MX}^\gamma +$$

$$+3m^2\left[\frac{2\,|z_M z_X|^{1/2}\,(\nu_M\nu_X)^{3/2}}{\nu}\right]C_{MX} \tag{6.19}$$

where B_{MX}^γ is given by eqn. (6.18) with $p=1$.

Eqn. (6.19) describea the activity coefficient of aqueous electrolytes over a wide range of concentration, that may reach $m>1$ mole/kg, or near the solution saturation.

For NaCl in aqueous solution, at 298 K, the values of the coefficients in eqn. (6.18) are $\beta^{(0)}=0.0765$ kg/mole, $\beta^{(1)}=0.2664$ kg/mole, and $C=0.000635$ kg^2/mole^2, adjusted in the concentration interval up to $m=6$ mole/kg. Using these values over that concentration interval, the activity coefficient of aqueous NaCl was calculated and the results are illustrated in figure 6.7. For the sake of comparison, the values predicted by the extended Debye-Hückel equation (eqn. (6.13)) with $Ba\neq0$, $a=0.4$ nm, and $C=0.055$ kg/mole, adjusted in the same concentration interval, are also plotted in the same figure. Figure 6.7 clearly reveals that Pitzer's model gives a very precise description of the thermodynamic properties of the solution over a wide concentration range.

6.8   Partial Phase Equilibrium or Osmotic Equilibrium

For a system having two phases in equilibrium, and in one of them a solute is dissolved, the phase equilibrium will be broken and the system will evolve toward a new equilibrium situation. This is because the chemical potential of the solvent is different in both phases. The easiest way of regaining the phase equilibrium would be for the solute to diffuse from one phase to the other. However, if the solute is not able to diffuse (i.e., when the equilibrium is established between pure solvent in the vapor phase and the solution, or between a crystalline solvent phase and the solution), the only way to reach phase equilibrium will be for another thermodynamic

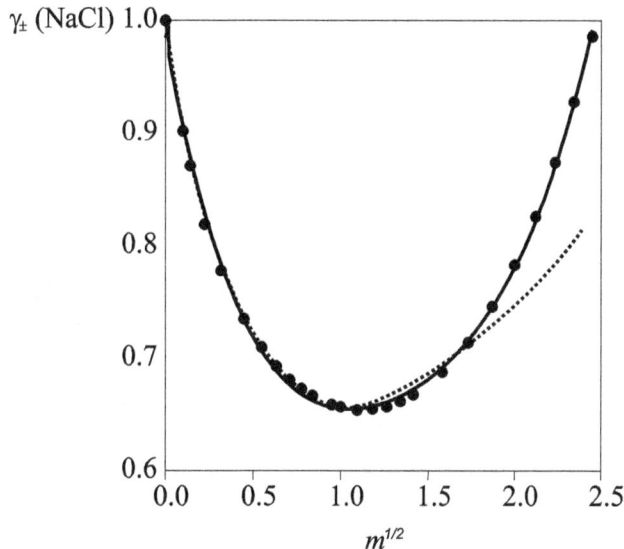

Figure 6.7: Mean ionic activity coefficient for an aqueous solution of NaCl at 25 ^{o} as a function of the square root of the concentration. • experimental values;——— Pitzer's eqn. (6.19); $\cdots\cdots$ Debye-Hückel's extended equation (eqn. (6.13)).

variable to change its value in the phase containing the pure solvent. Thus, it will be possible that this effect can compensate the change in the chemical potential of the solvent, which is in the solution due to the added solute, as indicated by Raoult's law. In this case, the equilibrium state is established through a single component of the solution, the solvent, and at least one component, the solute, cannot be in the two phases. That is why in section 1.6 we denominated *osmotic* or *membrane* equilibrium to this type of equilibrium. The real situations more frequently found are the following:

- A non-volatile solute in solution and in equilibrium with the solvent's vapor: The solute cannot go to the vapor phase.

- A solution in equilibrium with a solid solvent: The solute cannot go into the solvent crystals.

- A solution in equilibrium with a pure liquid solvent through a semipermeable membrane: The solute cannot pass through the membrane.

These situations of phase equilibria between solution and pure solvent exist frequently in nature or in applications, especially for dilute solutions. We will see that

184   *Matter and Molecules*

they are associated with the colligative properties of the solutions. Traditionally, the most often use of a colligative property is for the determination of the molecular weight of solutes, especially when the condition of ideality exists in the system. Nevertheless, there are other important applications, such as the determination of excess chemical potentials, allowing also the analysis of systems in osmotic equilibrium, in both natural or laboratory situations.

Figure 6.8 shows the chemical potential change of a pure solvent with temperature, and also that of a dilute solution at constant pressure.

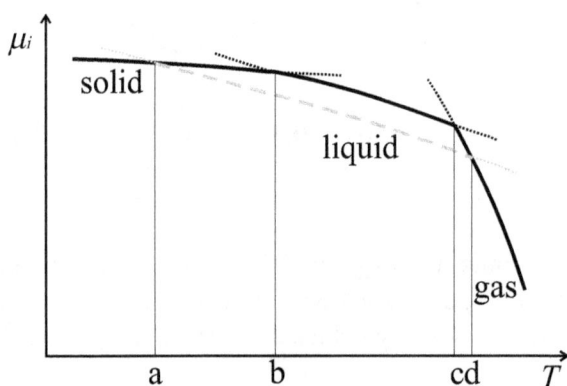

Figure 6.8: Chemical potential change of a pure substance ———, and of the same substance dissolving a non-volatile solute − − −. The presence of the solute causes the lowering of the solvent's melting point and the increasing of the solvent's boiling point. Vertical lines indicate the melting temperature of the solution, a, and of the pure solvent, b. Boiling temperature of the pure solvent, c, and of the solution, d. The dotted lines denote metastable states.

It has been mentioned that when there is a solution containing a non-volatile solute, the chemical potential of the solvent decreases with respect to that corresponding to the pure solvent, μ_1^*. The liquid-vapor equilibrium curve will be displaced toward lower pressures in an almost parallel curve to that of the pure solvent (especially for dilute solutions). It may be seen that, at ambient pressure, the new phase equilibrium will be attained at a different temperature. For the liquid-solid equilibrium, a new equilibrium temperature will be lower (depression of the melting temperature), while for the coexistence of liquid and vapor phases, the equilibrium temperature will be higher (ebulloscopic elevation).

It is then obvious that colligative properties are very useful to determine activity coefficients of substances of known molecular weight. However, it must be taken into account that to maintain the phase equilibrium, it is the solvent's chemical potential that must vary, because this is the substance found in both phases. That is, the colligative properties allow the direct determination of the solvent's chemical potential in the solution, and after this, it is necessary to use the Gibbs-Duhem relation to determine the excess chemical potential of the solute in the dilute solution. First, we will analyze the colligative properties in general, as a way to determine the components' activities in the solutions, and then we will discuss the particular case of electrolytes and the isopiestic method, which is of great importance to establish the excess chemical potential of electrolytes, although it can also be applied to solutions of non-ionic solutes.

6.9 Depression of the Melting Point and Ebulloscopic Elevation

These phenomena are related to the study of dilute solutions of non-volatile solutes, and the derivation of their thermodynamic relations allows the description of both properties: Melting temperature depression and boiling temperature elevation. Here, we will deduce how the depression of the melting temperature depends on the solvent's activity. Let's assume that there is equilibrium between the pure solid solvent and the solution at T, thus the chemical potentials of those species that can be transferred between both phases will be equal; that is,

$$\frac{\mu_1(\mathrm{sln}, T)}{T} = \frac{\mu_1^*(\mathrm{s}, T)}{T}$$

Since in this case equilibrium can only occur if the change of the solvent's melting temperature compensates the difference in the solvent's chemical potential in the solution, due to the presence of solute, it will be interesting to relate the solvent's chemical potential at T with that corresponding to its natural melting temperature, T_m. For the solid phase, we have

$$\frac{\mu_1^*(\mathrm{s}, T)}{T} - \frac{\mu_1^*(\mathrm{s}, T_m)}{T_m} = \int_T^{T_m} \frac{H_1^*(\mathrm{s})}{T^2} \mathrm{d}T$$

Now, the equilibrium established between the solid solvent and the solution at T implies that

$$\mu_1^*(\mathrm{s}, T) = \mu_1(\mathrm{sln}, T, x_1) = \mu_1^*(\mathrm{l}, T) + RT \ln a_1^{\mathrm{R}}$$

Since

$$\frac{\mu_1^*(\mathrm{l}, T)}{T} - \frac{\mu_1^*(\mathrm{l}, T_m)}{T_m} = \int_T^{T_m} \frac{H_1^*(\mathrm{l})}{T^2} \mathrm{d}T$$

and, remembering that $\mu_1^*(\mathrm{s}, T_m) = \mu_1^*(\mathrm{l}, T_m)$, one finally arrives to,

$$-\ln a_1 = \int_{T_m}^{T} \frac{H_1^*(\mathrm{s}, T)}{RT^2} \mathrm{d}T - \int_{T_m}^{T} \frac{H_1^*(\mathrm{l}, T)}{RT^2} \mathrm{d}T \tag{6.20}$$

From the previous equation, it results that

$$\ln a_1 = -\int_{T}^{T_m} \frac{\Delta_m H_1^*}{RT^2} \mathrm{d}T \tag{6.21}$$

where $\Delta_m H_1^*$ represents the melting enthalpy change of the pure solvent, given by $H_1^*(\mathrm{l}) - H_1^*(\mathrm{s})$. This derivation of the depression of the melting temperature has the advantage of introducing clearly a simple and exact thermodynamic relation, which may be applied to the resolution of different problems. Assuming that $\Delta_m H_1^*$ is constant between T and T_m, the following expression is obtained for ideal solutions,

$$\ln x_1 = \frac{\Delta_m H_1^*}{R}\left(\frac{1}{T_m} - \frac{1}{T}\right)$$

If the solution is dilute, it is possible to expand the logarithm and, remembering that $x_2 \approx M_1 m$, where M_1 is the solvent's molecular weight expressed in (kg/mole) and m the molality of the solution, one obtains the well-known linear relationship between the depression of the melting temperature, θ, and the concentration of the ideal dilute solution

$$\theta \equiv (T_m - T) = \frac{RT_m^2 M_1}{\Delta_m H_1^*} m = k_{cr1} m$$

where the proportionality constant k_{cr1} is usually called cryoscopic constant.

Once the solvent activity has been determined, as indicated in eqn. (6.21), it is possible to calculate the solute's activity coefficient using the Gibbs-Duhem relation (5.39). It must be remembered that in the case non-volatile solutes it is usual to adopt Henry's standard state (infinite dilution) for the solute and Raoult's standard state (pure liquid) for the solvent. This is not a problem at all; on the contrary, this flexibility is an advantage, but one has to be careful to derive the corresponding relations on the basis of eqn. (5.39). Measurements of the depression of the melting temperature have very often been used to determine the activity coefficients of solid solutes dissolved in water, especially electrolytes. In these cases, it should be taken into account that strong electrolytes dissociate into ions and, therefore, should be considered the concentration of all the dissolved species in the solution. It is interesting to contrast that statement and what happens in the following equilibrium:

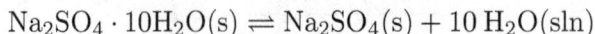

$$\mathrm{Na_2SO_4 \cdot 10H_2O(s)} \rightleftharpoons \mathrm{Na_2SO_4(s)} + 10\,\mathrm{H_2O(sln)}$$

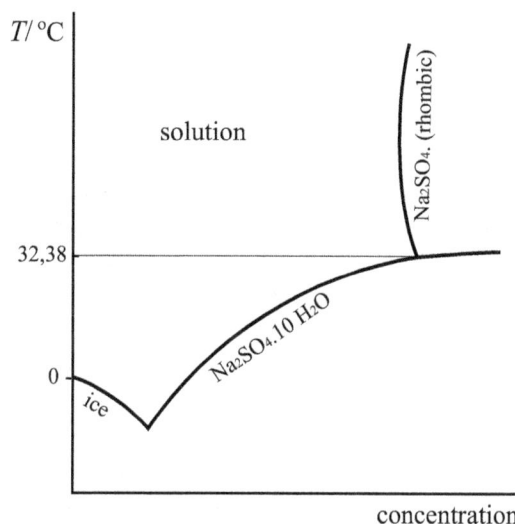

Figure 6.9: Phase diagram of the system $Na_2SO_4 - H_2O$.

In this case, the transition phenomenon is between two crystalline forms of sodium sulfate, in equilibrium with the saturated aqueous solution (see figure 6.9). The transition temperature at 32.384 ^oC ($p = 1$ bar), is also affected by the presence of other solutes dissolved in the solution. The ions Na^+ and SO_4^{2-} are some of the species found in the three-phase system and are sensible to the thermodynamic changes produced in the aqueous solution when solutes are added. These species translate the changes of chemical potential between the solid phases and the solution in equilibrium. When the added solutes are ionic and contain the cation Na^+ or the anion SO_4^{2-}, the relevant concentration to produce a modification of the transition temperature is that of all the ions, except Na^+ and SO_4^{2-}. Effectively, the cation Na^+ is in the crystalline lattice of both solids and also exists in the solution. That is why if you add an ionic solute containing the sodium cation, its contribution to the concentration of Na^+ will, in general, be negligible with respect to that in the saturated sodium sulfate solution (*ca.* 3.7 molal). That is, the small quantity of sodium cation of the added solute will only slightly modify the sodium cation chemical potential of Na^+, already present in the system. Conversely, if the added ionic solute does not contain sodium cations, then the relevant concentration of cations will be that of all the cations present.

6.10   Osmotic Pressure

The osmotic phenomenon is of great importance for living beings because it is present practically in all cells, where the cellular membrane is the semipermeable barrier (always permeable to water, but selectively permeable to some solutes). The process of osmosis also is important in technology, such as in the case of reverse osmosis to obtain pure water from brine and other aqueous media. Also, it is important to determine the water concentration in food and fluids.

A representation of an osmotic pressure measuring device is shown in figure 6.10, in which a vessel containing a pure solvent is separated by a membrane from another vessel containing the same substance and a dissolved solute. If we consider that the solute cannot go through the membrane (semipermeable membrane), the unique way of achieving equilibrium between both liquid phases is by increasing the pressure difference between both phases. This will produce and enhancement of the water chemical potential in the solution (cf. eqn. (1.12)). In this way, it will compensate the decrease of the water's chemical potential in the solution due to the presence of solute. The increase of water's chemical potential is produced by hydrostatic pressure, $p - p_o = \rho g h$, where p_o is the atmospheric pressure, ρ is the density of the solution, and h is the height of the liquid column. The condition for equilibrium is

$$\mu_1^*(p_o, T) = \mu_1(p, T, x)$$

where $\mu_1(p, T, x)$ can be written in terms of the Raoult scale as

$$\mu_1(p, T, x) = \mu_1^*(p, T) + RT \ln a_1$$

Finally, the chemical potential of pure water at the two different pressures (p and p_o) can be accounted for, as already done when treating the Poynting effect, yielding

$$-\ln a_1 = \int_{p_o}^{p} \frac{V_1^*}{RT} dp = \frac{p - p_o}{RT} V_1^* = \frac{\pi V_1^*}{RT} \tag{6.22}$$

where π is the osmotic pressure at the temperature T. This expression is the known as the van't Hoff equation, for the case of dilute ideal solutions.

If two solutions of the same solvent have the same osmotic pressure, they have the same osmolality. Let us consider a living cell, for instance an erythrocyte immersed in water; since the cell membrane is semipermeable, water will penetrate into the cell in order to dilute the solution inside and achieve equilibrium. Since the osmotic pressure of the intracellular medium is high, water transfer into the cell will produce the membrane disruption (lysis) because the equilibrium is never attained. That is why, when medicines are directly injected into the blood system, it is necessary that the solution in which they are dissolved have the same osmotic pressure as the intracellular liquid, as is the case of the so-called physiological solution. However, one should be aware that in this case it is not enough to know that both solutions have identical osmolalities, because the cell membrane is impermeable only for some

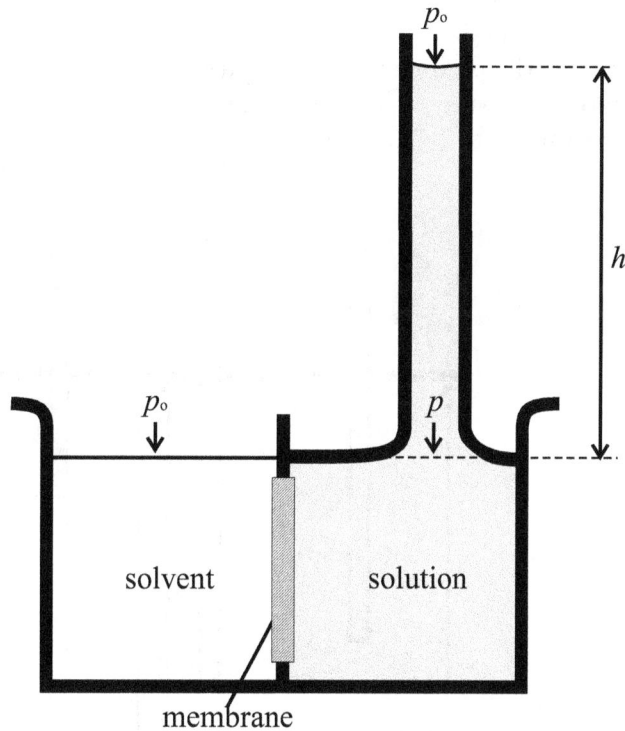

Figure 6.10: Osmotic pressure measuring device.

solutes (e.g., salts), but not for others. Even for an iso-osmotic solution with the intracellular medium, lysis will occur upon injection when it contains permeable solute. When two solutions have the same effect on the cell, they are called isotonic solutions, and this can be confirmed when the erythrocytes suspended in these two solutions have the same volume as the serum.

The Osmotic Pump

Around 1970, it was common in chemistry departments of universities to discuss a situation where a pipe is introduced into the ocean with the lower end, covered by a semipermeable membrane. We will analyze three immersion steps:

- At the begining, the pipe is empty. This situation is maintained, while the end closed by the membrane is at a depth above 231 m.

This depth, produced a hydrostatic pressure that is equal to the osmotic pressure of sea water, and, when that depth is reached, pure water may be able to penetrate inside the pipe (see figure 6.11 (a)).

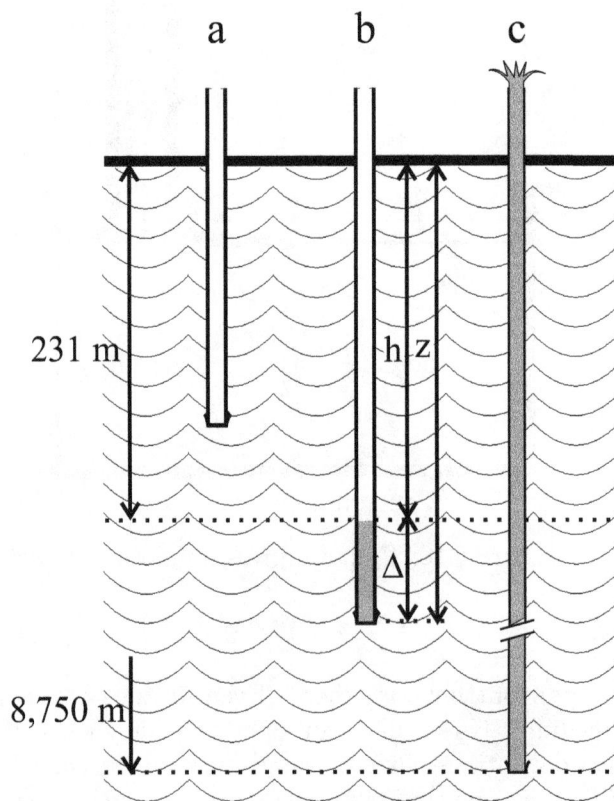

Figure 6.11: Osmotic pump.

- As the pipe is immersed beyond 231 m, reaching for instance a depth z, as illustrated in figure 6.11 (b), more pure water will enter the pipe. Pure water density inside the pipe, ρ, is lower than the density of the ocean water, ρ_m. So, in order to maintain the same pressure on both sides of the membrane, the height of the column of pure water inside the pipe, given by $(z - h)$ in the figure, must be larger than the depth of the pipe beyond 231 m. This difference, noted by Δ in the figure, will increase as the pipe is immersed deeper in the ocean.

- When the pipe reaches a great depth, it will be possible that the pure water inside the pipe raises above the ocean's surface. Then, it would be possible to let the excess water fall on top of a rotating device (like a wind mill) that can produce electricity and return to the ocean again, as illustrated in figure 6.11 (c). That would imply that it is possible to obtain energy without doing any work.

Is this possible? If it were, would not that violate the second principle of thermodynamics?

O. Levenspiel and N. de Nevers published a detailed study of this hypothetical problem[3]. We will make the same analysis, using the descriptive methodology of solutions that we have developed in this and previous chapters.

In order to describe the thermodynamic situation of the problem, it must be considered that, in an equilibrium state, the water's chemical potential in the internal and external sides of the membrane must be identical. Thus, at depth z, it should be valid that[4]

$$\mu^*_{\text{in}}(p_{\text{in}}, z) = \mu_{\text{ex}}(p_{\text{ex}}, z, x_1)$$

where x_1 represents the water concentration of the ocean. The chemical potential of water inside the pipe will be

$$\mu^*_{\text{in}}(p_{\text{in}}, z) = \mu^*(p^{\circ}, z = 0) + \int_{p^{\circ}}^{p_{\text{in}}} V_1^* \mathrm{d}p$$

were p° is the atmospheric pressure. Outside the pipe, in the ocean, the chemical potential of water is given by

$$\mu_{\text{ex}}(p_{\text{ex}}, z, x_1) = \mu^*(p^{\circ}, z = 0) + \int_{p^{\circ}}^{p_{\text{ex}}} V_1^* \mathrm{d}p + RT \ln x_1$$

Since both chemical potentials of water must be identical, and considering the hydrostatic internal and external pressures due to the respective columns of water, we have

$$V_1^* \rho g(z - h) = V_1^* \rho_m g z + RT \ln x_1 \tag{6.23}$$

---

[3]O. Levenspiel, and N. de Nevers, "The Osmotic Pump", *Science*, 183 (1974): 157-160.

[4]In this example, all the chemical potentials refer to the water component.

where ρ_m is the density of the ocean water, g the acceleration of gravity, and h the depth of the water meniscus inside the pipe, as indicated in figure 6.11 (b). Calculating h we get,

$$-V_1^* gh = V_1^* gz(\rho_m - \rho) + RT \ln x_1$$

Since $\rho_m > \rho$, it is possible that pure water flows through the upper end of the pipe $(h < 0)$, if z is sufficiently large. Putting the values of the quantities in the previous expression, we get that this effect would occur if $z > 8{,}750$ m.

From this analysis, the generation of work by means of the osmotic pump seems feasible. Does it mean the second principle does not apply to this case? It should be observed that in the equations used, several approximations have been made, and they are easily detected. However, they do not change the essence of the problem, with one exception. It has been assumed that the concentration of salt in the sea is independent of depth. Is that so? We will see that, if the system is in equilibrium, that assumption is wrong. To show this, let's imagine we have a very long closed pipe that is buried in the ground, full of water, at a constant temperature through all the pipe. That implies that the chemical potential of water is the same along the tube. Now, let us consider two points of this pipe, separated by a distance z, measured from the surface of the Earth, and let p° be the atmospheric pressure. We have

$$\mu^*(p, z) = \mu^*(p^\circ, z = 0) + \int_{p^\circ}^p V_1^* \mathrm{d}p = \mu^*(p^\circ, z = 0)$$

Being the pressure p at the base of the pipe given by the atmospheric pressure, plus the hydrostatic pressure due to the water column, the last equation can be written as

$$\mu^*(p, z) = \mu^*(p^\circ, z = 0) + V_1^* \cdot \rho gz = \mu^*(p^\circ, z = 0) \qquad (6.24)$$

If the previous two equations are valid, the term $V_1^*(p - p^\circ)$ or the equivalent term $V_1^* \rho gz$ must be zero.

Since $(V_1^* \rho)$ is equal to the mass of one mole of water, M_1, one gets that $M_1 gz = 0$, which is an absurd and impossible result. This wrong conclusion results from having omitted the contribution of the gravitational energy, which for the long pipe (imagine a pipe of 2 km in length) is important. In general, this contribution will not be appreciable in the majority of processes, because they occur in the surface of the

Earth; however, the gravitational effect is responsible, for example, for the change of density and composition of air when we are above the planet's surface.

Going back to the pipe filled with water, the liquid in the upper part can be made to fall down and work may be obtained from it; hence, this gravitational work, U_g, must be taken into account and added to the chemical potential that the excess of water in the upper part of the pipe has. Thus, the first equality of eqn. (6.24) must be replaced by

$$\mu^*(p, z) = \mu^*(p^\circ, z = 0) + U_g + V_1^* \rho g z$$

with $U_g = -M_1 g z$, and the problem is solved.

Whenever in the pipe's interior there is a solution, we will have

$$RT \ln \frac{x_1(z)}{x_1(z = 0)} = V_1^* g z (\rho_m - \rho) = M_1 g z \left(1 - \frac{\rho_m}{\rho}\right) \qquad (6.25)$$

That is, the composition of a system that is at equilibrium changes with the depth, and x_1 will be smaller (the salt solution will be more concentrated) the bigger z is. Then, if we apply to the problem of the osmotic pump the criterion of equal equilibrium chemical potentials, taking account the gravitational energy (water on the surface of the ocean has a gravitational energy greater than at a given depth), instead of eqn. (6.23), we get

$$RT \ln x_1(z) = V_1^* \rho g (z - h) - V_1^* \rho_m g z \qquad (6.26)$$

Expressing $x_1(z)$ as a function of $x_1(z = 0)$, according to eqn. (6.26), it finally gives

$$-V_1^* \rho g h = RT \ln x_1(z = 0)$$

That is, h is independent of z and has the constant value 231 m, (if the system is effectively in equilibrium), Hence, work cannot be obtained with the osmotic pump if the ocean is a system in equilibrium.

As epilogue of this analysis, it should be pointed out that real ocean is not at equilibrium; chemical analysis shows that the content of salt is quite independent of the depth. That is, the density of the ocean is very uniform and has a weak dependence on depth. Evidently, then, there is a possibility to obtain work with the osmotic pump without violating the second principle of thermodynamics, because the system from which energy can be obtained is not really at equilibrium.

6.11   Determination of Mean Activity and Osmotic Coefficients

The methods used to determine $\ln \gamma_{\pm}$ for electrolyte solutions are of two types: (a) those measuring the properties directly related to the solute's chemical potential and, (b) those that measure properties related to the solvent's chemical potential. Type (a) includes the determination of the solubility of a slightly soluble salt in presence of another electrolyte having no ions in common with that salt, and also the determination of the electric potential of galvanic cells. Methods of type (a) are applicable to a restricted number of ionic systems, while those of type (b) are among the most important methods to determine the activity coefficient of electrolytes, and will be analyzed in chapter 9. The latter type directly determines the value of the solvent's chemical potential and usually consists in measuring a colligative property of the solution. Also, type (b) includes methods that measure other properties related to the colligative properties, like the isopiestic method, where the solvent's chemical potential in the solution being studied is equal to that of the solvent in equilibrium in a reference solution. The equilibrium is established through the solvent vapor phase, which is in equilibrium with both solutions. This method is applicable to all non-volatile solutes, and especially used for the study of electrolytes. It is convenient to remember that, according to eqns. (5.15) and (5.16), the activity of the solvent (in Raoult's scale) is $a_1 = p_1/p_1^*$.

In the case of methods depending on the determination of the solvent's activity, a_1, based on the determination of colligative properties, it is very usual to calculate the so-called *osmotic coefficient* of the solution (ϕ) to express the deviations from ideality. For a binary solution, ϕ is defined by

$$- \ln a_1 = \nu m M_1 \phi \tag{6.27}$$

where ν is the number of particles produced in solution per mole of solute (for the case of non-ionic solutes $\nu = 1$), m is the solute's molality, and M_1 is the molecular weight of the solvent expressed in kg/mole. In the case of solutions having more than one solute, the previous expression is written as the sum over the concentration of all the independent species. In this situation, $\ln a_i = \sum m_i M_1 \phi$, where m_i is the molality of species i, which can be an ion or a molecule having no charge.

In table 6.1, the numerical advantage of using, for some solutions, the osmotic coefficient to express the deviations from ideality can be seen. In the three cases, the deviations from ideality of solutes are appreciable and the value of the solvent's activity coefficient (Raoult's scale) do not seem to reflect it. However, the values of the osmotic coefficient, which is unity for ideal solutions, reflect more clearly the deviations.

Table 6.1: Osmotic coefficients of selected aqueous solutions

Solute	$m/(\text{mole/kg})$	γ_2^{H}	γ_1^{R}	ϕ
Sucrose	0.5	1.085	0.9996	1.041
NaCl	2.0	0.614	1.004	0.912
CaCl$_2$	0.1	0.518	1.0008	0.854

In order to calculate the solute's mean activity coefficient from the osmotic coefficient of the solution, Gibbs-Duhem's relation must be used; for example, starting from eqn. (5.37) we have, for a binary electrolyte solution,

$$-\frac{1}{M_1}\mathrm{d}\ln a_1 = \nu m\,\mathrm{d}\ln(m\gamma_\pm)$$

according with the definition of ϕ (eqn. (6.27)), we obtain

$$\mathrm{d}\phi + \phi\frac{\mathrm{d}m}{m} = \mathrm{d}\ln(m\gamma_\pm) \tag{6.28}$$

Upon integration, this equation yields

$$\ln\gamma_\pm = (\phi - 1) + \int_0^m (\phi - 1)\,\mathrm{d}\ln m \tag{6.29}$$

On the other hand, if one wishes to calculate the osmotic coefficient from the eletrolyte's mean activity coefficient, then

$$\phi = 1 + \frac{1}{m}\int_0^m m\,\mathrm{d}\ln\gamma_\pm \tag{6.30}$$

An alternative expression of practical use may be obtained integrating by parts the previous expression, giving

$$\phi - 1 = \ln\gamma_\pm - \frac{1}{m}\int_0^m \ln\gamma_\pm\,\mathrm{d}m \tag{6.31}$$

Here, the deviation from ideality at concentration m is given by $\phi - 1$, which is equal to the value of $\ln\gamma_\pm$ at the same concentration minus the mean value of $\ln\gamma_\pm$ over the interval $[0, m]$.

We have already presented the isopiestic method as an important method to measure the solvent's activity. The device to perform isopiestic measurements is schematized in figure 6.12.

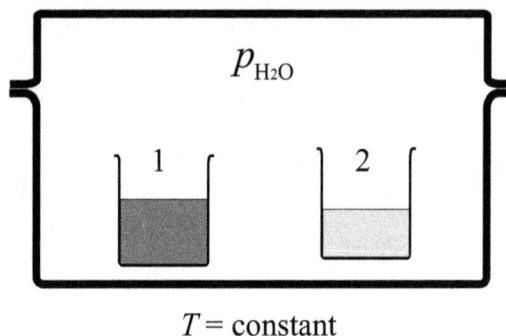

Figure 6.12: Scheme of a device to perform isopiestic measurements and calculate the solvent's activity.

The solution being studied and one reference solution prepared in the same solvent are put inside two beakers (or better, metal cups), which are at the same temperature, and are separated from the environment in a closed vessel (i.e., the two liquids are in contact with a single vapor phase). At the beginning of the experiment, the two solutions will have different vapor pressures. For instance, if the vapor pressure of the test solution (1) were bigger than that in the reference solution (2), the solvent will distill from (1) to (2), and the equilibrium will be reached when both solutions have the same solvent chemical potential. If the solutes dissolved in both solutions are non-electrolytes, or electrolytes dissociated in the same number of ions ν per mole, then

$$m\phi = m_{\mathrm{ref}}\phi_{\mathrm{ref}}$$

where the subscript ref indicates the value in the reference solution, for which the value of the osmotic coefficient is known for different values of m. Once equilibrium between the two solutions is attained, the molality of each one is determined analytically and, since we know $\phi_{\mathrm{ref}} = f(m_{\mathrm{ref}})$, the osmotic coefficient of the solution being studied is calculated. An inconvenience of the isopiestic method for the determination of $\gamma_{\pm}$ is that it is necessary to use the Gibbs-Duhem relation, as described by eqn. (6.29). Therefore, it requires knowing the dependence of ϕ with m for very low molalities. The isoptiestic method cannot be used for concentrations below 0.1 molal, because otherwise the equilibration of the dilute solutions is excessively slow. In order to complete the integration required by the Gibbs-Duhem relation, as described in eqn. (6.29), it is necessary to use an adequate theory for concentrations below 0.1 molal, Debye-Hückel being the most frequent one. The requirement of relying on a theoretical expression to establish $\phi = f(m)$ for very dilute solutions is only applicable to electrolytes, because it is there where one observes appreciable deviations from ideality, even at high dilution.

Problems

Problem 1

The determination of the cryoscopic descent of cyclohexanol (2) solutions in benzene (1) allowed to establish the following relation between the osmotic coefficient, ϕ, and the cyclohexanol molality, m, according to the relation

$$\phi = \frac{1 - 0.237m}{1 + 0.940m}$$

which is valid over the 0.02 and 0.27 molal range.

a) Calculate the solute's activity coefficient for $m = 0.25, 0.15, 0.05$, and 0.03 molal, at the melting point of benzene (278.7 K).

b) If it is assumed that the ideal deviations are due only to dimerization of cyclohexanol, due to the formation of a hydrogen bond, and that the cyclohexanol monomer dissolved in benzene behaves ideally, calculate the value of K_{dim} at the four molalities given in (a).

c) Discuss the validity of the hypothesis made in (b).

Answers: a) $\gamma_2 = 0.605, 0.802, 0.893$, and 0.933 (for the solutions 0.25, 0.15, 0.05, and 0.03 molal, respectively); b) $K_{\text{dim}} = 3.48, 1.74, 1.42$, and 1.30 (kg/mole)

Problem 2

Table 6.2 records data of different properties of caffeine (I) and theobromine (II), as well as the solubility of these solutes in supercritical CO_2 at different solvent temperatures and pressures.[5]

---

[5]S. Li, G. S. Varadarajan, and S. Hartland, "Solubilities of Theobromine and Caffeine in Supercritical Carbon Dioxide: Correlation with Density-Based Models", *Fluid Phase Equilibria*, 68 (1991): 263-280.

Table 6.2: Properties of caffeine and theobromine

Property	Caffeine	Theobromine
Molecular weight / (g.mole^{-1})	194.2	180.18
Melting point / ^{o}C	238	351-357
Sublimation point / ^{o}C	178	290
$\Delta_m H$ / (cal.mole^{-1})	5,044	9,819
Density at 25 ^{o}C / (g.cm^{-3})	1.349	1.453
Dipole moment / D	3.83	3.11
Boiling point / ^{o}C	6,009	7,235
Sublimation pressure at 40 ^{o}C / bar	3.717×10^{-9}	
Sublimation pressure at 60 ^{o}C / bar	4.769×10^{-8}	1.542×10^{-12}
Sublimation pressure at 80 ^{o}C / bar	4.583×10^{-7}	5.445×10^{-11}
Sublimation pressure at 95 ^{o}C / bar	2.129×10^{-6}	5.695×10^{-10}

Caffeine

Temperature ^{o}C	Pressure bar	Density mol.dm^{-3}	Solubility g.dm^{-3}
60	106	7.465	2.74×10^{-6}
	148	13.55	5.05×10^{-5}
	200	16.45	1.80×10^{-4}
	242	17.69	3.06×10^{-4}
	298	18.83	4.14×10^{-4}
80	105	5.427	1.94×10^{-6}
	148	9.510	2.77×10^{-5}
	198	13.39	1.51×10^{-4}
	236	15.11	2.82×10^{-4}
	298	16.90	4.82×10^{-4}

Theobromine

Temperature ^{o}C	Pressure bar	Density mol.dm^{-3}	Solubility g.dm^{-3}
60	150	13.73	6.29×10^{-7}
	200	16.45	6.90×10^{-7}
	240	17.64	7.93×10^{-7}
	300	18.87	1.13×10^{-6}
80	150	9.707	6.76×10^{-7}
	200	13.50	1.07×10^{-6}
	240	15.25	1.20×10^{-6}
	268	16.14	1.34×10^{-6}
	300	16.95	1.77×10^{-6}

a) Calculate the values of the incremental factor I for the conditions of p and T given in table 6.2.

b) Calculate Φ_2^∞ for the conditions reported in the tables.

c) Calculate the ideal solubility of (I) and (II) for the conditions reported in table 6.2.

d) Calculate $\Delta_{dis}H_2^\infty = H_2^\infty(sln) - H_2^\theta$, at 200 and 300 bar.

e) Explain the differences of solubility between (I) and (II).

Answers: c) x_2^{id} (60°C) = 0.070 (I), 6.01×10^{-4} (II); x_2^{id} (80°C) = 0.108 (I), 2.21×10^{-3} (II); d) $\Delta_{dis}H$ (200 bar) = 8.59 kJ/mole (I), 21.44 kJ/mole (II); $\Delta_{dis}H$ (300 bar) = 7.43 kJ/mole (I), 21.93 kJ/mole (II)

Problem 3

What is the thermal effect produced by the dilution of an aqueous NaCl solution, 0.01 molal, to take it to infinity dilution at 298.2 K? Compare the value obtained with $\Delta_{dis}H^\infty = 3.88$ kJ/mole.

Data: The slope in the limiting Debye-Hückel equation is $A_H = 0.81448\ RT$

Answer: $\Delta_{dil}H^\infty(NaCl) = 20.2$ J/mole

Problem 4

Calculate the solubility of NaCl in water at 298 K, using the data reported in the table, and the Pitzer equation for the activity coefficient of the saturated solution. Compare the obtained result with the experimental value ($m_{sat} = 6.145$ mole/kg).

species	Na^+(aq)	Cl^-(aq)	NaCl(s)
μ^o/RT	−105.651	−52.955	−154.99

Note: Since the activity coefficient depends on the ionic force, and this depends on the dissolved quantity of salt, find the answer for the previous problem using and iterative method, assuming that initially $\gamma_\pm = 1$. Calculate the equilibrium molality and, with it, the ionic strength; next, recalculate $\gamma_\pm$ and the new molality. Repeat the procedure until the molality does not change

Answers: $m_{sat} = 6.135$ mole/kg, $\gamma_\pm = 0.994$

Problem 5

Two closed vessels of 10.00 cm^3, at 298.0 K, are communicated to each other in the lower part of the vessels through an osmotic membrane, only permeable to water, as shown in figure 6.13.

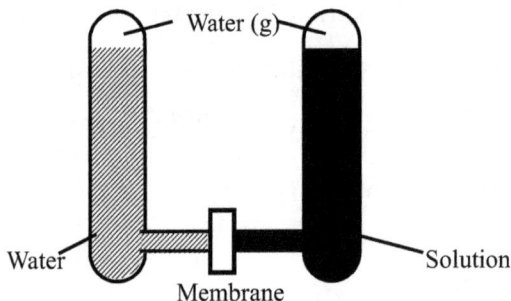

Figure 6.13: Two closed vessels communicated by an osmotic membrane.

Initially, the vessel on the left contains 9.00 cm^3 of water (in equilibrium with its vapor), and the vessel on the right contains 9.00 cm^3 of an aqueous solution that contains 0.001 mole of a non-volatile non-ionic solute and 0.490 mole of water (in equilibrium with its vapor). After attaining equilibrium, it may be observed that water moves from one vessel to the other.

a) In which direction does water go through the osmotic membrane? What mass of water moves from one side to the other, once equilibrium is established?

b) What is the pressure in each of the vessels?

Data: Neglect the effect of hydrostatic pressure. At 298.0 K, $p^*_{water} = 0.0317$ bar and $\delta^*_{water} = 1.000$ g/cm^3 (assume that p does not change)

Answers: a) 1 g of water is transferred from the left vessel toward the right one; b) $p_{left} = 0.0317$ bar, $p_{right} = 2,512$ bar

Chapter 7

Surface Phenomena

7.1 Introduction

In this chapter, we deal with systems in which the surface energy plays an important role. To make clear what we are talking about, it is important to contrast them with the systems that we have already studied in the previous chapters. These were constituted by homogeneous infinite phases and, up to now, when analyzing phase equilibria, we did not ask what happened in the regions of the system separating the phases. For instance, if we have the system composed by isobutiric acid and water, where there is partial miscibility at temperatures below 25.6 ^oC, at atmospheric pressure, a dense homogeneous phase rich in water and, over it, a phase rich in the organic acid, which is also homogeneous, is observed. Between the two phases, a meniscus will be seen separating them. This situation is illustrated in figure 7.1. The composition of the dense phase is the same from the bottom of the vessel up to the region close to the meniscus; from there on, and until the upper homogeneous phase is encountered, the system's composition suffers a strong change, going from the characteristic value of the dense phase (β) to that corresponding to the less dense phase (α). This region is denominated *interface*, and its width is not defined in a rigid manner. It can only be said that all change in composition, and of other intensive thermodynamic properties of the system, takes place in that zone. It can be assumed that there are two planes separating the two phases in equilibrium, and that they define the interface (cf. figure 7.1); their positions are only conditioned by the need that all Inhomogeneity is contained between both planes (they are totally inside the interface).

Also, one can think of a drop of liquid (far from its critical temperature) in equilibrium with its vapor. Near the surface of the drop the density changes drastically from the large value characteristic of a liquid phase to a value corresponding to the

Phase α

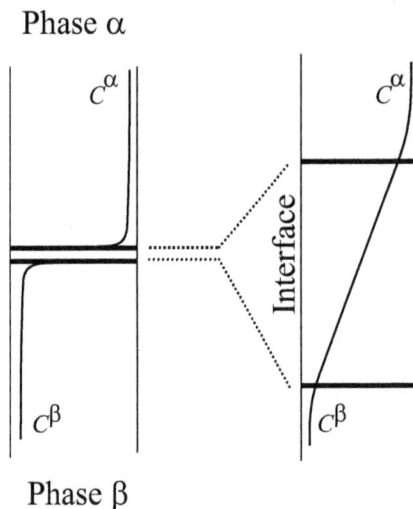

Phase β

Figure 7.1: Schematic representation of the change in concentration of the mixture composed by isobutiric acid in water through the interface separating the two phases α and β.

vapor in equilibrium. It is not a question, here, to discuss in detail how this occurs; nevertheless, it is important to understand that in all surfaces that limit portions of matter, the molecules are, specially in what concerns to energy and spatial arrangement of atoms, very different to those prevailing in a bulk phase. The molecules in the surface are subjected to a force field different than that existing in its bulk phase. For instance, in a solid, its surface clearly establishes a limit where its molecules can exist and, consequently, neighboring molecules on the surface of the solid do not have neighboring molecules in half of the space surrounding them. Also, it must be considered that the transference of matter between phases in equilibrium always goes through an interface; hence, it is very important to describe its characteristics.

At the molecular level, all pieces of matter constituting a phase are somewhat different at the surface than in the bulk of the phase; hence, it is convenient to separate the description of the interface from that of the bulk phase. However, it must be clear that only when there is an appreciable number of molecules at a surface the properties of the superficial molecules affect the global behavior of the system. Let's take, for example, a cubic sample of solid matter having a length L at each side. Its surface will be the sum of the areas of the six cube's faces, that is $6 \times L^2$. If the material being considered is constituted by spherical molecules having a diameter t, the number of molecules on its surface will be $N_s = 6 \times (L/t)^2$, and the total number of molecules contained in the cube will be $N_v = (L/t)^3$. The ratio between these two quantities is $(N_s/N_v) = 6 \times (t/L)$. Taking 0.3 nm as a typical value for t, when the cube has $L = 1~\mu$m, there will be approximately two molecules on the surface for every 1,000 inside the cube; if $L = 0.1~\mu$m = 100 nm, there will be 100 molecules inside the cube for two molecules on the surface. This example shows that only for very small samples of matter the surface properties will appreciably modify the behavior of the whole solid sample. In general, for particles of the size of $1~\mu$, or bigger, the contribution of atoms and molecules that are on the surface will be small. On the other hand, it is convenient to say that the energy of surface molecules will be larger (less negative) than in the bulk solid. There are a series of processes that, although they start with a very small number of molecules (like crystallization), the change of molecular surface energy cannot be neglected.

There are other processes where the properties of the surface are of interest even when ratio of the number of surface molecules and those in the bulk is small. This is the case of processes in which the surface molecules acquire special importance, because it is only through them that the molecules forming the bulk matter are able to interact with those in a different phase or with external systems; molecules of different phases can only communicate each other through the interface. This is of special interest in the important field of chemical recognition, and it is also the case of transport through membranes and interfaces. Also, for heterogeneous catalysis to occur, surface atoms and molecules are fundamental for the global reaction to take place.

7.2   Thermodynamic Description of the Role of Surfaces

When the first principle of thermodynamics was presented in chapter 1, we realized that the changes of internal energy of a system are determined by the exchange of thermal energy and by the work made over, or by, the system. It was made clear that the term δw includes all the types of work: Mechanical, magnetic, electric, gravitational, as well as changes of surface size. This last work, δw_{sup}, is the one we are now dealing with. Like other types of work, δw_{sup} is given by the product of an intensive variable and the differential of an extensive variable. The change in the system's energy due to surface work will imply a change in the specific area exposed, $\mathcal{A}$, which is the extensive variable corresponding to this type of work. Then, for a reversible process including surface work, the change in internal energy is written, considering that it is subjected to mechanical and surface work [see eqn.(1.18)], by

$$dU = TdS - pdV + \gamma d\mathcal{A} + \sum_i \mu_i dn_i \tag{7.1}$$

where $\gamma = (\partial U/\partial \mathcal{A})_{V,S,n_i}$, the interfacial tension, is the natural intensive variable for surface work $\delta w_{\text{sup}} = \gamma \mathcal{A}$.

Likewise, if the system is modified increasing the interfacial area and keeping constant the number of moles of all components, the temperature as well as the pressure, there will be an increase in the system's Gibbs energy. The inclusion of superficial work in eqn. (1.22) provides a definition for the interfacial tension in terms of the Gibbs energy, given by $\gamma = (\partial G/\partial \mathcal{A})_{T,p,n_i}$.

According to the discussion in the previous section, surface phenomena will manifest themselves in the region designated as the interface of the system, which is separating two adjacent bulk phases. In order to give a thermodynamic description of it, let's start analyzing how the concentration of a component i may be described in the surface. A convenient way of doing this is to write the superficial concentration as if it were an excess quantity, that is, the difference between the total number of moles of the component i and the number of moles in the bulk phases α and β. Hence, it is possible to write it as

$$n_i^s = \Gamma_i \mathcal{A} = n_i - n_i^\alpha - n_i^\beta = n_i - c_i^\alpha V^\alpha - c_i^\beta V^\beta \qquad (7.2)$$

where Γ_i represents the surface excess of component i in the interface. It must be noted that this quantity may be positive (i.e., an excess) or negative (i.e., a defect).

Let us imagine for a moment an interface separating two fluid phases, like that depicted in figure 7.1. In equilibrium, the concentrations of component i in both phases are denoted by c_i^α and c_i^β, both being independent of the volume element that is considered. However, when the considered volume element includes part of the interface, as we have already seen, the concentrations will change. The region of change, the interface, where concentrations go from the values corresponding to the first bulk phase to those in the second bulk phase, has a width of a few molecular diameters; that is, it is extremely thin. It is obvious that the calculation of Γ_i, like for any other thermodynamic property of the interface, will depend on the positions where the virtual geometric planes delimiting the interface are considered (cf. figure 7.1).

In order to clarify this apparent arbitrariness, it is necessary to clearly identify the position and the width of the interface. One possibility is to consider that the interface has no volume (it is a geometric plane having only two dimensions) and assign all the available space to the bulk phases α and β; that is, assume that $V = V^\alpha + V^\beta$. Then, the interface is restricted to a plane localized at a given position y, as shown in figure 7.2. If the plane defining the position of the interface is displaced to a new position y', one of the phases will increase its volume in $\Delta V = \mathcal{A}(y' - y)$ and the volume of the other phase will decrease by the same ΔV, thus no physical change will occur in the system. In spite of this, the displacement of the plane that defines the interface will modify the surface excess of component i, according to the equation

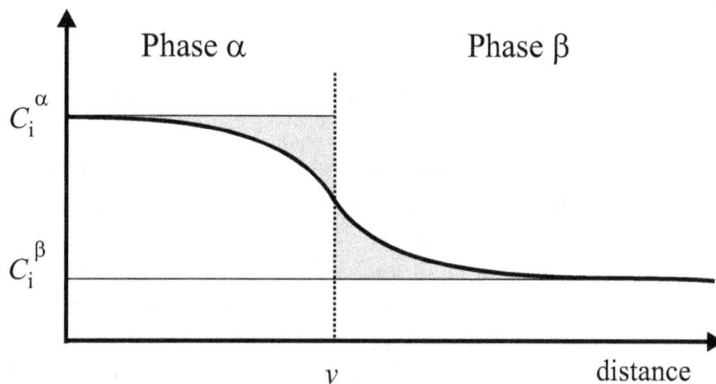

Figure 7.2: Position of the interfacial plane. Once the position y is fixed, the thermodynamic surface properties can be univocally defined.

$$\Delta\Gamma_i = (c_i^\beta - c_i^\alpha)(y' - y)$$

obtained from ΔV and the definition of surface excess of component i, given by eqn. (7.2).

When the inhomogeneous interface is replaced by a simple plane, it is really assumed that the concentration maintains the value c_i^α up to the surface plane, and from there adopts discontinuously the value c_i^β. According to figure 7.2, this implies overestimating the value of the concentration of one component until reaching the interface plane, and from there on underestimating it. Depending on the position of the geometric plane, the concentration will present an excess, a defect or no effect at all (when the excess before reaching the surface is equal to the defect, after it). For solutions, and especially in the case of binary systems, it is usual that the plane's position (frequently denoted as the Gibbs surface) is fixed so that the excess of solvent is made zero (i.e., $\Gamma_1 = 0$ when component 1 refers to the solvent). Under these conditions, the quantities $\Gamma_{i(1)}$ refer to excesses or defects with respect to the Gibbs surface where the solvent has no excess (or defect). In this way, the thermodynamic surface properties can be defined without any ambiguity.

A very simple case is that of the vapor-liquid interface for a one-component system, i, where $(c_i^\beta - c_i^\alpha)$ differs substantially from zero, except for the region close to the critical point. In such a system, it is easy to choose the position of the Gibbs plane dividing both phases so that $\Gamma_i = 0$. In that situation, $n_i = n_i^\alpha + n_i^\beta$ (cf.

eqn. (7.2)) and, consequently, $c_i^\alpha \, dV^\alpha + c_i^\beta \, dV^\beta = 0$, which corresponds to the plane placed in a position so that the areas indicated in figure 7.2 are equal. Near to the vapor-liquid interface, the fluid molecules' density changes very abruptly (in fact, discontinuously) and the interface profile has a very thin width (really, it is a line), which can be considered to define a geometric plane. Hence, it is quite reasonable in this case that the excess should be measured from that plane–the curve in figure 7.2 would be a step function.

In order to make this argument clearer to the reader, it will be formulated once more, but now for systems with more than one component. When the system has two or more components, the procedure to determine the position of the interfacial plane, as presented, will allow a zero superficial excess for only one component. Usually, the solvent is chosen as the reference component having a zero surface excess. On the other hand, the other components will present finite values of surface excess or defect, relative to the solvent. This is indicated with the symbol $\Gamma_{i(j)}$, where the subscript, which is between parenthesis, indicates the component chosen to place the interfacial plane, j, which designates in general the solvent, being $\Gamma_{j(j)} = 0$.

Once the interfacial plane has been defined, the thermodynamic properties of the surface per unit area of the interface, f^s, can be defined in a similar manner to Γ, according to the equation

$$F^s = f^s \mathcal{A} = F - \sum_i n_i^\alpha F_i^\alpha - \sum_i n_i^\beta F_i^\beta \qquad (7.3)$$

where F represents any extensive thermodynamic function, F_i^α and F_i^β are the partial molar quantities of F in phases α and β, and n_i^α and n_i^β are the number of moles of component i in the phases α and β, respectively. The quantity v^s (interface's volume per unit of area) is zero because it was assumed that the interface is a geometrical plane (the Gibbs plane) and, hence, it has no volume.

Analyzing the total process of equilibrium between bulk phases α and β, separated by the interface s, in terms of eqn. (7.3), it is observed that the extensive interfacial quantities can be described based on the properties of the total system, and of the bulk phases α and β. Thus, for instance

$$G^s = g^s \mathcal{A} = G - G^\alpha - G^\beta \qquad (7.4)$$
$$S^s = s^s \mathcal{A} = S - S^\alpha - S^\beta \qquad (7.5)$$
$$U^s = u^s \mathcal{A} = U - U^\alpha - U^\beta \qquad (7.6)$$

where the quantities having no superscripts are those corresponding to the global system. If the presence of an interface had no effect over the system, the values of all the thermodynamic superficial functions would be zero.

Now, we will derive an important relationship between the solution interfacial tension and the superficial excess of its components. The starting point is the Gibbs energy of the interface. Taking into account the fact that the volume of the interface is considered negligible, the relation between the internal energy and the interfacial Gibbs energy can be written as

$$G^s = U^s - TS^s \tag{7.7}$$

On the basis of eqns. (7.7) and (7.1), we get

$$\mathrm{d}G^s = -S^s\mathrm{d}T + \gamma\mathrm{d}\mathcal{A} + \sum_{i=1}^{C}\mu_i\mathrm{d}n_i^s \tag{7.8}$$

This equation assumes constant temperature in all over the system and uses the fact that the two phases are in equilibrium, so that the chemical potentials of all the substances in the system are equal in both phases and also at the interface.[1]

Integrating eqn. (7.8) according to the Euler method (chapter 3) we have

$$G^s = \gamma\mathcal{A} + \sum_{i=1}^{C}\mu_i n_i^s \tag{7.9}$$

Now, $\mathrm{d}G^s$ can be calculated using eqn. (7.9) and this expression can be compared with eqn. (7.9). This yields the equivalent Gibbs-Duhem relation for interfaces, given by

$$S^s\mathrm{d}T + \mathcal{A}\mathrm{d}\gamma + \sum_{i=1}^{C}n_i^s\mathrm{d}\mu_i = 0 \tag{7.10}$$

Then, dividing the Gibbs-Duhem equation given in eqn. (7.10) by the interface area, we get

$$\mathrm{d}\gamma = -s^s\mathrm{d}T - \sum_{i=1}^{C}\Gamma_{i(j)}\mathrm{d}\mu_i \tag{7.11}$$

The previous equation is the starting point to the study of thermodynamic properties of solution's interfaces. It is interesting to note that, if the geometric plane is placed

---

[1]It must be pointed out that the G^s definition adopted in eqn. (7.7) can include the term $\gamma\mathcal{A}$; this is done, for example, by Guggenheim. The form of defining G^s depends on which thermodynamic function one is interested to calculate. The form of G^s adopted here is particularly useful for processes at constant $(T, \mathcal{A})$, while the other one is suitable for processes at constant (T, γ).

in a position where there is no solvent excess or defect, the Gibbs-Duhem relation for the interface looses one term of the series, namely that corresponding to the solvent, because $\Gamma_{j(j)} = 0$.

If we consider a pure liquid substance in equilibrium with its vapor,[2] the surface tension only depends on temperature (cf. eqn(7.11)), because there is no surface excess ($\Gamma = 0$). Therefore,

$$s^s = -\frac{\mathrm{d}\gamma}{\mathrm{d}T} \tag{7.12}$$

Hence, the heat necessary to generate a surface of unit area will be given by the surface enthalpy

$$h^s = g^s + Ts^s = \gamma - T\left(\frac{\mathrm{d}\gamma}{\mathrm{d}T}\right) \tag{7.13}$$

For a solution's interface, the change of interfacial tension at a given temperature and pressure, according to eqn. (7.11), is

$$(\mathrm{d}\gamma)_{p,T} = -\sum_{i=1}^{C} \Gamma_i RT \mathrm{d}\ln\left(\frac{a_i}{\mathcal{C}^{\ominus}}\right) \tag{7.14}$$

where $\mathcal{C}^{\ominus}$ is defined on page 112. From this expression, it is possible to obtain the surface excess $\Gamma_{i(j)}$ for an ideal system as

$$\Gamma_{i(j)} = -\frac{1}{RT}\left(\frac{\partial\gamma}{\partial\ln c_i}\right)_{T,\mu_{k\neq i}} \tag{7.15}$$

This last equation, and its analogue for non-ideal systems (in which case c_i must be replaced by the activity), are known as Gibbs adsorption isotherm. They allow us to show that, for solutions where liquid and vapor are in equilibrium, the solutes that decrease the interfacial tension of the solution (e.g., detergents and many other organic compounds in aqueous solution), will have an excess of concentration at the interface with respect to that in the bulk solution. The opposite, a surface defect, occurs with those substances that increase the interfacial tension when their concentrations increase; this is the case of some ethers and of inorganic salts in aqueous solutions. Eqn. (7.15) illustrates the possibility of changing the interfacial tension of solutions by changing the composition of solutes, at constant temperature.

---

[2]Despite the surface tension, γ, defined for a pure liquid substance, refers to the interface appearing when the liquid-vapor equilibrium is established; the term *surface tension* is usually applied to the single liquid phase as well.

7.3   Curved Surfaces. Bubbles and Drops

If the restriction of geometric planes as interfaces is released, there will appear effects caused by the curvature of the interfaces. In the analysis of this problem, it will be assumed that the surface curvature does not affect the value of the interfacial tension. This supposition can be sustained by the fact that the curvature radius of the systems which we deal with is much larger than the distances where the surface effect manifests itself, which is the interfacial thickness.

Internal Pressure of Bubbles and Drops

Let us consider a liquid system having inside a vapor bubble of radius r. Imagine that the bubble's volume increases dV_b and that the corresponding volume change in the liquid is dV_l. The change of the total internal energy will be

$$dU = dU_l + dU_b = T dS_l - p_l dV_l + T dS_b - p_b dV_b + \gamma d\mathcal{A}$$

In order to establish the equilibrium condition on the basis of the internal energy, $dU = 0$, the process must be considered occurring at constant S and V (eqn. (1.8)), that is $dS_b = -dS_l$ and $dV_b = -dV_l$; hence, if the temperature is constant

$$(p_l - p_b) dV_b + \gamma d\mathcal{A} = 0$$

The bubble's volume change, $dV_b = 4\pi r^2 dr$ is equal to $-dV_l$, and the area change is $d\mathcal{A} = 8\pi r dr$. Replacing these changes in the previous expression, it is obtained the Young-Laplace equation, given by

$$(p_b - p_l) = \frac{2\gamma}{r} \tag{7.16}$$

Thus, under equilibrium, the pressure inside the bubble will be larger than the pressure in the bulk liquid, and that pressure difference increases when the bubble is smaller. A similar argument lets us deduce that the pressure inside a drop of liquid is also $2\gamma/r$ larger than the external pressure.

It is interesting to derive for a drop of radius r, the expression for equilibrium vapor pressure $p_v(r)$. The vapor will have a pressure that allows its equilibration with the liquid's drop. The chemical potential of the drop, μ_d, will be that of the pure liquid at the pressure existing inside the drop, p_d. Hence, according to Poynting's equation

$$\mu_d = \mu_d^* + V^*(l)(p_d - p_v^*)$$

and, when the state of equilibrium exists between liquid and vapor, we will have to a good approximation (cf. Problem 3 on page 232)

$$\mu^\ominus(T) + RT \ln \frac{p_v(r)}{p^\ominus} = \mu_d^* + V^*(l)(p_d - p_v^*)$$

Since $\mu^* = \mu^\ominus + RT \ln(p_v^*/p^\ominus)$, using the Young-Laplace equation to express the pressure differences, one obtains

$$\ln \frac{p_v(r)}{p_v^*} = \frac{2\gamma V^*(l)}{rRT} \tag{7.17}$$

Consequently, the vapor pressure that equilibrates both phases will be larger the smaller the drop's diameter is. This point will be discussed again in section 7.5.[3]

On the other hand, if we take as example a soap bubble, which is a system having a thin liquid film separating two gaseous phases, the internal pressure will be twice the value obtained in the previous example. This is due to the fact that in this system, both the internal as well as the external liquid film surfaces contribute to the surface work.

It is possible to perform a more general approach considering two volume phases and one interfacial region. In this case (cf. figure 7.3), R^α and R^β are the radii of curvature of the two menisci limiting the region we name as interface.

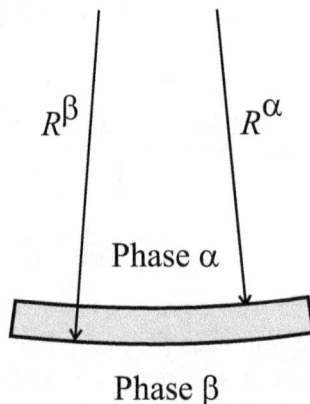

Figure 7.3: Diagram of an interface limited by two curved menisci.

In general, one assumes that

$$R^\beta - R^\alpha \ll \bar{R}$$

---

[3] A similar equation explains the condensation of liquids at $p < p_v^*$, known as Kelvin's equation.

where $\bar{R} = (R^\beta + R^\alpha)/2$. If this were not the case, the very concept of interfacial tension has no sense, practically all methods for the determination of γ assume that the thickness of the interface is small compared with $\bar{R}$. If the interface were a plane, the radius of curvature would be infinite, and in this example the pressure inside a bubble, p_b, will be identical to the pressure outside it.

Capillarity

The previous analysis can explain the ascent (descent) of a liquid inside a thin capillary tube. Figure 7.4 illustrates the case of the ascent (descent) of a liquid column inside a capillary tube of small diameter when immersed in liquids.

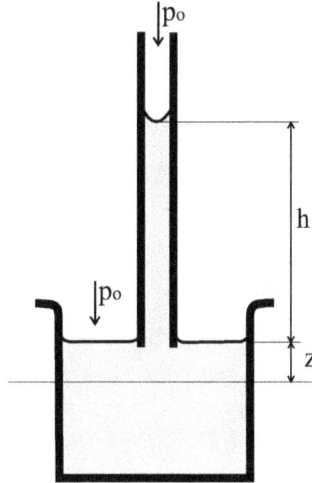

Figure 7.4: Capillarity phenomenon.

According to eqn. (7.16), in order that the external pressure (p_{ex}) at a bulk liquid surface be equal to the pressure inside the capillary tube at the same level, there must be a pressure difference at the meniscus, as dictated by the Young-Laplace equation, to compensate the hydrostatic pressure caused by the capillary liquid column. Hence,

$$\rho g h - \frac{2\gamma}{r} = 0$$

where r is the curvature radius of the meniscus. Finally,

$$h = \frac{2\gamma}{r\rho g}$$

Thus, the resulting height, h, inside the capillarity tube will be positive (capillary ascent) when the curvature radius is negative (i.e., a concave meniscus). On the other hand, if $h < 0$ (capillary descent), there will be a convex meniscus. Considering the meniscus to be (approximately) part of the surface of a hemisphere of radius r, the ratio between the radius of the sphere and that of the capillary is given by $-\cos\phi$, where ϕ is the contact angle between the meniscus and the capillary wall.

The contact angle is a well-known property that is present at the liquid-solid-gas interfaces, and is related to the wetting of solids or hydrophobic materials in aqueous media. The contact or wetting angle develops between a solid surface and the plane tangent to the liquid surface, measured from the liquid phase (cf. figure 7.5). The figure also shows, as examples, two liquid-solid systems that present different wetting and, naturally, they evidence different contact angles.

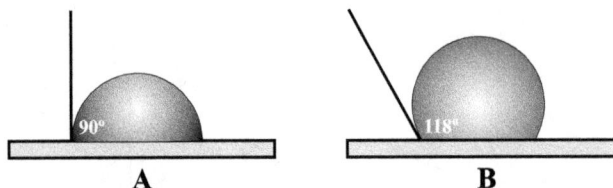

A   B

Figure 7.5: Contact angle between a drop of liquid and a solid surface in contact with it. For example, two liquid-solid systems showing distinct wetting are shown, with contact angles: $\phi_A = 90^o$ and $\phi_B = 118^o$.

Considering a liquid drop situated over a solid surface, the contact angle that is formed is the result of the balance of three forces, originated in the following interfacial tensions: i) γ_{SL}, between the solid and the liquid phases; ii) γ_{SV}, between the solid and the vapor phases; and iii) γ_{LV}, between the solid and the vapor phases. The contact angle can be modified by adding surfactants to the liquid solution, and therefore changing the values of γ_{SL} or γ_{LV}.

7.4   Solid-Gas Interface. Adsorption

So far, we have discussed the effect of having interfaces in the systems, without analyzing the specific kind of substance being part of it, or the type of phases that coexist. The thermodynamic principles presented in previous sections are still valid, with some small modifications, for all types of interfaces. The nucleation process, which will be discussed in another section, can be easily generalized so that it covers specifically the processes of precipitation, sublimation, boiling, and freezing, besides the condensation process that will be discussed more extensively. However, the study of the solid-gas interface deserves particular treatment due to the great importance of the physical phenomena associated to it, especially the adsorption phenomenon.

Molecular Aspects of Adsorption

The interaction between gas molecules and those on a solid surface generally lead to the adsorption process. We denominate adsorption the process by which one molecule (adsorbate) is retained by attractive interactions over a solid surface (adsorbent). On the other hand, absorption implies that molecules of adsorbate penetrate into the solid. Sometimes, the difference between the two phenomena is not clear (e.g., when a piece of cotton wool is wet with water, do we have adsorption or absorption of H_2O?).

When a solid surface is exposed to gas molecules, the latter will collide against the surface and, in many cases, some of them will be retained over the surface of the solid. Often, this situation implies that the surface contains sites of atomic or molecular size, where the gaseous molecules remain attached. These sites will usually correspond to the positions of some atoms in the crystalline lattice, or to defects in the lattice. We call cover fraction, θ, to the ratio between the number of surface sites occupied by the adsorbate, N_s, and the number of total sites on the solid's surface, N_T. Thus,

$$\theta = \frac{N_s}{N_T} \tag{7.18}$$

Two types of adsorption can occur: Associative, which refers to molecules that are adsorbed on the surface as a whole (there is no molecular fragmentation), and dissociative, when adsorbate molecular fragmentation occurs on the surface. For example, the adsorption of N_2 over alumina is an associative adsorption, but that of H_2 over platinum is typically dissociative.

Adsorption Isotherms

Even when a system has a negligible contribution of surface's molecules to the values of its properties, there may be great interest to know the processes that take place on its surface. That is the case of solid catalysts, where molecules absorbed on them, for instance gases, may suffer a chemical transformation. Catalysts need to have a large surface per unit mass, so that the number of sites per unit mass of the catalyst is very large, because is there where the adsorbate molecules will stick.

The isotherms of adsorption are curves that describe the dependence of θ on the adsorbate pressure at constant temperature, $\theta = f_T(p)$. A simple model considers that the adsorption process of gaseous adsorbate molecules occurs through a molecular bombing of a surface, and that every molecule that impinges on the surface is retained there. The degree of coverage will be determined by the balance between the adsorption and desorption processes. When the surface is clean (i.e., $(\theta \simeq 0)$), the degree of coverage will be proportional to the gas pressure, which is a relation analogous to Henry's law. The degree of coverage described by the adsorption isotherms is that of equilibrium, corresponding to the situation where the rate of adsorption is equal to the rate of desorption. The equilibrium condition can be represented by the equality of the chemical potential of the adsorbate in the gaseous phase and that on the adsorbent surface or, in other words, by the equality of the rates of the adsorption and the desorption processes. Starting with the latter process, in the case of associative adsorption, the process can be described by

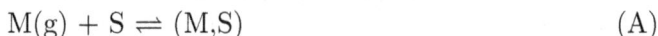

$$M(g) + S \rightleftharpoons (M,S) \tag{A}$$

where S represents the adsorbent surface and M the gaseous adsorbate.

The rate of adsorption, v_a, will be proportional to the gas pressure and to the fraction of the surface that is still not covered, free of adsorbate. That is

$$v_a = k_a(1 - \theta)p \tag{7.19}$$

The rate of desorption, v_d, will be independent of the gas pressure and proportional to the number of sites covered by adsorbate molecules, according to

$$v_d = k_d\theta \tag{7.20}$$

When $v_a = v_d$, the process achieves a state of equilibrium, and in that situation

$$\theta = \frac{N_s}{N_T} = \frac{K^*p}{1 + K^*p} \tag{7.21}$$

where $K^* = k_a/k_d$. The adsorption isotherm represented by the eqn. (7.21) is known as the Langmuir's isotherm. The quantity K^* is related to the equilibrium process (A) (for an associative adsorption) through the standard pressure, $p^{\ominus}$, according to the relation $K^* = Kp^{\ominus}$. It can be said that the model is very similar to that used to explain kinetically the equilibrium between a liquid and its vapor. Figure 7.6 shows the type of curve derived from eqn. (7.21) for a process like (A).

Considering the chemical potential of a species M that participates in an equilibrium of type A, we have

$$\mu(M, g) + \mu(S) = \mu(M, S)$$

Assuming that the activity of the free adsorbent and of the absorbate are given by the fraction of free sites and of occupied sites, respectively, one gets

$$\mu^{\ominus}(M, g) + RT \ln \frac{p}{p^{\ominus}} + \mu^*(S) + RT \ln(1 - \theta) = \mu^{\infty}(M, S) + RT \ln \theta$$

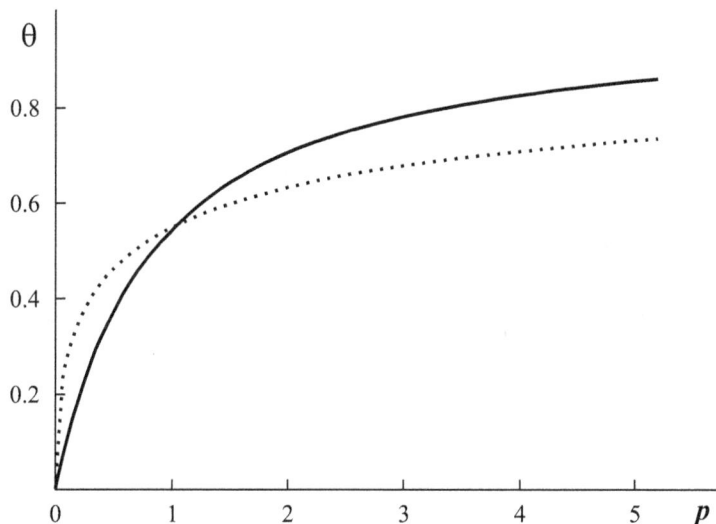

Figure 7.6: Degree of coverage, θ, as a function of gas pressure, for a non-dissociative (also called associative) (——) and a dissociative ($\cdots$) adsorption.

where it has been assumed that the free and occupied sites behave ideally; that is, they do not interact among themselves. In this way, we can obtain the equilibrium constant

$$K = \frac{\theta p^{\ominus}}{(1 - \theta)p} = K^{*} p^{\ominus}$$

arriving to the same result as given in eqn. (7.21).

In the case of dissociative adsorption, the process for a diatomic homonuclear molecule can be represented by

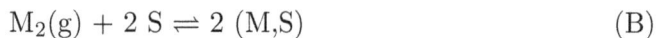

$$M_2(g) + 2\,S \rightleftharpoons 2\,(M,S) \tag{B}$$

The kinetic argument, as well as that based on using the thermodynamic equilibrium between the phases, leads to

$$\theta = \frac{(K'p)^{1/2}}{1 + (K'p)^{1/2}} \tag{7.22}$$

and the ratio between the rate of the adsorption and the desorption processes is equal to K'.

The plot of θ as a function of pressure for the dissociative adsorption (D) is very similar to that of the associative process (A), as illustrated in figure 7.6. However, for low coverage, the curve corresponding to the dissociative process does not have a constant slope; in particular, it is a parabolic curve with infinite slope for $\theta \to 0$.

Molecular Models to Describe Adsorption Isotherms

It is convenient to consider two extreme possibilities to model, at the molecular level, the adsorption of monoatomic gases over solids,

- Localized model (loc): The adsorbed atoms are localized in the sites existing over the surface of the adsorbent.

- Non-localized model (nl): Adsorbate atoms are free and can move over the surface plane.

In the localized model, the adsorbed molecules can vibrate on the adsorption site in the three degrees of freedom, and their partition function will be

$$q_{\text{loc}}(T) = q_x q_y q_z \exp\left(-\frac{\epsilon_o}{kT}\right)$$

where q_i is the vibrational partition function of the molecular coordinate i. On the other hand, in the non-localized model, the absorbed molecules vibrate only on the coordinate x, perpendicular to the surface, and maintain two translational degrees of freedom on the surface. For a surface of area $\mathcal{A}$, the partition function is

$$q_{\text{nl}}(T) = q_x \frac{2\pi mkT}{h^2} \mathcal{A} \exp\left(-\frac{\epsilon_o}{kT}\right)$$

The energy ϵ_o is the adsorption energy of one molecule at 0 K, not including the zero point contribution to the vibrational degrees of freedom.

In these two equations, it has been assumed that the internal degrees of freedom of the adsorbate molecules, characterized by frequencies that are very different compared with the vibration in the adsorption coordinate, are not modified in the adsorption process.

Let us assume that N molecules are adsorbed on the surface of area $\mathcal{A}$. If on the surface there are M sites for a localized adsorption, so that $M = c\mathcal{A}$, where c is the surface concentration of adsorption sites, the partition functions for the system having N adsorbed molecules will be

$$(Q_N)_{\text{loc}} = \frac{M!}{N!(M-N)!} [q_{\text{loc}}(T)]^N$$

and

$$(Q_N)_{\text{nl}} = \frac{[q_{\text{nl}}(T)]^N}{N!}$$

where the number of distinguishable microconfigurations have been taken into account for each value of ϵ_o, by eliminating the permutations of molecules that do not introduce new microconfigurations.

According to eqn. (7.8), at constant p and V, we can write

$$\mathrm{d}A = \mathrm{d}G = -S\mathrm{d}T + \gamma\mathrm{d}\mathcal{A} + \mu\mathrm{d}N = -S\mathrm{d}T + \frac{\gamma\mathrm{d}M}{c} + \mu\mathrm{d}N = -kT\ln Q_N$$

and hence,

$$-\left(\frac{\partial \ln Q_N}{\partial N}\right)_{\mathcal{A},T} = \frac{\mu}{kT} = \frac{\mu^{\ominus}}{kT} + \ln\frac{p}{p^{\ominus}}$$

Using the Stirling approximation to express the factorials, it yields for the localized model

$$\left(\frac{\partial \ln(Q_N)_{\mathrm{loc}}}{\partial N}\right)_{\mathcal{A},T} = -\ln\frac{\theta}{(1-\theta)q_{\mathrm{loc}}}$$

where $\theta = N/M$. Likewise, in the case of the non-localized model, we arrive to the following expression

$$\left(\frac{\partial \ln(Q_N)_{\mathrm{nl}}}{\partial N}\right)_{\mathcal{A},T} = \frac{N}{q_{nl}}$$

If all the quantities depending on the temperature are gathered in a function $a(T)$, the localized adsorption isotherm is then given by

$$\theta = \frac{a_{\mathrm{loc}}(T)p}{1 + a_{\mathrm{loc}}(T)p}$$

which is the Langmuir's isotherm.. For the non-localized adsorption, the expression of the isotherm is

$$\frac{N}{\mathcal{A}} = a_{\mathrm{nl}}(T)p$$

which is analogous to the general gases equation, but now refers to a two-dimensional gas. The functions of temperature $a_{\mathrm{loc}}(T)$ and $a_{\mathrm{nl}}(T)$ are given respectively by

$$a_{\mathrm{loc}}(T) = \frac{q_{\mathrm{loc}}(T)\exp(\mu^{\ominus}/kT)}{p^{\ominus}}$$

and

$$a_{\mathrm{nl}}(T) = \frac{q_{\mathrm{nl}}(T)\exp(\mu^{\ominus}/kT)}{\mathcal{A}p^{\ominus}}$$

Adsorption Energies

The equilibria considered to obtain Langmuir's isotherm assumes, as already seen, that the adsorption of one molecule does not alter the probability that another molecule will be absorbed. This requires that there are no interactions between adsorbate molecules. A simple model that can be used to explain this fact must assume that adsorption will not be so large that the adsorbate exceeds a monolayer on the adsorbent. If this is not true and two or more layers of adsorbate molecules are absorbed, it would imply that the molecules on these layers would be interacting in a different way than those on the first layer (i.e., they would not have direct access to the sites on the surface of the adsorbent).

It can be appreciated that adsorption equilibria are similar to that existing when a fluid condenses. That is, the change of pressure with temperature in the adsorbate gas, at constant coverage, is formally similar to the Clausius-Clapeyron equation, namely

$$\left(\frac{\partial \ln K}{\partial T}\right)_\theta = \frac{\Delta_{ad} H^\ominus}{RT^2} = -\left(\frac{\partial \ln p}{\partial T}\right)_\theta \qquad (7.23)$$

The quantity $\Delta_{ad} H^\ominus$ is the isosteric heat of adsoption; that is, the change of enthalpy of adsorption at constant coverage. Integration of eqn. (7.23) gives the expression

$$\ln\left(\frac{p_1}{p_2}\right)_\theta = \frac{\Delta_{ad} H^\ominus}{R}\left(\frac{1}{T_1} - \frac{1}{T_2}\right) \qquad (7.24)$$

that allows the calculation of the heat of adsorption from the adsorption's isotherms at two temperatures.

Mono and Multilayers

Physisorption denotes the adsorption in which the adsorbate and the adsorbent interact through multipoles, induced multipoles, and dispersion forces. In these cases, there is practically no transfer of electronic density between molecules of adsorbate and the surface sites. On the other hand, when a strong electronic exchange is observed, the adsorption is denoted as chemisorption. It can be said that, typically, $-\Delta_{ad} H^\ominus < 40$ kJ/mole in physisorption, while chemisorption is characterized by values of $-\Delta_{ad} H^\ominus > 40$ kJ/mole. If atom X in the interaction scheme illustrated in figure 7.7 were an O atom, it might refer to the following process

$$O_2 \text{ (g)} \rightleftharpoons O_2 \text{ (prec)} \rightleftharpoons 2\,O \text{ (ad)}$$

where O_2(prec) denotes the molecules of oxygen that are precursors of adsorbed oxygen atoms. The curves in figure 7.7 illustrate the changes in the energy of interactions as physisorption of oxygen molecules become a chemisorption of oxygen atoms.

It can also happen that the molecules are adsorbed forming more that one layer. It is evident that in this case the enthalpy of adsorption will change according to

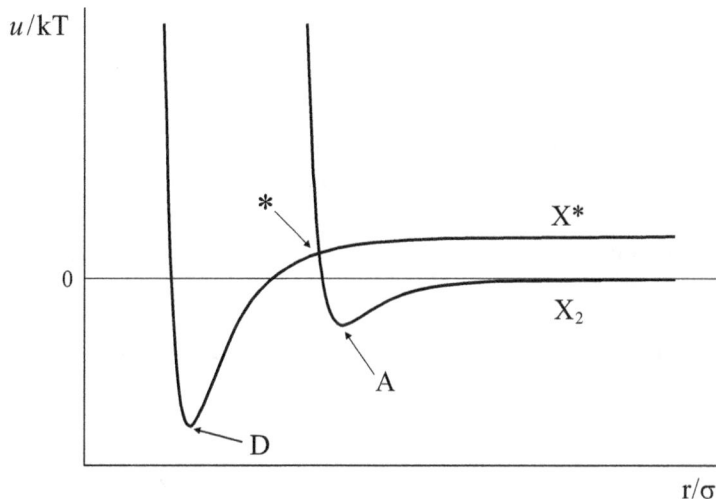

Figure 7.7: Interaction energy between a molecule and the surface. A: Associative interaction of the molecule X_2 with the surface. D: Dissociative interaction of the atom $X\cdot$ with the surface. Symbol $*$ indicates the transition state.

the layer where the molecule is adsorbed. Molecules in the first layer will have a predominant interaction with the surface, while the second layer will be adsorbed on top of the first monolayer, and the adsorption enthalpy will be different than that on the first one. If one goes on adding molecules in successive layers, there will be a moment in which the enthalpy of adsorption becomes equal to the enthalpy of condensation of the gas, when it becomes a liquid. The multilayer adsorption was studied by S. Brunauer, P. H. Emmet, and E. Teller (BET), who assumed that beyond the second adsorbed layer the enthalpy of adsorption is equal to that involved gas condensation. They derived the following equation

$$\frac{x}{V(1-x)} = \frac{1}{V_m C} + \frac{(C-1)x}{V_m C} \tag{7.25}$$

where $x = (p/p_v)$ is the pressure relative to the vapor presure, p_v, V is the gas volume (under ambient conditions of p and T) adsorbed over the surface at p, V_m is the volume of adsorbate in a monolayer, and C is a constant given by

$$C \approx \exp\left(\frac{\Delta_{ad}H + \Delta_v H}{RT}\right)$$

where the argument of the exponential is the difference between the absolute magnitudes of adsorption and gas liquefaction enthalpies.

The determination of the BET isotherm is one of the most frequently used methods to determine the specific area of adsorbents. Effectively, starting from eqn. (7.25), it may be observed that if the lhs member is plotted as a function of x, from the value of the ordinate at the origin C is obtained, and from the curve's slope, V_m can be calculated. Then, by assigning a coverage area per molecule of adsorbate, it is possible to calculate the total area of the adsorbent.

7.5 Microheterogeneous Systems: Phase Formation and Molecular Aggregation

In the first part of this section, the formation of phases in one-component systems will be discussed, such as the case of the formation of bubbles or drops and the crystallization of a compound. Afterwards, the formation and properties of microheterogeneous systems will be analyzed (i.e., systems containing molecular aggregates inside a phase with different properties). These systems are related with the formation of a new phase, where the contribution of surface properties is very important. In the second part of this section, systems having molecular aggregates will be discussed. In general, the aggregates will be inside a liquid, formed by self-assembly of its molecules due to different types of interactions existing between solute and solvent molecules.

Nucleation and Condensation Phenomena

Nucleation is a non-equilibrium process related to the incipient formation of new phases in material systems; the formation of interfaces plays a fundamental role in it. The processes of condensation, precipitation, crystallization, sublimation, and boiling are related to nucleation, a phenomenon of great importance for various branches of science and technology. Nucleation is the process by which individual molecules and molecular aggregates come together to give origin to germinal and nuclei of condensed mater. Once they are formed, they increase their size in a fast process to form a new macroscopic phase. In the following paragraphs, the basic rudiments of the classical theory of homogeneous nucleation, for the particular case of the condensation of liquids, will be developed.

The nucleation process of a drop happens under conditions of supersaturation of the vapor. This is due to the large surface Gibbs energy of small aggregates, which, in this case, is always positive and then the drop growth is not favored (cf. eqn. (7.17)). The supersaturation S at temperature T is defined as the ratio between the vapor pressure of the system and the saturation vapor pressure of the liquid, p^*, according to

$$S = \frac{p(T)}{p^*(T)} \tag{7.26}$$

If $S < 1$, the vapor is subsaturated; if $S = 1$, there is equilibrium between liquid and vapor at T; finally, if $S > 1$, the vapor is supersaturated.

In the microscopic scale, the collisions between molecules in the gas phase result in the formation of molecular aggregates, which then grow to form germs or nuclei. According to the classical theory of nucleation, molecular aggregates are clusters of particles formed by only a few molecules; these are formed transiently and they may disintegrate spontaneously to produce individual molecules. The clusters are then described as particles in the gas phase. On the other hand, the properties of germs, having a sufficiently large number of molecules to be described as a continuous medium, can be considered equivalent to those of the macroscopic liquid. Vapor condensation occurs when the germs grow beyond the critical size; these germs are called critical condensation nuclei . The size of the critical nuclei depends on the system and the experimental conditions, having in general hundreds of molecules. Once the nuclei are formed, they grow rapidly and spontaneously until condensation into a liquid occurs. The classical theory of nucleation is based on the idea that a germ of a phase grows or decreases its size by addition of loss of single molecules, and the process can be represented by the scheme

$$A_i + A_1 \rightleftharpoons A_{i+1}$$

This reaction proceeds to the right when the germ increases its size, or to the left when it decreases. Moreover, the following additional assumptions are made: i) the gaseous phase behaves as an ideal gas, ii) nucleation is an isothermal process, and iii) the Gibbs energy for a liquid germ formation can be written in terms of the interfacial macroscopic tension, even if the i molecules that form it are less than one hundred.

The expression for the change of Gibbs energy for the formation of one germ of radius r is

$$\Delta G_g = -\frac{4\pi\, r^3}{3}\, \frac{(\mu_V - \mu_L)}{V_L} + 4\pi r^2 \gamma \tag{7.27}$$

where μ_V and μ_L are the chemical potentials in the vapor and liquid phases, respectively, at pressure p and temperature T; $(4\pi r^3/3)$ represents the volume of each germ, V_L is the liquid molar volume, and γ denotes the macroscopic interfacial tension.

The difference of chemical potential of the substance in both phases can be represented in terms of supersaturation S, using eqn. (7.26), as

$$\Delta G_g = -\frac{4\pi r^3}{3}\frac{RT \ln S}{V_L} + 4\pi r^2 \gamma \qquad (7.28)$$

In figure 7.8, ΔG_g is qualitatively plotted as a function of the germ's size.

The second term in the rhs of eqns. (7.27) and (7.28) take into account the increase in Gibbs energy due to the surface work involved in the liquid germ formation (cf. curve (a) in figure 7.8). The supersaturation contribution (first term in the rhs of eqn. (7.28)) is negative and is given by curve (b) in figure 7.8. Both contributions have opposite signs, and that leads to a maximum for $r = r^*$ and then decreases monotonously as the size of the nucleus increases.

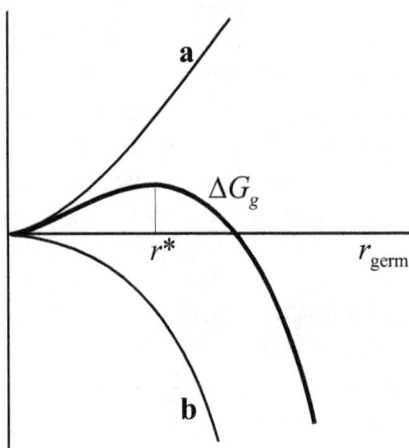

Figure 7.8: Gibbs formation energy of a liquid germ as a function of its radius, for a constant supersaturation in the vapor: (a), free energy increase due to the surface work; (b), free energy decrease due to the vapor condensation; r^*, condensation radius of the nuclei.

Hence, starting from r^*, the aggregation process of a new molecule to the germ will be spontaneous. That is, those liquid germs reaching the critical radius r^* become condensation nuclei that will grow spontaneously. The value of r^* may be obtained by differentiating the expression (7.28) with respect to r and making the result equal to zero. Thus, we get

$$r^* = \frac{2\gamma V_L}{RT \ln S} \qquad (7.29)$$

The effect of vapor supersaturation S over the number i^* of molecules forming the critical nucleus of condensation and, consequently, over the size of the nucleus, is very marked. An increase in supersaturation diminishes the Gibbs energy to form the germs and this also reduces the mean value of the critical radius of the condensation nuclei. For instance, an increase from 2 to 10 in water vapor supersaturation at room temperature reduces the value of i^* from approximately 1,000 to 20 molecules of water. Although germs are formed for $\mathcal{S} < 1$ due to fluctuations in the vapor density, the number of them is relatively small and disintegrates easily. Only when the vapor is supersaturated ($\mathcal{S} > 1$), germs growth will be favorable statistically as well as thermodynamically. In other words, only those germs that exceed the critical barrier of $\Delta G_g(r^*)$ will be able to reach the critical size, represented by r^*, and become condensation nuclei.

If the supersaturation condition is not achieved ($\mathcal{S} \leq 1$), the present clusters in the vapor phase can be considered part of an average distribution, maintaining on average the fraction of germs of each size. Under those conditions, the number of germs A_i formed by i molecules A, n_i can be expressed using the Boltzmann distribution function in terms of their energy of formation, given by eqn. (7.28), as

$$n_i = n \exp\left(\frac{-\Delta G_g(i)}{kT}\right) = n \exp\left(i \ln \mathcal{S} - \frac{4\pi r^2 \gamma}{kT}\right) \qquad (7.30)$$

where $n = \sum_i n_i$ is the total number of germs. On the other hand, the number of molecules i is considered equal to $4\pi \, r^3 N_A / 3V_L$. From the analysis of eqn. (7.30), it is possible to conclude that for subsaturated vapors ($\mathcal{S} < 1$) n_i is a decreasing function of the number of i molecules present in the germ. For instance, studies with water vapor at 273 K show that many clusters have more that 5 water molecules, but the number of clusters with more than 10 molecules is very small.

The previous equations provide a simple but clarifying description of the nucleation phenomenon, which is exclusively controlled by the Gibbs energy of formation of germs. More specifically, the classical theory highlights the two opposite effects acting over ΔG_g, namely the formation the condensation nuclei surface and the consequent formation of the tridimensional phase.

Crystalization

Another thermodynamic consequence of the surface effects is the relation between a solid solubility and the size of its crystals. Let us consider a solution in equilibrium with the solute's solid crystals. If the crystals' size are denoted by r, the volume of $n(\mathrm{s})$ moles of these crystals is $V(\mathrm{s}) = n(\mathrm{s})k_v r^3$, and the total area of the surfaces of all these crystals will be $\mathcal{A}(\mathrm{s}) = n(\mathrm{s})k_a r^2$; k_v, k_a being the proportionality of constants, which depend on the geometry of the crystal. The equilibrium condition between the crystalline solute and that dissolved in the solution is $\mu_2(\mathrm{sln}) = \mu_2(\mathrm{s})$, and then

$$\mu_2^{\infty}(\text{sln}) + RT \ln \frac{a_2}{C^{\circ}} = \mu_2^{*}(\text{s}) + \gamma \mathcal{A}_m(\text{s})$$

where $\mathcal{A}_m(\text{s})$ is the area per mole, which may also be expressed by

$$\mathcal{A}_m(\text{s}) = k_{\text{s}} \frac{V_m(\text{s})}{r}$$

where $k_{\text{s}} = k_a/k_v$ also depends on the crystalline geometry. Now, we denote as a_2° the solute's activity in equilibrium with the solid crystal when the crystals' size is very large (i.e., when the surface energy contribution is negligible). Hence,

$$RT \ln \frac{a_2}{a_2^{\circ}} = \mathcal{A}_m(\text{s}) \gamma$$

Let's suppose that the solution is ideal, so that it is possible to replace activities by concentrations; in this case, we easily verify that the solubility of small crystals is larger than that of bigger crystals. This result confirms the fact that the surface energy is quite appreciable when the crystals are small. If the saturated solution is left standing for some time, the system having small crystals will spontaneously grow to form larger crystals, which will have a smaller surface Gibbs energy per mole of substance.

Another phenomenon related with the previous one, which is convenient for the reader to elaborate, can also be observed when a solid precipitates from a solution. The size of the formed crystals upon precipitation are smaller whenever the supersaturation of the solution at the moment when precipitation starts to happen is higher. That is, whenever the activity of this substance in the solution phase is greater, before the crystallization process begins, the smaller the crystals formed by crystallization will be. As already mentioned, if we let the crystals stand for some time, their size will increase spontaneously to reduce the surface Gibbs energy and, in consequence, their chemical potential. This process is usually referred to as Ostwald's ripening.

Thermodynamics of Aggregates' Self-Assembly

Now, we will look into the phenomenon of molecular aggregation in liquid solutions, where the surface properties are important. These phenomena are of great interest in the fields of biology, chemical synthesis, and for a large number of industrial processes.

The principles of thermodynamics will be used in these cases to predict the type of supramolecular structures formed when molecular aggregation of solutes occurs, in aqueous solutions. It is convenient to make it clear that similar aggregates are produced in non-aqueous systems, as is the case of micelles in supercritical fluids, or of reversed micelles dispersed in organic solvents that contain water in their interior.

Molecules that tend to form aggregates in aqueous solutions are amphiphilic (i.e., molecules with one polar or ionic part), and another having a non-polar part that is hydrophobic. Examples of these molecules are the fatty acids of long chain alkanes, like the stearic and oleic acids, or molecules like phosphatydylcholine. Acid and basic groups, which may be highly polar and also may be ionized, form the hydrophilic head of the molecule. The less polar part of the chain, generally hydrocabonated, forms the monomer's tail. The molecular aggregates of amphiphilic molecules will self-assemble spontaneously in aqueous solutions forming micelles, lipidic layers, vesicules, liposomes, or membranes similar to the biological membranes. All these supramolecular structures have sizes, approximately between 10 nm and 100 μm. Amphiphilic molecules tend to self-assemble, a property that has in general an important effect on the interfacial properties of the solutions where they are dissolved, and they are usually denoted as surfactants, like detergents. Self-assembly of molecular aggregates is a consequence of the different type of interactions: Between the polar part of an amphiophilic solute and a solvent molecule, hydrogen bonds, (screened) electrostatic interactions, and van der Waals forces among non-polar chains.

These amphiphilics can be depicted as spheres that indicate the polar part of the molecule and a tail that represents the hydrophobic chain. The association process produces aggregates of different sizes, as those depicted in figure 7.9. It is convenient to imagine it as a series of coupled chemical equilibria to form aggregates of 2, 3, ..., N molecules, as it was the case of drop germs;

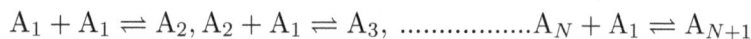

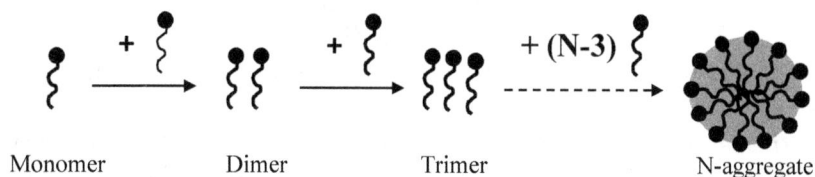

$$A_1 + A_1 \rightleftharpoons A_2, A_2 + A_1 \rightleftharpoons A_3, \ldots\ldots\ldots A_N + A_1 \rightleftharpoons A_{N+1}$$

Figure 7.9: Aggregate formed by N monomers.

Since the systems are in equilibrium, the thermodynamic condition imposes the equality of chemical potentials for all the species. It is useful to assume the condition of ideality for the solutions to isolate the phenomenon of self-assembly from the effect of interactions among aggregates, and of these with monomers and ions that can be present in the system; rigorously, both effects need to be accounted for when describing the process. Hence, one gets the following relations

$$\mu = \mu_1^\infty + kT \ln c_1 = \mu_2^\infty + \frac{kT}{2} \ln\left(\frac{c_2}{2}\right) = \mu_3^\infty + \frac{kT}{3} \ln\left(\frac{c_3}{3}\right) = \ldots$$

where μ is the chemical potential of one molecule in one of the aggregates, remembering that all of them are in equilibrium, μ_N^∞ is the standard chemical potential of one molecule in an aggregate of N molecules, and c_N its concentration, is expressed as the number of molecules of the monomer in the aggregate. The total number of molecules in the system will be

$$c = \sum_{N=1}^{\infty} c_N$$

The equilibrium constant for the formation of an aggregate having N monomer molecules, as shown in figure 7.9, is

$$K = \frac{(c_N/N)}{c_1^N} = \exp\left(-\frac{N(\mu_N^\infty - \mu_1^\infty)}{kT}\right) \tag{7.31}$$

from which, the following expression can be obtained

$$c_N = N\left[c_1 \exp\left(-\frac{(\mu_N^\infty - \mu_1^\infty)}{kT}\right)\right]^N \tag{7.32}$$

It is important to note that the value of c_N depends on the difference between the chemical potentials of the non-associated molecules and those forming aggregates of N molecules, $(\mu_N^\infty - \mu_1^\infty)$. The condition for spontaneous assembly of aggregates is $\mu_N^\infty < \mu_1^\infty$; thas is usually ascribed to cohesiveness between the aggregating molecules. If $\mu_{N+1}^\infty > \mu_N^\infty < \mu_{N-1}^\infty$, the distribution of aggregates reaches a maximum for those containing N monomers, so that it is important to know how μ_N^∞ varies with N.

At this moment, it is convenient to the reader to verify the similarity between eqn. (7.32) and eqn. (7.30), which was deduced for the formation of condensation nuclei. In the first of these equations, two terms contribute to the difference of chemical potentials $(\mu_N^\infty - \mu_1^\infty)$. One term accounts for the different chemical potentials of individual molecules in their respective states (as monomer or as aggregate), and the other is due to the surface work, as discussed in the page 222. Thus, eqn. (7.32) lets us write the surface term that contributes to μ_N^∞ in terms of the molecule's chemical potential of an aggregate containing an infinite number of monomers, which due to its large size will not be influenced by the surface work. We will denote $\mu_{N\to\infty}^\infty$ the aggregate's chemical potential, as having $N \to \infty$ which, and as already argued, will have a negligible surface contribution. Hence, according to eqns. (7.30) and (7.32), μ_N^∞ per monomer in the N-aggregate may be written as

$$\mu_N^\infty = \mu_{N\to\infty}^\infty + \frac{4\pi r^2 \gamma}{N} \tag{7.33}$$

Assuming that the aggregate with N molecules is spherical, its radius is given by $r = [3Nv/4\pi]^{1/3}$ where $v = V_L/N_A$ is the volume of each molecule. The surface contribution in the previous equation yields

$$\mu_N^\infty = \mu_{N\to\infty}^\infty + \frac{[\pi(6v)^2]^{1/3}\gamma}{N^{1/3}} = \mu_{N\to\infty}^\infty + \frac{\alpha kT}{N^p} \tag{7.34}$$

where $p = 1/3$, for tridimensional aggregates, particularly spherical ones, and in such conditions

$$\alpha = \frac{[\pi(6v)^2]^{1/3}\gamma}{kT}$$

Eqn. (7.34) can be generalized for aggregates having a different dimensionality. In those cases it is necessary to use the last equality on the rhs of this equation. For aggregates of different dimensionality, $(\alpha\, kT/\gamma)$ represents the area of the surface per monomer. For unidimensional aggregates as rigid chains or rods, $p = 1$, and for bidimensional aggregates as disks or lamellas, $p = 1/2$. These exponents result from the relation between the aggregate size (its radius diameter or its longitude) and the size of the corresponding monomer.

Let's see now which is the monomer concentration that forms aggregates. Using eqns. (7.32) and (7.34), and replacing $(\mu_N^\infty - \mu_1^\infty)$ by the value obtained from the previous model, for a given geometry, one gets

$$c_N = N \left\{ c_1 \exp\left[\alpha \left(1 - \frac{1}{N^p} \right) \right] \right\}^N \approx N \left[c_1 \exp(\alpha) \right]^N \tag{7.35}$$

If the monomer concentration is very low, $c_1 \exp(\alpha) \ll 1$, and then c_1 is bigger than the concentration of the aggregates (i.e., the solution is mainly formed by non-associated molecules). The experiments show that for surfactants or detergents dissolved in water, once the monomer concentration reaches a limit value denominated by critical micelle concentration (c_{cmc}), abrupt changes in the solution properties occur. Frequently, the c_{cmc} value for aqueous solutions at room temperature is between 0.1 and 10 milimolal. For larger concentrations, c_1 is constant and, when the monomer concentration increases, aggregates are formed following a quite narrow distribution around the value of N having a maximum number of aggregates. Figure 7.10 illustrates schematically how two properties of sodium dodecylsulfate solutions change with concentration. It is clearly appreciated that the osmotic and

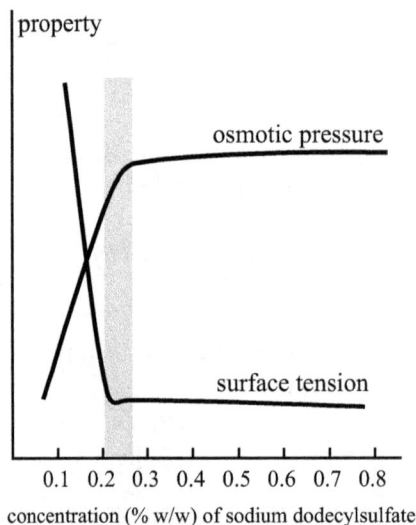

Figure 7.10: Change of the osmotic pressure and the surface tension of a solution of sodium dodecylsulfate, as a function of the concentration.

the interfacial tension of these solutions show a sharp variation, almost a discontinuity, when their concentration reaches c_{cmc}. For concentrations larger than c_{cmc}, these two properties change very little.

The features of the aggregation phenomenon or micellization are singular when compared with ordinary chemical equilibria. In many senses, it is more like a phase transition, something very similar with what has been described in the previous section with reference to the nucleation phenomena.

In order to observe the effect of a large increase of the surfactant concentration, we will consider the case of unidimensional aggregates ($p = 1$). It can be seen from eqn. (7.35) that above c_{cmc}, c_N increases with N since c_1 is constant and, consequently, the size of the aggregates increases. When the value of N is very large, the term $(c_1\,e^{\alpha})^N$ dominates and c_N tends to zero. This means that c_N reaches a maximum value for $N_{max} = (c_1\,e^{\alpha})^{1/2}$. The system of unidimensional aggregates, however, is now polydispersed; that is, aggregates exist with values of N bigger and smaller than N_{max}.

For aggregates of another dimensionality, where $p < 1$, one observes that a phase transition occurs; that is, an infinite aggregate ($N \to \infty$) is formed. Those aggregates containing a large number of molecules, which might be considered as macroscopic phases different of the medium where they are dispersed, adopt various forms. They may be appreciated in figure 7.11. The resulting systems are

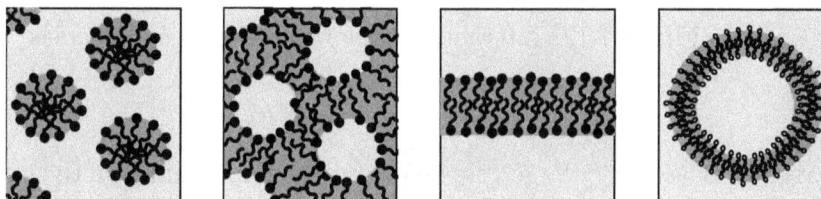

Figure 7.11: Structures formed by ampiphiles: a) micelles; b) inverted micelles; c) bilayers; d) vesicles. Adapted from figure 16 of Israelachvili's book (appendix C).

micro-heterogeneous (i.e., the size of the dispersed aggregates in the liquid phase are not greater than a few micrometers). Moreover, these aggregates of different forms can transform one into others when the temperature, the pH, or the concentration changes.

In order to decide what determines the form of the self-assembled phase, it is necessary to analyze with more detail the nature of the interaction forces among amphiphilic molecules. An important feature of these molecules is that μ_N^∞ reaches a minimum value for some finite N. This is a clear result, taking into account that the forces that operate on the molecules that aggregate are mostly attractive, in relation to the interfacial tension γ, and mostly repulsive among the monomers. This repulsive force is difficult to describe because there are steric, hydration, and electrostatic contributions. It is not necessary to know explicitly the nature of these contributions; we only need to assume that they are inversely proportional to the area, a, of the polar head of every aggregated molecule (cf. figures 7.12 and 7.9). This can be summarized on the basis of eqn. (7.33), leading to

$$\mu_N^\infty = \mu_{N\to\infty}^\infty + \gamma a$$

where $a = (4\pi r^2 / N)$ represents the area occupied by each monomer's polar head in the aggregate. The term $\mu_{N\to\infty}^\infty$ does not contain any surface contributions because

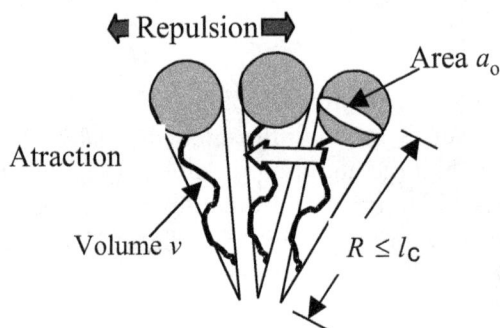

Figure 7.12: Interactions among anphiphilic molecules.

this aggregate has a very large volume ($N \to \infty$), but includes all the interactions between chains, monomeric polar heads, and between them and the solvent molecules, which can also contain dissolved salts. This repulsive term increases its value the smaller the space between polar heads is, and it may be represented by the relation

$$\mu_{N \to \infty}^{\infty} = \frac{C_0}{a}$$

where C_0 is a proportionality constant whose characteristics are unknown. Despite that, it is interesting to describe the principal features of the phenomenon. It is obvious that the attractive and the repulsive interfacial forces will make the system to adopt an optimal area per monomer, a_0, which makes μ_N^{∞} be a minimum. From the minimum condition ($\partial \mu_N^{\infty}/\partial a = 0$), it is possible to eliminate the unknown constant C_0, and finally obtain

$$\mu_N^{\infty} = 2\gamma a_o + \frac{\gamma}{a}(a - a_o)^2$$

where $a_0 = (C_0/\gamma)^{0.5}$ is the polar head group area at the minimum. This expression, in spite of being a very primitive one, can represent the main features of the interactions between amphiphilic molecules in the aggregates, micelles, films, and membranes that are formed in the solution. The parameters C_0 and a_0 contain information about the optimal chain packings, which is fundamental to establish the form of the aggregates. This will depend on a_0, the monomers' volume, v, and the maximum length of the hydrophobic chains l_c. The last parameter corresponds to the effective length of the non-polar chain, which allows the surfactant to behave as a fluid; it is considered an adjustable parameter.

   Let us analyze, as an example, the case of molecules that self-assemble, forming a spherical micelle of radius r. It is evident that the condition necessary to obtain such a system is given by $r \leq l_c$. If N is the average number of molecules in the micelle, its area will be $Na_0 = 4\pi r^2$ and its volume $Nv = 4\pi r^3/3$, so that $r = 3v/a_o$. The previous condition is valid only if the form factor, defined by $F = v/a_o l_c$, is less than $1/3$. If F exceeds $1/3$, even slightly, the micelle will deform and become not spherical. If $1/3 < F < 1/2$, self-assembly will adopt the form of cylindrical micelles with the aspect of rods.[4] As seen before, these structures, basically unidimensional, are polydisperse, and the size of the aggregate increases with the concentration of monomer.

When the molecules posses charged groups, like sodium dodecylsulfate, the repulsion also depends on the ionic strength of the solution. Increasing the salt concentration there is some screening of the charges, which leads to a reduction of the repulsion (lower a_0) and, hence, the number of monomers in the micelle increases.

For F values between $1/2$ and 1, the edges curve themselves, leading to vesicular structures, as observed in figure 7.11d. If $F \approx 1$, as in the case of fosfatidylcholine (lecitin), planar structures are formed, like the bilayers shown in figure 7.11c. When lecithin, a molecule having a double hydrocarbonate strand, is subjected to the action of the enzyme phospholipase that cuts one of the chains, the value of v is reduced approximately to the half and, hence, F is reduced to a value close to 0.5. This is the reason why isolecithins, with a single hydrocarbonated chain, form micelles instead of bilayers.

Problems

Problem 1

a)Calculate the ascent of water at 25 °C inside a capillary of radius 0.01 μm. Discuss if the result is possible, remembering that the trees cannot be taller than 130 m so that water can reach their top.

b) Calculate the vapor pressure of a water drop 0.01 μm in radius.

c) Calculate the relative humidity[5] at which water condensate in a capillary 0.01 μm in radius.

Data: γ(H$_2$O, 298 K) = 0.0728 N/m; p_v(H$_2$O, 298 K) = 31.7 mbar

---

[4]The reader could get a deeper insight about the relation between the form factor and the structure of the self-assembled aggregates by consulting chapter 17 of Israelachvili's book (appendix C).

[5]Ratio of the partial pressure of water vapor to the equilibrium vapor pressure of water, at a given temperature.

Answers: a) $h \simeq 1500$ m; b) $p_v(\text{drop}) = 35.2$ mbar; c) relative humidity = 90 %

Problem 2

Explain why bubbles or drops levitate (that is, they are not on a surface) have spherical form.

Problem 3

a) The equilibrium pressure at constant T, p, for transferring mass between vapor and a drop of liquid of radius r, or between vapor and liquid inside a capillary having the same radius, can be calculated with the general expression

$$dG = -TdS + Vdp + \gamma d\mathcal{A} + \sum_{i=1}^{C} \mu_i dn_i$$

Remembering that for a pure substance $\mu(\text{v}) = \mu(\text{drop or capillary})$, show that the exact form of eqn. (7.17) is

$$RT \ln \left(\frac{p}{p^*} \right) = \left[(p - p^*) + \frac{2\gamma}{r} \right] V^*(l)$$

b) Explain why $p < p^*$ is required for condensation to occur inside a capillary.

Problem 4

Butler and Wightman determined the vapor pressure of ethanol for different ethanol(2)-water(1) solutions. The data at 298 K are reported in the following table

x_2	$(\gamma/10^2)$ (N m^{-1})	p_2 (Torr)
0.12	3.442	20.70
0.10	3.672	18.03

For a solution with $x_2 = 0.11$:

a) Estimate the surface excess, $\Gamma_{2(1)}$, considering: i. ideal solution, and ii. the vapor pressure data

b) What is the relative error in the calculation of $\Gamma_{2(1)}$ if you assume that the solution is ideal?

c) Which is the sign you expect for $\Gamma_{1(2)}$? Justify your answer

d) Peter, a young researcher somewhat too eager, bought a humidity detector for which the instruction's manual says that it was highly sensitive to water, and he measured with the detector a $\Gamma_{1(2)}(x_2 = 0.11, 298 \text{ K})$ value of -5×10^{-10} mole/m^2. Did Peter make a good choice of detector? Justify your answer.

Answers: a) i. 5.09×10^{-6} mole/m^2, ii. 6.73×10^{-6} mole/m^2; b) 28 %; c) opposite sign compared with $\Gamma_{2(1)}$; d) no

Problem 5

As stated in problem 6 of chapter 5, phenol is not fully soluble in water. At 293.0 K, the interfacial tension, γ, measured for phenol in water as a function of the concentration, decreases when the total concentration of phenol increases. When a concentration limit 0.496 mole/dm^3 is reached, it is observed that, with further increase of concentration, γ becomes constant.

c(phenol,water) / (mole/dm^3)	0.450	0.496
γ / (N/m)	46.16×10^{-3}	44.97×10^{-3}

a) How do you interpret such observation?

b) From the data in the table, estimate the excess surface of phenol in the interface, $\Gamma_{\text{phenol(water)}}$, for concentrations close to the limit of 0.496 mole/dm^3.

c) Estimate the area occupied by each phenol molecule under those conditions

Answers: a) a monolayer of phenol has been fully established at 0.496 mole/dm^3; b) $\Gamma_{\text{phenol(water)}} = 5.58 \times 10^{-6}$ mole/m^2; c) 0.3 nm^2 per molecule

Problem 6

The solubility of cubic crystals of $BaSO_4$ is observed to change with the size of the crystals, so that for 1.8 μm crystals it is 1.778×10^{-5} mole/dm^3. Calculate the value of the interfacial tension (water/crystal) at 298.2 K, knowing that the molar volume of crystalline $BaSO_4$ is 46.5 cm^3/mole.

Answer: $\gamma = 3.35$ J/m^2

Problem 7

Calculate the relative increase of the vapor pressure of drops of radii 10^{-5} mm, for the three liquids in the following table.

Liquid	γ mN/m	V cm^3/mole
Hg	460.6	14.72
H$_2$O	69.85	17.95
CCl$_4$	25.02	976.43

Answers: $\Delta p/p$: 0.735 for Hg, 0.108 for H$_2$O, and 0.216 for CCl$_4$

Problem 8

Sodium dodecylsulfate (SDS) forms micelles in aqueous solution with a critical length $l_c = 1.67$ nm, having a chain of volume $v = 0.350$ nm^3.

a) What will be the number of SDS molecules so that the micelle is spherical? Compare this value with $N = 74$, found experimentally, and discuss the difference.

b) Taking into account that the molecule of SDS is ionic, and that the degree of dissociation depends on the pH of the solution, predict how should the pH be modified so that the micelle becomes larger (i.e., so that N increases).

c) Indicate how the value of N modifies for SDS micelles in a 0.06 M NaCl solution.

Answers: a) $N = 56$; b) pH should increase; c) N increases in the presence of salt

Chapter 8

Chemical Equilibrium

8.1 Its Characteristic Features

In physical chemistry textbooks, a separate chapter is often dedicated to the problem of chemical reactions and chemical equilibrium, and here we follow the same traditional presentation. There is really no fundamental reason for dealing with chemical processes and their equilibrium states have to be treated separately from other processes, like phase equilibrium. It is simply a didactic reason, since chemical equilibria are ruled by the same principles of thermodynamics. It is convenient to deal separately with chemical equilibria because of the following:

- Chemical reactions and equilibrium are always related to a stoichiometric relation, as results naturally from the fact that they are based on the ponderal laws that rule chemical processes.

- It allows to emphasize temporal (kinetic) aspects together with the chemical thermodynamic features of chemical processes.

- It enables to center the attention in the capacity to determine many thermodynamic quantities using the characteristics of the chemical reactions and chemical equilibrium.

- It is relatively simple to link thermodynamic quantities characterizing chemical equilibrium with molecular properties of the intervening substances; this relation is made using statistical mechanics.

As we have seen, the equilibrium states at molecular level are dynamic. The system constituted by molecules continuously adopt several microconfigurations and all these microconfigurations are part of the macroscopic state of the system. This already shows the relation existing between chemical equilibrium and statistic mechanics, in particular, how the velocity of conversion of reactants into products and the inverse process must be equal when the chemical reaction is at equilibrium, which implies that the microconfigurations corresponding to the state of equilibrium must contain reaction products as well as reactants, and that the probability that they can convert into each other must be equal. We must begin by considering the stoichiometry relation of chemical reactions.

It is essential to establish clearly which is the chemical reaction we are referring to. Thus, for instance, if we are dealing with the ammonia synthesis, the reaction

$$N_2(g) + 3H_2(g) \rightleftharpoons 2NH_3(g)$$

is not the same as the reaction

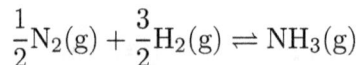

$$\frac{1}{2}N_2(g) + \frac{3}{2}H_2(g) \rightleftharpoons NH_3(g)$$

We will use the term *mole of reaction* to characterize the amount of matter involved in the process. This concept will be clearly understood if we consider the previous two reactions of synthesis of ammonia. One mole of the first reaction will be equal to two moles of the second, and because of that, the value of any extensive property characterizing the first reaction will double those of the second one.

The best way to express the stoichiometry of chemical reactions consists in defining the *extent of reaction*; this implies establishing a single variable to measure how much the chemical process has advanced. Let us consider the general process

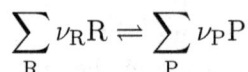

$$\sum_R \nu_R R \rightleftharpoons \sum_P \nu_P P$$

In this general process, R represents all the reactants and ν_R, the stoichiometric numbers corresponding to each one of them. Likewise, P stands for the products and ν_B for the stoichiometric coefficients of products. The extent of reaction, ξ, is a variable that determines the moles of reactants that have been transformed into products, starting from reactants, according to the chemical reaction that has been written. For 1 mole of reaction, the values that may adopt the variable ξ lie between 0 mole (no reaction yet) and 1 mole (all the reactants have turned into products). This value indicates what fraction of the considered reaction has been produced. When n moles of reaction are produced, ξ will change from 0 (no products) to n (no reactants remaining).

When the number of moles of reactants n_R has decreased by $dn_R = -\nu_R \, d\xi$, the products will increase in the amount $dn_P = \nu_P \, d\xi$, which may be expressed by

$$d\xi = -\frac{dn_R}{\nu_R} = \frac{dn_P}{\nu_P} \tag{8.1}$$

or, in general, one can write

$$d\xi = \frac{dn_i}{\nu_i}$$

where the stoichiometric coefficient of reactants and products have been noted with the same symbol, ν_i, being ν_i negative and equal to $-\nu_R$ for reactants and positive and equal to ν_P for products. It is important to emphasize here that the extent of a reaction has the dimensions of quantity of matter.

If equation (8.1) is integrated between 0 and ξ, when the process has advanced to n_R/ν_R mole of reaction, then $\xi = (n_R^0 - n_R)/\nu_R$, where n_R are the number of moles of reactant remaining when the reaction has advanced ξ. The previous equation is closely related to that used to define the rate of reaction, v, in chemical kinetics, given by $d\xi/dt$. As a result of the stoichiometric relation, in a chemical process the reactants and products' concentration cannot vary arbitrarily (restriction usually added to the phase rule as well). Hence, the kinetics of a chemical reaction and the state of equilibrium will depend on a single thermodynamic variable of concentration, the extent of reaction, the definition of which does not relay in the reactant or product chosen to write eqn. (8.1).

Let us assume that a gaseous system is being studied where a chemical process takes place and that, at a certain point of the reaction, its extent is ξ. What is important is to express the change of the system's Gibbs energy per mole of reaction. At constant p and T we have

$$dG = \sum \mu_i dn_i = \left(\sum_P \nu_P \mu_P - \sum_R \nu_R \mu_R \right) d\xi \tag{8.2}$$

Now, the quantity between parentheses in the previous expression is the change of Gibbs energy of the reaction $\Delta_r G$; hence, we have

$$\left(\frac{\partial G}{\partial \xi} \right)_{p,T} = \Delta_r G \tag{8.3}$$

When the chemical process is under equilibrium conditions, $(\Delta_r G)_{p,T} = 0$; if the process is occurring spontaneously, $(\Delta_r G)_{p,T} < 0$; and on the other hand, if $(\Delta_r G)_{p,T} > 0$, the chemical reaction does not take place spontaneously or naturally; that is, it will be necessary to make work for it to occur. This is illustrated in figure 8.1.

Frequently, the affinity $\mathbb{A}$, is used to characterize the chemical process; this is defined by

$$\mathbb{A} = \sum_R \nu_R \mu_R - \sum_P \nu_P \mu_P \tag{8.4}$$

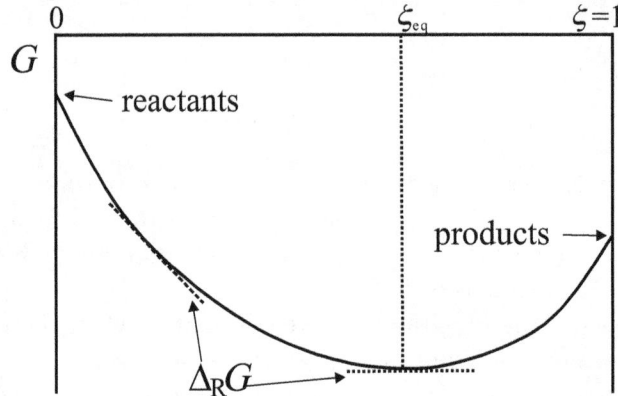

Figure 8.1: Gibbs energy as a function of the extent of one mole of the chemical reaction. The slope is $\Delta_r G = f(\xi)$.

and, according to eqn. (8.3), we can write

$$\mathbb{A} = -\left(\frac{\partial G}{\partial \xi}\right)_{p,T} \tag{8.5}$$

At this time, it is convenient for a short digression about the relation between the thermodynamic and the kinetic aspects of a chemical processes. According to eqn. (1.18), the change of entropy when there is only mechanical work is given by

$$dS = \frac{dU}{T} + \frac{pdV}{T} - \sum \frac{\mu_i dn_i}{T}$$

and remembering the first principle (eqn. (1.1)) we have

$$dS = \frac{\delta q}{T} - \sum \frac{\mu_i dn_i}{T}$$

The first term of the rhs is related with the change of entropy due to the environment, produced by exchanging heat with the system. The second term is the change of energy due to the internal process when the reaction takes place (eqn. (1.2)). Hence, these equations yield

$$d_i S = \frac{\mathbb{A}}{T} d\xi$$

Thus, the production of entropy, which gives an indication of the degree of irreversibility due to the occurrence of a process, becomes

$$\frac{d_i S}{dt} = \frac{\mathbb{A}}{T} \frac{d\xi}{dt} \geq 0 \tag{8.6}$$

where the factor $d\xi/dt$ represents the rate of the reaction. This means that the irreversibility of a process, measured by the production of entropy, depends on the product of a thermodynamic factor and a kinetic one. The reactions occurring with a large rate will be more irreversible than those which are slower, obviously depending also on the value of $\Delta_r G = -\mathbb{A}$. This conclusion is very important when dealing with processes that do not occur under equilibrium.

8.2   Homogeneous Equilibrium and Variation of G With the Extent of Reaction

This section will be devoted to a detailed analysis of a chemical reaction in gaseous phase that attains equilibrium. The participant gases will be considered ideal and the process will occur at constant p and T. The reaction we will take as an example is the dissociation of oxide of dinitrogen N_2O, according to

$$N_2O_4(g) \rightleftharpoons 2NO_2(g)$$

The Gibbs energy of the mixture of gases will be

$$G = n_D\mu_D + n_M\mu_M = n_D\left[\mu_D^\ominus + RT\ln\left(\frac{x_D p}{p^\ominus}\right)\right] +$$

$$+ n_M\left[\mu_M^\ominus + RT\ln\left(\frac{x_M p}{p^\ominus}\right)\right]$$

where the subscript D denotes the dimer and subscript M the monomer. We can define an adimensional quantity, called *degree of conversion*, as $\xi' = \xi/n_D^\circ$, which adopts values between 0 and 1 and is analogous to the dissociation degree of the dimer. Assuming that the reaction started with n_D° ($\xi = 0$) moles of dimer, the number of moles of D and M expressed as a function of the dissociation degree will be

$$n_D = n_D^\circ(1 - \xi') \quad \text{and} \quad n_M = 2\xi' n_D^\circ$$

Hence, the mole fractions of the reactant and the product as a function of ξ' are given by

$$x_D = \frac{1 - \xi'}{1 + \xi'} \quad \text{and} \quad x_M = \frac{2\xi'}{1 + \xi'}$$

Since the system is at pressure $p = 1$ bar, which is identical to the standard pressure $p^{\ominus} = 1$ bar, the Gibbs energy of the gaseous system will be

$$G = n_D^0 \left[(1 - \xi')\mu_D^{\ominus} + 2\xi'\mu_M^{\ominus} + RT(1 + \xi')\ln\frac{p}{p^{\ominus}} + \right.$$
$$\left. + RT(1 - \xi')\ln\frac{1 - \xi'}{1 + \xi'} + RT2\xi'\ln\frac{2\xi'}{1 + \xi'} \right]$$

For the dissociation reaction at $T = 300$ K, $(\mu_D^{\ominus} - 2\mu_M^{\ominus}) = 5435$ J/mole and, since $p = p^{\ominus}$, we get

$$\frac{G}{n_D^0} - \mu_D^{\ominus} = 5435\,\xi'\text{ J/mole} + RT(1 - \xi')\ln\frac{1 - \xi'}{1 + \xi'} + RT\,2\xi'\ln\frac{2\xi'}{1 + \xi'} \tag{8.7}$$

Figure 8.2 shows the curve of G as a function of the degree of conversion, for the dissociation of dinitrogen tetroxide given by the previous reaction.

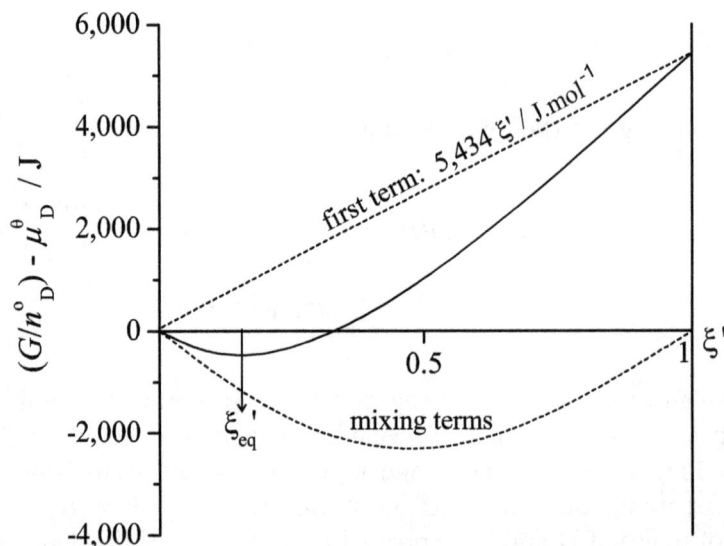

Figure 8.2: Gibbs free energy associated to the reaction $N_2O_4(g) \rightleftharpoons 2\,NO_2$. The solid curve represents $(G/n_D^0 - \mu_D^{\ominus})$ as a function of ξ'.

The minimum in the curve, according to eqn. (8.3), corresponds to the position of the equilibrium for the chemical reaction being considered. For this state, the dissociation degree ξ'_{eq} is equal to 0.166. The figure shows the contribution of the two main contributions to G: The first, accounts for the difference of standard states (notice that it is linear with ξ'); and the second, gives the contribution of the process of mixing of the two ideal gases for a dissociation degree ξ', under the condition that the total pressure remains constant–that is, if the reaction is made inside a cylinder having a piston that keeps the pressure p constant. If the reactions were done at constant volume, for instance in a closed reactor, the curve would be different because the change in total pressure as the reaction proceeds must be taken into account.

8.3   van't Hoff's Isotherm and Chemical Equilibrium in Real Systems

For a general chemical reaction, the variation of the Gibbs energy that takes place when the reaction occurs at constant p and T is given, according to eqn. (8.2), by

$$\Delta_{\mathrm{r}}G \;=\; \sum_{\mathrm{P}} \nu_{\mathrm{P}} \mu_{\mathrm{P}} - \sum_{\mathrm{R}} \nu_{\mathrm{R}} \mu_{\mathrm{R}} = \tag{8.8}$$

$$=\; \sum_{\mathrm{P}} \nu_{\mathrm{P}} \mu_{\mathrm{P}}^{\circ} - \sum_{\mathrm{R}} \nu_{\mathrm{R}} \mu_{\mathrm{R}}^{\circ} + RT \ln \frac{\prod_{\mathrm{P}}(a_{\mathrm{P}}/\mathcal{C}^{\circ})^{\nu_{\mathrm{P}}}}{\prod_{\mathrm{R}}(a_{\mathrm{R}}/\mathcal{C}^{\circ})^{\nu_{\mathrm{R}}}}$$

where $\prod$ denotes the product of the substances activities and $\mathcal{C}^{\circ}$ is the concentration corresponding to the standard state,[1] which depends on the state of aggregation of each substance that participates in the reaction and on the adopted activity scale. The first two terms of the rhs member yield the standard reaction Gibbs energy; thus, the previous equation may be written as

$$\Delta_{\mathrm{r}}G = \Delta_{\mathrm{r}}G^{\circ} + RT \ln \frac{\prod(a_{\mathrm{P}}/\mathcal{C}^{\circ})^{\nu_{\mathrm{P}}}}{\prod(a_{\mathrm{R}}/\mathcal{C}^{\circ})^{\nu_{\mathrm{R}}}} = \Delta_{\mathrm{r}}G^{\circ} + RT \ln Q \tag{8.9}$$

where Q represents the activities quotient raised to the corresponding stoichiometric numbers. Depending on the values of the activities of reactants and products and the value of $\Delta_{\mathrm{r}}G^{\circ}$, $\Delta_{\mathrm{r}}G$ will be either negative (the reaction will proceed spontaneously to products), zero (the reaction will be in equilibrium), or positive (the reaction that occurs will be the decomposition of products forming the reactants; that is, the reaction turns round and the extent of reaction will decrease). When the chemical reaction is at equilibrium we will have, using eqn. (8.9), that

$$\Delta_{\mathrm{r}}G^{\circ} = -RT \ln \frac{\prod_{\mathrm{P}}(a_{\mathrm{P}}^{\mathrm{eq}}/\mathcal{C}^{\circ})^{\nu_{\mathrm{P}}}}{\prod_{\mathrm{R}}(a_{\mathrm{R}}^{\mathrm{eq}}/\mathcal{C}^{\circ})^{\nu_{\mathrm{R}}}} \tag{8.10}$$

---

[1]The supraindex $^{\circ}$ is used to denote a standard state of any kind.

The rhs member in the previous expression contains the equilibrium constant of the reaction, K, since $Q(\xi'_{eq}) \equiv K$:

$$\Delta_r G^\circ = -RT \ln K \qquad (8.11)$$

This expression indicates that the equilibrium constant only depends on temperature, because it is only related with the corresponding $\mu_i^\circ(T)$ of reactants and products, standard thermodynamic quantities which, as already shown, depend only on temperature.

The relation (8.11) is very important and is the basis of one of the methods more frequently used to obtain the values of the thermodynamic quantities that characterize chemical processes. Nevertheless, its use requires some care. Eqn. (8.11) relates the change of molar Gibbs energy of reaction when the reactants are transformed into products, all of them being in their respective standard concentrations, with the equilibrium constant of the system corresponding to the equilibrium concentrations for all the substances participating in the equilibrium. That is, both members of the relation (8.11) refer to different states of the system, and the two members of that equation appear related only because in the state of equilibrium $\Delta_r G = 0$. This can be illustrated with the dissociation reaction of dinitrogen tetroxide that was analyzed in the previous section. There, we showed that $\Delta_r G^\ominus = 5435$ J/mole, and hence the products at the standard pressure will evolve spontaneously to reactants at the same pressure. On the other hand, the extent of reaction found in the previous section for the equilibrium condition leads to an equilibrium constant at 300 K, $K = 0.1132$ (bar) for the same reaction. In the equilibrium state, the partial pressures are $p_D = 0.713$ bar and $p_M = 0.287$ bar, which are different from the standard pressure $p^\ominus = 1$ bar, characterizing the definition of $\Delta_r G^\ominus$.

It is important to make clear that, in order to simplify the notation in what follows, the equilibrium constant will be written

$$K = \frac{\Pi(a_P^{eq})^{\nu_P}}{\Pi(a_R^{eq})^{\nu_R}}$$

That is, in the expressions similar to the latter, the concentrations in the standard state will be omitted, but it should be remembered that they are implicitly dividing the activities. Rigorously, the equilibrium constant has no units. This is because we are dealing with expressions that contain quotients of real and standard state concentrations. On the other hand, the equilibrium constant cannot have units because, otherwise, eqn. (8.11) makes no sense, since it is impossible to calculate the logarithm of a magnitude having dimensions. In order to see it even more clearly you can try to answer the following question: What is the value the logarithm of ten apples? However, in this book the values of K will be given followed by apparent dimensions, in parenthesis, in order to facilitate the understanding of the values; this is a way to remind the reader which standard states are being used.

8.4   Heterogeneous Chemical Equilibria

Usually, chemical equilibrium is established among substances that are not in the same phase; in this case, we talk of heterogeneous chemical equilibria. These types of equilibria do not present major problems, except for the fact that the standard states of the substances involved in the chemical processes will be different. The most frequently encountered heterogeneous equilibria are the thermal decompositions of carbonates, metal oxides, and hydrated salts. In all these cases the system has two pure solid phases and one gaseous one in equilibrium; this will be illustrated with the decomposition equilibrium of $CaCO_3(s)$, as indicated by the following chemical equilibrium

$$CaCO_3(s) \rightleftharpoons CaO(s) + CO_2(g)$$

The equilibrium constant for this reaction is given by

$$K = p(CO_2)/p^{\ominus}$$

because the pure solids have unity activity. In cases as simple as this it is easy to see the similarity of the chemical heterogeneous equilibrium between pure substances with phase equilibrium; for instance, in the fact that the equilibrium pressure of the gaseous product will only be function of temperature since, as already shown, the equilibrium constant is given by the relation between the standard chemical potentials defined at a fixed $T^{(\ominus)}$.

Frequently, it is important to know if in an equilibrium between solid phases, these are pure phases or they constitute a mixture. Let us consider the following chemical equilibrium

$$MY_3(s) \rightleftharpoons MY(s) + Y_2(g)$$

If n° is the initial number of moles of MY_3, the equilibrium quantities of MY_3, MY, and Y_2, will be $n^{\circ}(1 - \xi')$, $n^{\circ}\xi'$, and $n^{\circ}\xi'$ (expressed in moles), respectively. The adimensional quantity ξ' is the decomposition degree of MY_3.

If the two solid phases form a mixture, we can write the general equilibrium constant of the previous process as

$$K = \frac{x(MY)}{x(MY_3)} \frac{p(Y_2)}{p^{\ominus}}$$

where we have assumed that the solid mixture is ideal and x represents the mole fractions of the species MY and MY_3 in the solid phase. Finally, introducing the expressions for the degree of conversion of the reaction, when it has started with n° moles of MY_3, we have

$$K = \frac{\xi'}{1 - \xi'} \frac{p(Y_2)}{p^\ominus}$$

Figure 8.3 illustrates how the pressure of Y_2 varies with the decomposition degree of MY_3, at constant temperature. This case is analogous to the dissociation of a weak acid, where the concentration of protons is given by $K(1-\alpha)/\alpha$, where α represents the dissociation degree.

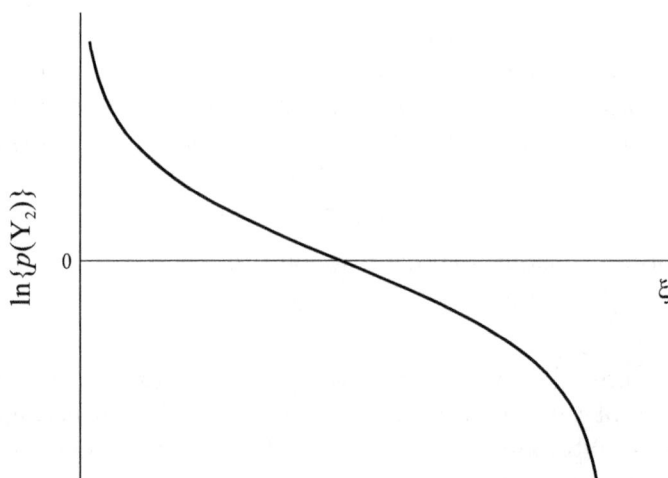

Figure 8.3: Heterogeneous chemical equilibrium $MY_3(s) = MY(s) + Y_2(g)$ with the formation of a solid solution. Changes in the partial pressure of Y_2 as a function of the decomposition degree of MY_3, at constant temperature.

On the other hand, if the solid phases MY and MY_3 are pure, their mole factions will be the unit, and we have

$$K = \frac{p(Y_2)}{p^\ominus}$$

Hence, at constant temperature, the pressure of Y_2 will not change during the reaction while both solid phases are present, analogous with the case of the vapor pressure of a pure solid or liquid. According to figure 8.3, it then becomes possible to distinguish if the two solids are distinct phases or it they form a mixture: In the first case $p(Y_2)$ is independent of ξ', while in the second one it will depend on the extent of reaction.

Figure 8.4 is another example of heterogeneous equilibrium. Here, $FeBr_2$ yields, upon equilibration with water vapor, different hydrates $FeBr_2 \cdot (H_2O)_n$, where the subscript n represents their mole-water-per-mole-salt content. At constant temperature, the anhydrous salt is allowed to stand in contact with an atmosphere where the humidity increases. Initially, a mixture of the anhydrous salt and $FeBr_2 \cdot (H_2O)_1$ is formed and, as long as this equilibrium persists, the water vapor pressure remains unchanged at a value $p(n = 0, 1)$. When no anhydrous salt is left, the water vapor pressure suddenly increases to a new value $p(n = 1, 2)$, and remains constant until the equilibrium between $FeBr_2 \cdot (H_2O)_1$ and $FeBr_2 \cdot (H_2O)_2$ ceases, which occurs when the monohydrate is totally consumed. A third equilibrium between $n = 2$ and $n = 4$ hydrates is also observed in the figure at a water vapor pressure $p(n = 2, 4)$.

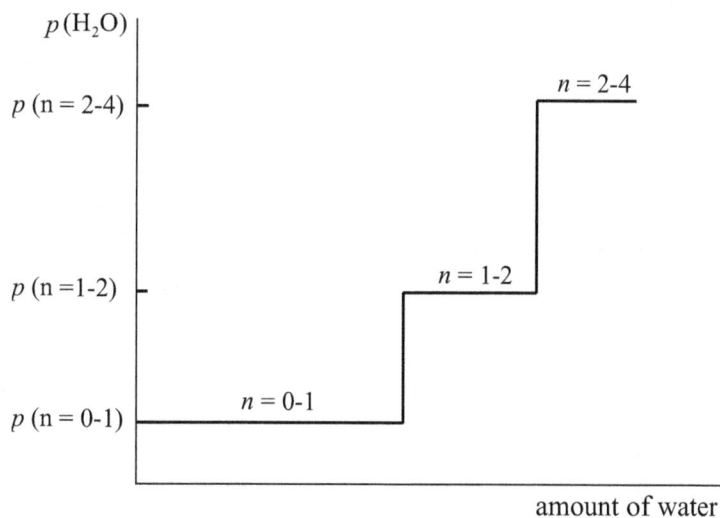

Figure 8.4: Water binding isotherm of $FeBr_2$. The partial pressure of water, $p(H_2O)$, is plotted as a function of the amount of water introduced in the system (see text). Each pressure step corresponds to the equilibrium reaction between two hydrates.

8.5   Le Chatelier's Principle and Thermodynamic Stability

Chemical equilibria play an important role in many chemical processes, natural and industrial. In 1884, H. L. Le Chatelier formulated his well-known principle to explain how the systems evolve when they are at equilibrium and how some of the variables that characterize it are perturbed. This principle can be stated simply by saying that, when a system at equilibrium is perturbed, the equilibrium state is modified in such way that the effect of the perturbation is minimized. This formulation must be applied with great care because, as we will see, there are situations where it looks as if did not apply; nevertheless, if the effect of the perturbation over the whole system is analyzed in detail, the principle is always correct. The principle of Le Chatelier is only a consequence of the principles of thermodynamics, but its great practical use makes it worthy to illustrate its consequences in a special form.

Eqn. (8.2) gives the change of Gibbs energy of a system when the extent of reaction varies at constant p and T. Now it can be extended to the case where the reaction can be also affected by changes of p and T. Then, for a chemical reaction in the gas phase

$$dG = V dp - S dT + \left(\sum_{\mathrm{P}} \nu_{\mathrm{P}} \mu_{\mathrm{P}} - \sum_{\mathrm{R}} \nu_{\mathrm{R}} \mu_{\mathrm{R}} \right) d\xi \qquad (8.12)$$

$$dG = V dp - S dT + \Delta_{\mathrm{r}} G d\xi \qquad (8.13)$$

This expression relates the variation of G per mole of the reacting system with pressure, temperature, and the extent of the reaction. It has already been shown that the condition of equilibrium at p and T constants is

$$\left(\frac{\partial G}{\partial \xi} \right)_{p,T} = \Delta_{\mathrm{r}} G = 0$$

It is convenient to make clear that if a change in the extent of the reaction occurs, at constant volume and temperature, then the previous expression is not applicable and it must be replaced by (starting from eqn. (8.13)),

$$\left(\frac{\partial G}{\partial \xi} \right)_{V,T} = V \left(\frac{\partial p}{\partial \xi} \right)_{V,T} + \left(\frac{\partial G}{\partial \xi} \right)_{p,T} = 0 \qquad (8.14)$$

The equilibrium condition is reached by adding the term of the change of pressure with ξ at constant V, and keeping always the system in chemical equilibrium. That is, when $\Delta_{\mathrm{r}} G = 0$, $(\partial G/\partial \xi)_{T,V}$ will be different from zero, unless the pressure is not modified with the extent of reaction, a situation where the pressure does *not* affect the chemical equilibrium. The curve of G as a function of ξ, at constant T and V, will be similar to that illustrated in figure 8.1, but the minimum *will not correspond* to the same ξ_{eq} as for constant p and T.

The quantity $\Delta_{\mathrm{r}} G$ is then a function of (p, T, ξ), being

$$d(\Delta_{\mathrm{r}} G) = -\Delta_{\mathrm{r}} S \, dT + \Delta_{\mathrm{r}} V dp + \frac{\partial^2 G}{\partial \xi^2} d\xi \qquad (8.15)$$

If, in spite of changing p and T, it is required to keep the condition of chemical equilibrium ($\Delta_r G = 0$), it will be necessary that $d(\Delta_r G) = 0$. In that case, if one considers that $\Delta_r S = \Delta_r H/T$, we have at constant pressure

$$\left(\frac{\partial \xi_{eq}}{\partial T}\right)_p = \frac{\Delta_r H}{T(\partial^2 G/\partial \xi^2)_{eq}} \tag{8.16}$$

and at constant temperature

$$\left(\frac{\partial \xi_{eq}}{\partial p}\right)_T = -\frac{\Delta_r V}{(\partial^2 G/\partial \xi^2)_{eq}} \tag{8.17}$$

These are the expressions that summarize the thermodynamic equilibrium and the Le Chatelier's principle. It can be seen that the eqns. (8.16) and (8.17) establish how the extent of the reaction changes when modifying a thermodynamic variable. If we deal with systems having other type of energies, like electric, elastic, etc., these will appear in the expression of G for the system and, using the same procedure as before, it is possible to establish how the equilibrium is modified when a variable that affects those energy types is perturbed. The quantity $(\partial^2 G/\partial \xi^2)_{eq} > 0$, will always appear; this is the quantity whose sign will be determined by the requirement of thermodynamic stability (cf. chapter 1). Thus, eqn. (8.16) indicates that $(\partial \xi_{eq}/\partial T)_p$ will have the same sign as $\Delta_r H$, while according to eqn. (8.17), $(\partial \xi_{eq}/\partial p)_T$ will have the opposite sign to $\Delta_r V$. To extend the Le Chatelier's principle away from the case of chemical equilibrium (cf. section 8.6), it is convenient to denote it in a more descriptive way, as proposed by I. Prigogine, and call it moderation principle.

Equilibrium State, Thermodynamics, and Kinetics of Chemical Reactions

We use as an example a chemical reaction in the gaseous phase that is very simple because its kinetics of reaction indicates that the process occurs in a single elemental stage, and we will assume that the conditions are such that the gases behave ideally. The reaction is

$$H_2(g) + I_2(g) \rightleftharpoons 2HI(g)$$

for which the overall rate of the reaction is $v = d\xi/dt$. Often, v is splitted in the contribution that generates products, rate v_+, and that consuming products and generating reactants, opposite rate v_-. Then,

$$\frac{dc_{HI}}{dt} \equiv 2\frac{d\xi}{dt} = v_+ - v_- = k_+ c_{H_2} c_{I_2} - k_- c_{HI}^2 \tag{8.18}$$

where k_+ and k_- are the rate constants of the direct and opposite reactions, respectively. Using the first term of the rhs as a common factor, we get

$$\frac{\mathrm{d}\xi}{\mathrm{d}t} = \frac{k_+}{2} c_{H_2}\, c_{I_2} \left(1 - \frac{k_- c_{HI_2}^2}{k_+ c_{H_2}\, c_{I_2}}\right) = \frac{v_+}{2}\left(1 - \frac{Q}{K}\right) \qquad (8.19)$$

In the previous expression, we used the nomenclature of eqn. (8.9) and (8.11), as well as the fact that in equilibrium $v_+ = v_-$, and $K = k_+/k_-$. Using the same equation (8.9), we obtain

$$\frac{\mathrm{d}\xi}{\mathrm{d}t} = \frac{v_+}{2}\left[1 - \exp\left(\frac{\Delta_r G}{RT}\right)\right] = \frac{v_+}{2}\left[1 - \exp\left(-\frac{\mathbb{A}}{RT}\right)\right] \qquad (8.20)$$

where we have also written the expression for the case one desires to use the affinity $\mathbb{A} = -\Delta_r G$.

Now, let us imagine that the system is not very far away from equilibrium (i.e., that $(\xi/\xi_{eq}) \simeq 1$). This situation is frequently encountered, and can be imagined that it occurs when chemical processes in equilibrium, are slightly perturbed by changes of temperature, pressure, or concentration. In these conditions, $\Delta_r G(\xi)$ will be very small (or close to zero), and then the value of the corresponding reaction Gibbs energy for the perturbed system can be expressed as a series around the equilibrium state, whose first term is the important one, according to

$$\Delta_r G(\xi) = \Delta_r G(\xi_{eq}) + \left(\frac{\partial \Delta_r G}{\partial \xi}\right)_{eq} (\xi - \xi_{eq}) + \dots \qquad (8.21)$$

Remembering that $\Delta_r G(\xi_{eq}) = 0$, the series expansion in eqn. (8.21) can be replaced in eqn. (8.20). Using the extent of reaction to express the concentration of reactants R, we get $c_R = c_{R,eq} - (\xi - \xi_{eq})$, thus the previous equation may be written

$$\frac{\mathrm{d}\xi}{\mathrm{d}t} \approx -\frac{v_+}{2}\left(\frac{\Delta_r G}{RT}\right) = -\frac{v_+}{2RT}\left(\frac{\partial \Delta_r G}{\partial \xi}\right)_{eq}(\xi - \xi_{eq}) =$$

$$= \frac{v_+}{2RT}\left(\frac{\partial \mathbb{A}}{\partial \xi}\right)_{eq}(\xi - \xi_{eq}) \qquad (8.22)$$

The negative sign in the first terms of the rhs in the previous expression denotes the restitution effect on the perturbation; that is, if the perturbation has displaced the equilibrium to larger values of ξ, when the system relaxes, the extent of reaction will decrease with time. Consequently, the opposite effect will occur when the perturbation tends to diminish the value of ξ.

It was shown that in the equilibrium state

$$k_+ \, c_{H2,eq} \, c_{I2,eq} = k_- \, c_{HI,eq}^2$$

In terms of the extent of reaction, this expression can also be written as[2]

$$k_+ \, (1 - \xi_{eq})^2 = 4k_- \, \xi_{eq}^2$$

Now, it is possible to calculate the rate of reaction for states slightly separated from the equilibrium state by differentiating $\Delta_r G$ with respect to the extent of reaction and using the value of the derivative for the equilibrium state, or simply writing the kinetics of the reaction for $\Delta\xi = \xi - \xi_{eq}$. In both cases we get

$$\frac{d\xi}{dt} = -[k_+ \, (1 - \xi_{,eq}) + 4k_- \, \xi_{eq}](\xi - \xi_{eq}) = -\frac{\xi - \xi_{eq}}{\tau}$$

using ξ as the variable or, using the concentrations of reactants and products as variables, it leads to

$$\frac{d\xi}{dt} = -[k_+ \, (c_{H2,eq} \, c_{I2,eq}) + 2k_- \, c_{HI,eq}](\xi - \xi_{eq}) = -\frac{\xi - \xi_{eq}}{\tau}$$

The denominator denoted by τ is known as the relaxation time of the reaction. It is possible to observe the relation between the kinetics of a chemical process and the change of Gibbs energy (or the affinity of the reaction) is given by

$$\frac{1}{\tau} = \frac{v_+}{2RT}\left(\frac{\partial \Delta_r G}{\partial \xi}\right)_{eq} = -\frac{v_+}{2RT}\left(\frac{\partial \mathbb{A}}{\partial \xi}\right)_{eq} = \frac{v_+}{2RT}\left(\frac{\partial^2 G}{\partial \xi^2}\right)_{eq}$$

an expression that takes us back to the discussion in section 8.1. It is an important relation because the differential in the last term of the previous equation emphasizes the importance of the thermodynamic stability of the equilibrium state (cf. discussion in section 8.6).

---

[2]In order to put the example in easier terms, we have to assume that the reaction was initiated with one mole of each reactant, in absence of products.

Replacing in eqn. (8.6), the rate of reaction according to eqn. (8.22), it yields

$$\frac{\mathrm{d}_i S}{\mathrm{d}t} = \frac{v_+}{2R}\left(\frac{\Delta_r G}{T}\right)^2 = \frac{v_+}{2R}\left(\frac{\mathbb{A}}{T}\right)^2$$

This expression is completely general, for small departures from the equilibrium state lead to

$$\frac{\mathrm{d}_i S}{\mathrm{d}t} \simeq \frac{v_+}{2RT^2}\left(\frac{\partial \Delta_r G}{\partial \xi}\right)^2_{\mathrm{eq}}(\xi - \xi_{\mathrm{eq}})^2 \geq 0$$

thus showing that the production of entropy depends of the thermodynamic stability of the chemical equilibrium state, given by $(\partial \Delta_r G/\partial \xi)_{\mathrm{eq}} = (\partial^2 G/\partial^2 \xi)_{\mathrm{eq}} > 0$ (cf. previous section).

8.6   A Few Consequences of the Le Chatelier's Principle

A trivial application of the Le Chatelier's principle consists in predicting the evolution of a system when, at constant temperature, the concentration of the substance participating in the chemical equilibrium is modified. In that case, the system will respond according with the requirement that the equilibrium constant does not vary; if the concentration of reactants is increased, the system will evolve toward the products (ξ_{eq} increases), and if the concentration of products is increased, the system will respond by regenerating the reactants (ξ_{eq} decreases). There are situations that appear as exceptions to the previous argument, but they are really not, as will be demonstrated.

i) Pressure effect in gaseous systems
Now we consider the equilibrium perturbation due to a change of pressure. This case will be illustrated with the homogeneous gaseous reaction of synthesis of ammonia, among other reasons, because it is a frequently used industrial process in which the adequate choices of the variables p and T to optimize the yield of products is based partially in the application of Le Chatelier's principle.

    If we refer to the reaction for the decomposition of ammonia

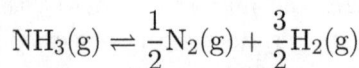

$$\mathrm{NH_3(g)} \rightleftharpoons \frac{1}{2}\mathrm{N_2(g)} + \frac{3}{2}\mathrm{H_2(g)}$$

and assume that the gaseous mixture behaves ideally at a temperature of 673 K, the following values of the properties that describe the equilibrium state are $\Delta_r G^\ominus =$

-24.38 kJ/mole, $\Delta_r H^\ominus = 51.47$ kJ/mole, and $K^{\text{syn}} = 78.74$ (bar). If we consider that the reaction started with n° moles of ammonia, the number of moles of NH_3 when equilibrium is established will be $n^\circ(1 - \xi'_{\text{eq}})$, that of H_2 will be $3n^\circ\xi'_{\text{eq}}/2$, and that of N_2 will be $n^\circ\xi'_{\text{eq}}/2$, where ξ' stands for the decomposition degree of ammonia.

Table 8.1 shows the values of ξ'_{eq} and also reports the mole fractions of the three substances at the total pressures of 10 and 50 bar.

Table 8.1: Total pressure effect on the synthesis and decomposition of NH_3 at 673 K. The values of the equilibrium constants refer to ideal gaseous mixtures.

p / bar	10	50
ξ'_{eq}	0.9260	0.7374
$x(NH_3)$	0.0384	0.1511
$x(N_2)$	0.2404	0.2122
$x(H_2)$	0.7212	0.6367
K_x^{syn}	0.128	0.646
K^{syn}	0.0128	0.0129
K_x^{dec}	7.820	1.549
K^{dec}	78.74	77.45

It is observed that the yield of the decomposition reaction of NH_3 is smaller as the total pressure increases (i.e., there is a larger quantity of ammonia in the equilibrium state at higher total pressure). The equilibrium constant for the ideal gaseous mixture is represented by

$$K^{\text{dec}} = \frac{[p/p^\ominus]^{3/2}(H_2)\,[p/p^\ominus]^{1/2}(N_2)}{[p/p^\ominus](NH_3)} = \frac{x^{3/2}(H_2)\,x^{1/2}(N_2)}{x(NH_3)}\,\frac{p}{p^\ominus} = K_x\,\frac{p}{p^\ominus}$$

Thus, the increase in pressure leads to a decease of K_x for the synthesis of ammonia (defined as the relation of the equilibrium for mole fractions instead of pressures), because it is necessary that K^{dec} remains constant.

The expression that yields the pressure effect on the chemical equilibrium, in ideal gaseous systems, can be generalized as

$$K(T) = K_x \left(\frac{p}{p^\ominus}\right)^{\Delta\nu} \tag{8.23}$$

where $\Delta\nu = \nu_P - \nu_R$ is the difference between the number of moles of products and that of the reactants, per mole of reaction. Hence, when the total number of moles of reactants is equal to the total number of moles of products, $\Delta\nu = 0$ and changes in the pressure do not affect the position of the equilibrium (i.e., on ξ'_{eq}).

The total pressure of the gaseous reacting system can also be modified by addition, at constant volume, of a gas that does not participate in the reaction. For the reaction involving the synthesis of ammonia, Ar gas might be added without changes in the equilibrium. Although adding Ar will increase the total pressure, this addition will not alter the equilibrium, if the gases behave ideally.

The indifference of the reaction to changes in total pressure due to the addition of ideal gases into the reacting system can be explained in terms of a compensation of two effects: i) the non-reacting gases decrease the mole fractions of the substances intervening in the equilibrium, and ii) they increase the total pressure in exactly the same factor. In this case, one must evaluate the quotient involving the quantities of matter of each substance, and not K_x, to figure out if the yield of the reaction is affected by the addition of an inert gas. Replacing the number of moles of the participating substances, given by the ideal gas equation, in eqn. (8.23) we obtain

$$K(T) = K_n\, n_t^{-\Delta\nu} \left(\frac{n_t RT}{p^{\ominus} V}\right)^{\Delta\nu} = K_n \left(\frac{RT}{p^{\ominus} V}\right)^{\Delta\nu} \qquad (8.24)$$

where $K_n = (\Pi n_{\mathrm{P}}^{\nu_{\mathrm{P}}} / \Pi n_{\mathrm{R}}^{\nu_{\mathrm{R}}})$ is the quotient of the moles of each species at equilibrium, elevated to the power equal to the respective stoichiometric number, and n_t represents the total number of moles in the ideal gas phase. It may be seen that K_n does not change with the total pressure in the system, if V is constant.

A situation related to the previous discussion appears in the case where we add reactants or products to chemical systems, in equilibrium in the gas phase, having certain stoichiometry, like that in the reaction among methanol, CO and hydrogen, represented by

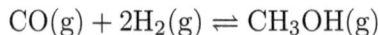

$$CO(g) + 2H_2(g) \rightleftharpoons CH_3OH(g)$$

or in the decomposition of ammonia in its elements, balanced using non-fractional coefficients,

$$2NH_3(g) \rightleftharpoons N_2(g) + 3H_2(g)$$

Both processes have great importance for chemical industry. In these cases, when one increases the number of moles of some substances, from those corresponding to the equilibrium state at constant p and T, the reaction will not always evolve in the direction where the number of moles of that substance decreases. The reason why this happens is that, in the gas phase, the addition of a reactant (or a product) will increase the concentration of those respective substances, but will simultaneously decrease the concentration of the rest of the gaseous species, when expressed using their partial pressures p_i.

Let us assume that n_j moles of j are added at constant pressure, taking the system out from the equilibrium state. Then, we write Q_p, which will be a relation of partial pressures of products divided by those of the reactants, all elevated to their corresponding stoichiometric numbers.[3] If Q_p is larger than K, the reaction will evolve so that it diminishes the value of Q_p, until it returns to be equal to K. On the other hand, if $Q_p < K$, Q_p will increase on addition of n_j moles so that again $Q_p = K$. Let us write again the general reaction

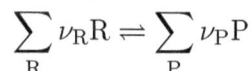

$$\sum_{R} \nu_R R \rightleftharpoons \sum_{P} \nu_P P$$

and let us use ν_i, as indicated in page 236 (i.e., $\nu_i = \nu_P$ and $\nu_i = -\nu_R$), and we have

$$Q_p = \left(\frac{p}{n_t}\right)^{\Delta\nu} Q_n$$

and the change of $\ln Q_p$ with n_j will be (taking into account the previous relations)

$$\left(\frac{\partial \ln Q_p}{\partial n_j}\right)_{p,T,n_{j\neq i}} = \frac{1}{n_j}(\nu_j - x_j \Delta\nu)$$

where x_j is the mole fraction of j.

For the case of the decomposition of NH_3, when nitrogen is added to a mixture having $x(N_2) < 0.5$, it will result in $(\partial \ln Q_p/\partial n_j)_{p,T,n_{j\neq i}}$ being larger than zero, according to the previous relation, since $[1 - 2x(N_2)]$ is positive. For $x(N_2) > 0.5$, the addition of nitrogen leads to a decrease of Q_p and the reaction will evolve toward the decomposition of ammonia in order to restore the equilibrium condition. The same effect, which appears to be a paradox, occurs for the reaction describing the synthesis of methanol when $x_{CO} > 0.5$; the addition of CO under those conditions increases the decomposition of CH_3OH.

The effect of the change in pressure, at constant volume, can also be derived from eqn. (8.14) because, for ideal gases

$$p = \frac{RT}{V}\left[\sum_{R} n_R + \sum_{P} n_P\right] = \frac{RT}{V}\left[\sum_{R}(n_R^\circ - \nu_R \xi) + \sum_{P}(n_P^\circ + \nu_P \xi)\right]$$

This equation was derived for the case in which the only gaseous substances are those involved in the chemical equilibrium; if there are other gases that do not participate

---

[3]See eqn. (8.9) for the meaning of Q.

in the chemical equilibrium, it will be necessary to add one more term. Hence,

$$\left(\frac{\partial p}{\partial \xi}\right)_{V,T} = \frac{RT}{V}\left[\sum_P \nu_P - \sum_R \nu_R\right]$$

and finally

$$\left(\frac{\partial G}{\partial \xi}\right)_{V,T} = RT\left[\sum_P \nu_P - \sum_R \nu_R\right] + \Delta_r G = RT\Delta\nu + \Delta_r G$$

This expression will be zero only at equilibrium (i.e., when $(\Delta_r G)_{p,T} = 0$), if the number of moles of reactant and products are equal; that is, no change in pressure occurs when there is a change in the extent of reaction.

If the gaseous mixture is not ideal, an increase in pressure will affect the fugacity of the substances participating in the chemical reaction. According to eqn. (8.10) applied to gaseous systems, the expression for the equilibrium constant is given in terms of a relation of fugacities that only can be replaced for a relation of partial pressures if the mixture behaves ideally. According to the discussion in chapter 4, the fugacity of each substance varies as a function of temperature, of pressure in the gaseous mixture, and of the molecular characteristics of its components; obviously, for a given temperature, the effect on non-ideality will increase with total pressure. As an example, we analyze the data for the synthesis of NH_3 at 723 K, which are listed in the table 8.2. It becomes clear that both, K_p and K, denoting the equilibrium constants expressed in partial pressures and fugacities, respectively, are related by

$$K = \frac{\Pi_B f_B^{\nu_B}}{\Pi_A f_A^{\nu_A}} = \frac{\Pi_B p_B^{\nu_B}}{\Pi_A p_A^{\nu_A}} \times \frac{\Pi_B \Phi_B^{\nu_B}}{\Pi_A \Phi_A^{\nu_A}} = K_p Q_\Phi$$

Table 8.2: Total pressure effect on the synthesis of $NH_3(g)$, at 723 K

p / bar	30	100	300	600	1000
K_p	0.00676	0.00725	0.00884	0.01294	0.02328
K	0.00659	0.00636	0.00608	0.00642	0.01010

The non-ideality effect on K is able to alter the position of equilibrium in chemical gaseous reactions having $\Delta\nu = 0$, when the pressure increases sufficiently. For instance, for the reaction denoted as water-gas shift reaction, which can be written as

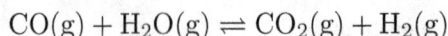

$$CO(g) + H_2O(g) \rightleftharpoons CO_2(g) + H_2(g)$$

the relation of fugacities Q_Φ, at 500 bar and 673 K, results 1.27 times greater than at 1 bar, where the equilibrium constant is practically equal to K_p.

ii) Pressure effect on condensed systems
According to eqn. (5.23), the expressions of the chemical potential for condensed phases have two terms: One expresses the dependence of the concentration through all the components' activities (when the phase is a mixture), and the other takes account of the effect of pressure on the standard chemical potential (Poynting effect). The latter phenomenon also affects the chemical equilibria involving systems having condensed phases. Really, eqn. (8.9) should be written by taking into account the general expression for the chemical potential of substance i, given in the eqn. (5.23), in order to include all the possible modifications of the chemical potentials. The complete expression is given by

$$\Delta_r G = \Delta_r G^\circ + RT \ln \frac{\Pi(a_P/\mathcal{C}^\circ)^{\nu_P}}{\Pi(a_R/\mathcal{C}^\circ)^{\nu_R}} + \int_{p^\ominus}^{p} \Delta_r V'^o dp \qquad (8.25)$$

In this expression, the quantity $\Delta_r V'^o = \sum'_P \nu_P V_P'^o - \sum'_R \nu_R V_R'^o$, gives the change of volume of reaction at the standard state, but considering only the intervening substances which are in a condensed phase; i.e. the symbol $\sum'$ indicates that the gaseous reactants and products are excluded from the sum. What we have just mentioned can be illustrated with the reaction

$$CuSO_4 \cdot 5H_2O(s) \rightleftharpoons CuSO_4(s) + 5H_2O(g)$$

for which
$$\Delta_r V'^o = V^*(CuSO_4) - V^*(CuSO_4 \cdot 5H_2O)$$

Eqn. (8.25) indicates that when there is chemical equilibrium and $\Delta_r G = 0$, the second term in the rhs becomes $-RT \ln K$. Since $\Delta_r G^\circ$ is an exclusive function of T, in agreement with the adopted convention (cf. section 5.2), and considering that the Poynting term depends in addition on pressure, it results that K must also be a function of (T, p) when there are condensed phases. This can be understood more clearly if one considers the following expression

$$\Delta_r G^\circ = -RT \ln K(T, p) - \int_{p^\ominus}^{p} \Delta_r V'^o dp$$

in which the two terms in the rhs depend on p. The fact that the equilibrium constant for a reaction where there are substances in condensed phase depends on temperature and pressure is an important conclusion that must be remembered.

Although the effect of pressure on the chemical equilibrium in presence of condensed phases is generally small, in some cases it is not negligible. For example, for the autoprotolysis reaction of water, at 298 K, given by the reaction

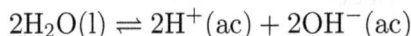

$$2H_2O(l) \rightleftharpoons 2H^+(ac) + 2OH^-(ac)$$

the product $c(H^+) \cdot c(OH^-)$ increases 10 percent when the pressure goes from 1 bar to 100 bar. This effect is important, for instance, in protolytic processes taking place in the deep oceans or inside the Earth's lithosphere.

iii) Temperature effect
According to eqn. (8.16), when the temperature increases (keeping p constant, and the chemical reaction at equilibrium) the extent of reaction will increase if $\Delta_r H^\circ > 0$; that is, if the process is endothermic. As a corollary, the extent of reaction will decrease when the process is exothermic.

It is useful to derive the equation that gives the change of $\ln K$ with T, which is done by differentiating eqn. (8.11), leading to

$$\left(\frac{\partial \ln K}{\partial T}\right)_p = \frac{\Delta_r H^\circ}{RT^2} \tag{8.26}$$

Since for chemical equilibria in gaseous phase, K is independent of pressure, it may be important to have an expression that allows us to know the value of K over a wide range of temperatures. For that purpose, the best is to begin with the expressions for the heat capacities of the intervening substances. These quantities are related to the effect of temperature over the enthalpy and the entropy of the system at equilibrium, and consequently with $\Delta_r G^\circ$. A usual expression that accounts for the temperature effect in the gaseous phase[4] is

$$\Delta_r C_p = \Delta_r a + \Delta_r b T + \Delta_r c T^2$$

where the operator Δ_r has the same meaning given in the section 1.3. $\Delta_r H^\circ$ may be calculated with the Kirchhoff equation, by integrating the equation given for $\Delta_r C_p$, leading to

$$\Delta_r H^\circ(T) = \Delta_r H^\circ_{T=0} + \int_0^T \Delta_r C_p dT =$$
$$= \Delta_r H^\circ_{T=0} + \Delta_r a T + \frac{\Delta_r b}{2} T^2 + \frac{\Delta_r c}{3} T^3$$

---

[4]This is valid for substances far away from the behavior observed for C_p at very low temperatures, discussed in chapter 2.

Using this expression for $\Delta_r H^\circ(T)$ in eqn. (8.26), and integrating from 0 to T, one arrives to the expression for K as a function of temperature

$$R \ln K = -\frac{\Delta_r H^\circ_{T=0}}{T} + \Delta_r a \ln T + \frac{\Delta_r b}{2}T + \frac{\Delta_r c}{6}T^2 + IR$$

where I is denoted as a chemical constant. If the previous expression is written separating its terms in the following way

$$\sum \equiv -R \ln K + \Delta_r a \ln T + \frac{\Delta_r b}{2}T + \frac{\Delta_r c}{6}T^2 = \frac{\Delta_r H^\circ_{T=0}}{T} - IR \qquad (8.27)$$

it is possible to obtain the values of $\Delta_r H^\circ_{T=0}$, the enthalpy of reaction at $T = 0$ K, and of I. The first is the slope of the straight line of $\sum$ as a function of $1/T$, and its ordinate at the origin is equal to IR. Hence, figure 8.5 gives a way of expressing the temperature dependence for the formation of gaseous H_2O,

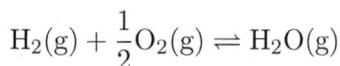

$$H_2(g) + \frac{1}{2}O_2(g) \rightleftharpoons H_2O(g)$$

resulting in the values indicated in the figure.

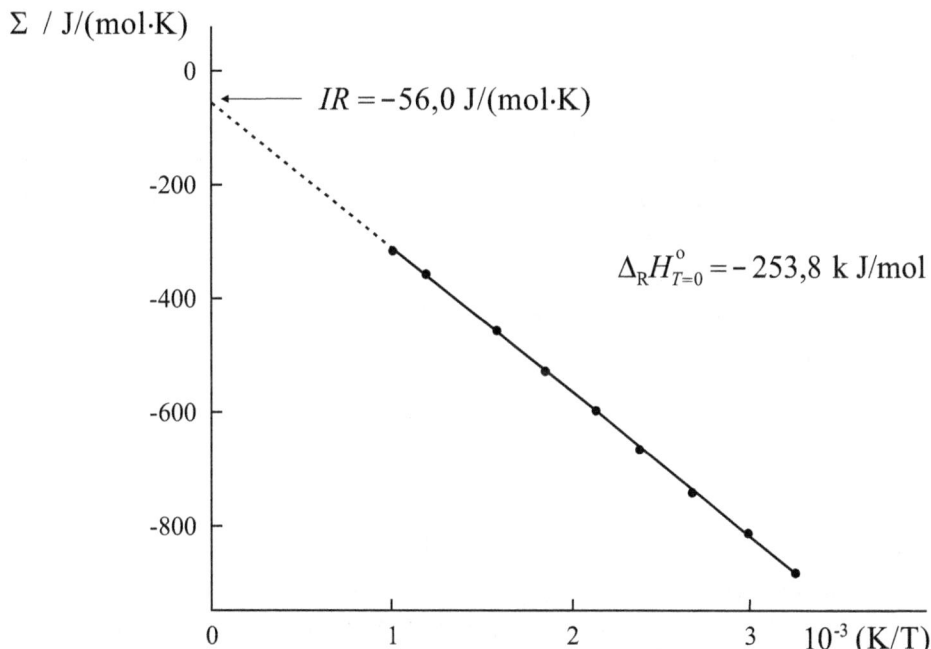

Figure 8.5: Temperature effect on the equilibrium of formation of $H_2O(g)$. The graph indicates the values of ordinates at the origin (IR) and the slope $(\Delta_r H^\circ)_{T=0}$ of $\sum$ as a function of $1/T$ (cf. eqn. (8.27)).

About fifty years ago, chemists were interested to know if it was possible to describe chemical equilibrium exclusively on the basis of calorimetric determinations. It is seen that, according to the example we have discussed, it is not possible to do it since it is necessary to have the value of the chemical constant for the reaction at a given temperature.

Geothermometers and Geobarometers

Chemical equilibria among minerals is important in geology. It is possible to use chemical equilibria between solids to extract information about the temperature and/or pressure prevailing in a given geological environment. It is possible to establish a relationship between the thermophysical features of some heterogeneous equilibria, and their relation with the extent of reaction. In particular, it would be possible to know the temperature or the pressure in a point inside the Earth's lithosphere from where the sample was obtained.

For the chemical reaction to be a good geothermometer, it is evident that it will be necessary for the extent of reaction to be very sensitive to temperature changes, but slightly sensitive to changes in pressure ($\Delta_r H$ very large and $\Delta_r V$ quite small). The opposite will be valid for the effect of p and T over those chemical equilibria that can be used as geobarometers. Let us see two examples of reactions that can be used: One as geobarometer, and the other as geothermometer.

i) Geobarometer. The following reaction describes the decomposition of anorthite, which forms quartz and $CaAl_2SiO_6$, a substance that remains dissolved in the mineral clinopyroxene. Anorthite and quartz are pure solids. The reaction is

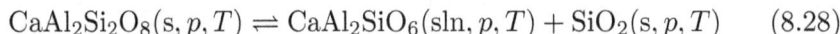

$$CaAl_2Si_2O_8(s, p, T) \rightleftharpoons CaAl_2SiO_6(sln, p, T) + SiO_2(s, p, T) \qquad (8.28)$$

and its equilibrium constant

$$K = a(CaAl_2SiO_6)$$

since $CaAl_2SiO_6$ is the only reactant that is not in a pure phase. In the state of equilibrium, the activity of $CaAl_2SiO_6$ is related to the change of reaction Gibbs energy and volume in the standard state (cf. eqn. (8.25)) according to

$$\Delta_r G^*(p_0, T) = -RT \ln a(CaAl_2SiO_6) - \int_{p_0}^{p} \Delta_r V^* dp \qquad (8.29)$$

At the laboratory, it is necessary to determine how the pressure and temperature affect the equilibrium of the mixtures of different activity of $CaAl_2SiO_6$ (which is taken to be equal to the mole fraction). As a result, a graph is obtained like the one illustrated in figure 8.6 for this example.

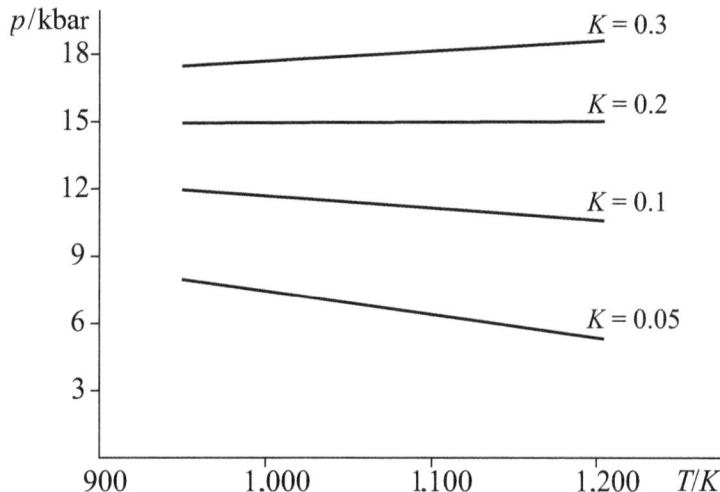

Figure 8.6: Geobarometer for the reaction (8.28).

It is clearly visible in this case, that the equilibrium constant varies very little with temperature for all the assayed solution concentrations. Hence, if in a sample obtained from the lithosphere it is possible to determine the mole fraction of $CaAl_2SiO_6$ present in the clinopyroxene, using the graph on figure 8.6 it is possible to know the pressure at the sampling point in the lithosphere. For the case of the geobarometers, the change in K with pressure is due to the Poynting effect, as it is deduced from eqn. (8.29).

ii) Geothermometer. The following chemical equilibrium, involving dissolved substances in clinopyroxene and in granate, can be used as a geothermometer:

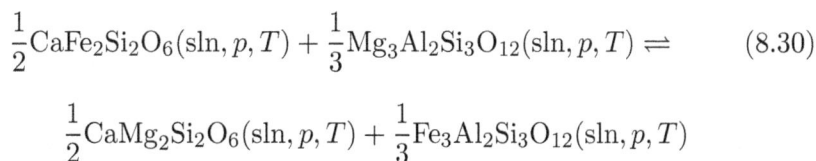

$$\frac{1}{2}CaFe_2Si_2O_6(\text{sln}, p, T) + \frac{1}{3}Mg_3Al_2Si_3O_{12}(\text{sln}, p, T) \rightleftharpoons \quad (8.30)$$

$$\frac{1}{2}CaMg_2Si_2O_6(\text{sln}, p, T) + \frac{1}{3}Fe_3Al_2Si_3O_{12}(\text{sln}, p, T)$$

The substances containing calcium are dissolved in clinopyroxene (cpx) and those containing aluminum in granate (gr). The equilibrium constant is therefore

$$K = \frac{a^{1/3}(\text{gr}, \text{Fe}) \cdot a(\text{cpx}, \text{Mg})}{a^{1/3}(\text{gr}, \text{Mg}) \cdot a(\text{cpx}, \text{Fe})}$$

Again, using the relation (8.29), it comes out that K is related to the effect of the variables p and T.

Figure 8.7 illustrates how the values of K change with p and with T for the reaction (8.30). It may be seen that this chemical equilibrium is an useful geothermometer because pressure has little effect of the values of K, when T varies.

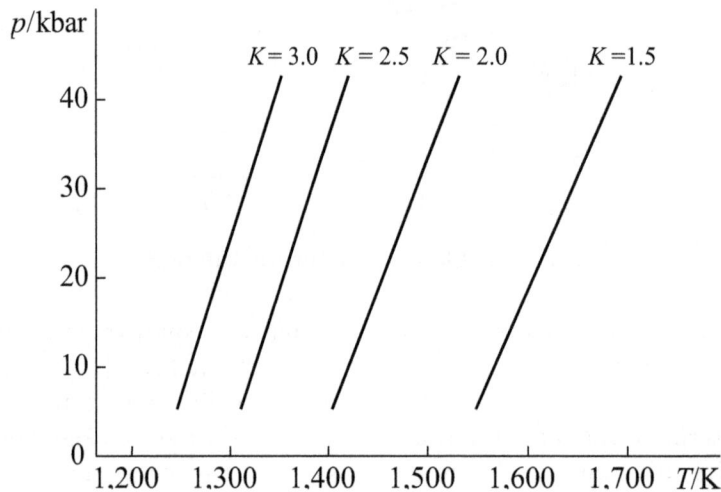

Figure 8.7: Geothermometer for the reaction (8.30).

It is interesting to observe how, in these cases, the concept of equilibration time results essential. Since the geological processes have characteristic time scales of thousand or millions of years, when the mineral samples are taken from the lithosphere at (p,T) and are then left in the laboratory under ambient conditions, it is reasonable to assume that the equilibrium will not be altered in the time necessary to accomplish the study (days).

8.7   Equilibrium Constants and Statistical Mechanics

Eqn. (2.5) is a fundamental relation between thermodynamic functions and the partition function, given by eqn. (2.11). From them it is possible to obtain eqn. (4.12) for the standard chemical potentials, $\mu_i^{\ominus}$. Then according to eqns. (8.8) and (8.9), the equilibrium constant can be related to the internal partition functions of products and reactants that are at equilibrium. As explained in chapters 2 and 4, the internal partition functions, as well as the $\mu_i^{\ominus}$, refer to isolated molecules (having no intermolecular interactions); that is, to systems hypothetically ideal. Then, it becomes fairly simple, on the basis of spectroscopic information, which gives values of the frequencies for the vibrations and rotations, from knowledge about the electronic state of the molecules, and of their geometric characteristic, to calculate the values of the equilibrium constants without measuring the concentrations at equilibrium. The advantage of this procedure can be seen, for instance, in the cases where the equilibrium considered involve species intrinsically unstable, like those participating as intermediates in reaction mechanisms, systems for which their equilibrium constants are difficult to assess by simple laboratory methods. Also, this strategy is used in chemical kinetics to calculate the thermodynamic properties of the transition complex.

To illustrate how the spectroscopic and molecular information may be used for the calculation of equilibrium constants, we shall only use here examples of homogeneous equilibria in gaseous phase. On the basis of eqns. (4.12) and (8.11), we get

$$\ln K = \ln \left[\frac{\prod_P (q_{int})_P^{\nu_P}}{\prod_R (q_{int})_R^{\nu_R}} \left(\frac{RT}{p^{\ominus}} \right)^{-\Delta\nu} \right] \tag{8.31}$$

Using eqn. (2.11), it is possible to replace in the previous equation the different contribution to the internal molecular modes.

In the following paragraphs we will calculate, for example, the equilibrium constant for the reaction

$$Na_2(g) \rightleftharpoons 2Na(g)$$

at 1000 K. To determine q_{int} of reactants and products, it is necessary to analyze each of the components in terms of eqns. (2.7), (2.8), (2.9) and (2.10). Both q_{vib} and q_{rot} are unity for the Na atoms. Instead, for the Na_2 molecule, a vibration mode exists having a wave number of $\bar{\nu} = 159.23 \text{ cm}^{-1}$ ($\bar{\nu}$ is defined as the reciprocal of its wavelength and is equal to the ratio between the vibration frequency ν and the speed of light), and its rotation will be characterized by a moment of inertia of $1.81 \times 10^{-38} \text{g cm}^2$ and a symmetry number $\sigma = 2$, because we are dealing with an homonuclear molecule.

The electronic contribution can be easily calculated by taking into account that only the fundamental electronic levels of the two species are involved, for which the degeneration of these states need be taken into account (cf. eqn. (2.7)), implying that $g(Na) = 2$ and $g(Na_2) = 1$. Consequently, it is possible to obtain the ratio of the electronic partition functions, if we know the interaction energy between the two sodium atoms in the fundamental state, as a function of the interatomic distance, as illustrated in figure 8.8.

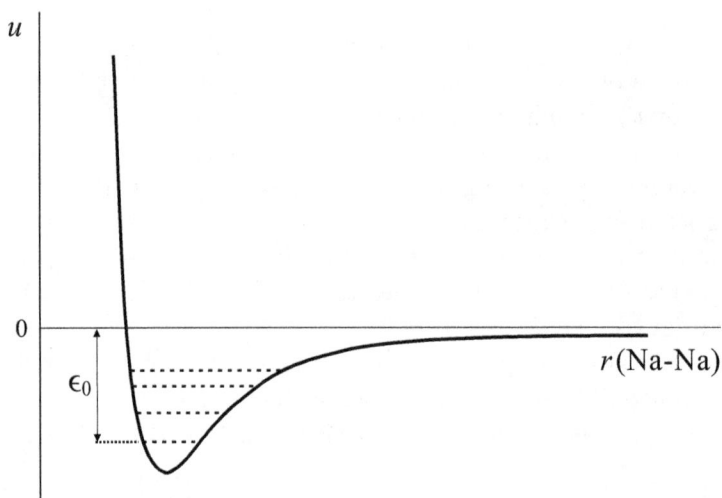

Figure 8.8: Variation of the interaction energy of two sodium atoms as a function of their separation; ϵ_o is the energy at 0 K, if the zero point energy is added.

The minimum of the potential energy corresponds to the equilibrium distance of the dimer Na-Na. Thus, the equilibrium distance of the two atoms that form the molecule Na_2 can be obtained. We must remark two points here. First, the energy of the minimum of the curve, $-\epsilon_0$, is the energy of the association reaction, at 0 K, and is related to the quantity $\Delta H^\circ(0\ K)$, defined in the previous section with reference to the effect of temperature on the chemical equilibrium. Second, the association energy, at 0 K, is normally added to the corresponding zero point vibrational energy, which is in the numerator of q_{vib} in eqn. (2.8); the energy ϵ_0 is not taken directly from the minimum of the curve in figure 8.8, but from the dashed line located $h\nu/2$ above it. In the case of the decomposition of Na_2, we get $\epsilon_0 = 70.4$ kJ/mole.

The final expression for the equilibrium constant results

$$K = \frac{\left[(2\pi m(\text{Na})\, kT/h^2)^{3/2}\right]^2}{\left[(2\pi m(\text{Na}_2)\, kT/h^2)^{3/2}\right]} \frac{p^{\ominus}}{RT} \frac{\sigma(\text{Na}_2)\, h^2}{8\pi^2 I(\text{Na}_2)\, kT} \left[1 - \exp(-\theta_v/T)\right]$$
$$\times \frac{g(\text{Na})^2}{g(\text{Na}_2)} \exp(-\epsilon_0/kT)$$

where $\theta_v = (h\nu/k)$ is the characteristic vibration temperature. The numerical values of each one of the factors is given in the following expression, remembering that the standard state is 1 bar = 0.1 MPa and the temperature 298 K,

$$K = (3.198 \times 10^7) \times (4.45 \times 10^{-4}) \times (0.205) \times 4 \times (2.10 \times 10^{-4})\ (\text{bar}) = 2.45\ (\text{bar})$$

It is interesting to analyze the factors that have the largest influence on the value of the equilibrium constant. The ratio between the translational partition functions is related to the reaction entropy, also referred to as an ideal gas contribution. On the other hand, it is quite important the term containing ϵ_0, which is related to the energy change when the reaction occurs at 0 K, is also related with the reaction enthalpy $\Delta_r H^{\circ}(0\ \text{K})$. Since we are dealing with the dissociation of a gaseous diatomic molecule, the entropy of reaction is positive because there is an increase in the number of gaseous particles. The entropic term will highly contribute to the spontaneity of the dissociation reaction at high temperature. On the other hand, the influence of the term containing ϵ_0 could be neglected at low temperature; for example, the dissociation of $\text{Na}_2(\text{g})$ going from the value 2.10×10^{-4} at 1000 K to the value 1.5×10^{-12} at 300 K.

Another example of the use of a molecular description to calculate the equilibrium constant is the simple reaction between two isotopologues

$$^{16}\text{O}_2 + ^{18}\text{O}_2 \rightleftharpoons 2\,^{16}\text{O}^{18}\text{O}$$

where we will call A to the lighter isotopologue, and B to the heavier one, and C to the product having an intermediate molecular weight. In this case, the equilibrium constant is given by the differences in i) the translational contributions, since there are differences in the mass of the participating substances, ii) the rotational partition functions, for symmetry reasons and having different inertia momenta (different reduced masses) and, finally, iii) the reaction energy of the reaction at $T = 0$ K. This last factor is small, but it includes small differences in the vibrational frequencies (zero point energy), and in the electronic energy of the isotopologues. Hence, we have

$$K = \left(\frac{m_C^2}{m_A m_B}\right)^{3/2} \left(\frac{\mu_C^2}{\mu_A \mu_B}\right) \left(\frac{\sigma_A \sigma_B}{\sigma_C^2}\right) \exp(-\epsilon_0/kT)$$

where μ_A, μ_B, and μ_C are the reduced masses[5] of the molecules A, B, and C, and σ represents the symmetry factors.

Considering a unity symmetry factor for the heteronuclear molecule, and $\sigma = 2$ for the homonuclear ones, the previous expression yields

$$K = 1.0035 \times 0.9965 \times 4 \times 0.971 = 4 \times 0.971 = 3.884$$

This result shows that the isotopic exchange is, in this case, essentially a statistical process (entropic), due to the larger probability that two different atoms are in the same molecule, compared with the probability that they form an homonuclear species. This fact is equivalent to the ratio of the probability of getting in two consecutive trials of a coin, two heads, or two tails. Only the term ϵ_0 contributes to the modification of the statistic result (i.e., by a factor 4). This example has been selected for its simplicity, and because it illustrates the way to obtain correct values of the equilibrium constants only by considering molecular features.

Problems

Problem 1

The crystals of the mineral ortopirocene have two types of cationic sites (M_1 and M_2) that can be occupied by Fe^{2+} or by Mg^{2+}. The exchange equilibrium (in the solid phase) is represented by

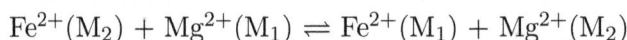

$$Fe^{2+}(M_2) + Mg^{2+}(M_1) \rightleftharpoons Fe^{2+}(M_1) + Mg^{2+}(M_2)$$

a) Calculate the equilibrium constants of the cationic exchange process at 600 $^{\circ}$C and at 800 $^{\circ}$C, using for the activity coefficients the expression corresponding to simple mixtures

$$\ln \gamma(\text{ion}, M_i) = \frac{B(M_i)}{RT}[1 - x(\text{ion}, M_i)]^2$$

b) Calculate ΔH^{θ}.

---

[5]For a molecule P having n atoms of mass m_i, its reduced mass is μ_P, which is defined by $(\mu_P)^{-1} = \sum_{i=1}^{n} m_i^{-1}$.

Data:

T °C	$B(M_1)$ kJ/mole	$B(M_2)$ kJ/mole	$x(Fe, M_1)$	$x(Fe, M_2)$
873	10.0	6.60	0.23	0.80
1073	6.87	4.42	0.20	0.60

Answers: a) $K(873 \text{ K}) = 0.279$, $K(1073 \text{ K}) = 0.288$; b) $\Delta H^\theta = 1.24$ kJ/mole

Problem 2

Gaseous acetic acid dimerizes by forming two hydrogen bonds. At 40 °C and 0.016 bar of pressure, the equilibrium is attained when 80.0 percent of the molecules of CH_3COOH are in the form of dimers. Calculate ΔH for the formation of one mole of hydrogen bonds, knowing that when the temperature increases 10 °C the fraction of dimers is 0.700, at a pressure of 0.0135 bar.

Answer: $\Delta H = -29.2$ kJ/(mole of hydrogen bond)

Problem 3

In HCl aqueous solutions in equilibrium with the vapor, the following equilibrium exists

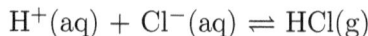

$$H^+(aq) + Cl^-(aq) \rightleftharpoons HCl(g)$$

It is known that the distribution constant (the standard state is 1 molal in both phases) as a function of temperature and density of the solvent, is given by

$$\log K_D = -13.4944 - \frac{934.466}{T/K} - 11.0029 \log \rho(l)/(g\,cm^{-3}) + 5.4847 \log(T/K)$$

a) Calculate $\Delta_D G$, $\Delta_D H$ and $\Delta_D S$, for this equilibrium at 373.2 K.

b) Using eqn. (6.13) for $\gamma_\pm$, with $A = 1.8244 \times 10^6/(\epsilon T)^{3/2}$ (kg K^3/mole)$^{1/2}$, and $C = 0.30$ kg/mole, calculate the molality of HCl in the vapor, at 373.2 K for i) a solution 1 molal in HCl, and ii) a solution 1 molal in HCl and 2 molal in $CaCl_2$.

Data: At 373.2 K, we have $p(H_2O) = 0.1$ MPa; $\rho = 0.9576$ g/cm^3; water expansivity $= 0.000857$/K; $\epsilon(H_2O) = 55.57$

Answers: a) $\Delta_D G = 12.04$ kJ/mole; $\Delta_D H = 45.83$ kJ/mole; $\Delta_D S = 90.34$ J/(mole.K); b-i) $m(HCl,g) = 0.0162$ mole/kg; b-ii) $m(HCl,g) = 0.553$

Problem 4

Often, for biochemical processes that occur in solutions involving protons, a standard state for the reaction, different from that used in processes that have been analyzed in this chapter, is defined. The new standard state corresponds to standard concentrations (1 molar) for all the solutes, pressure 1 bar for the gases, and concentration of H^+ equal to 10^{-7} molar.

Using as an example the reaction

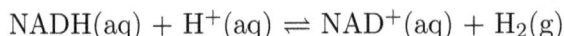

$$NADH(aq) + H^+(aq) \rightleftharpoons NAD^+(aq) + H_2(g)$$

that represents the equilibrium between two species of the dinucleotide nicotinimide-adenosine, and knowing that for this reaction $\Delta_r G^{\ominus} = -21.84$ kJ/mole, at 298 K, calculate:

a) The difference between $\Delta_r G^{\ominus}$ and $\Delta_r G'$, the last term denoting the change of Gibbs energy in the new standard state

b) K and K' for the previous equilibrium, and for the following concentrations: $[NADH] = 0.029$ molar; $[NAD] = 0.0040$ molar; $[H^+] = 2 \times 10^{-5}$ molar; and $p(H_2) = 0.015$ bar

Answers: a) $\Delta_r G^{\ominus} - \Delta_r G' = -38.96$ kJ/mole; b) $K = 103$, $K' = 1.03 \times 10^{-5}$

Problem 5

The equilibrium of esterification between benzoic acid and n-butanol

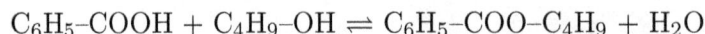

$$C_6H_5-COOH + C_4H_9-OH \rightleftharpoons C_6H_5-COO-C_4H_9 + H_2O$$

has been studied in a supercritical fluid. This equilibrium is described in terms of $K = K_x \Pi \gamma_i$, and it was found that at low concentration of reactants, K_x increases when the pressure decreases. Since K must be independent of pressure (because the process occurs at constant volume), the evidence was interpreted by the researchers by assuming that $\Pi \gamma_i$ must decrease.

a) Discuss this statement and indicate what is the correct expression for K.

b) Describe the equilibrium in the following situations (to express μ_i of the components at equilibrium): i) the supercritical solution is a mixture of compressed gases, and ii) the compounds are dissolved in the supercritical solvent.

c) Compare b-i with b-ii, and discuss both results.

Problem 6

Methane is very abundant in the lands that are north of the polar arctic circle, where it is found in the form of clathrates, or aqueous hydrates. Since there are non-stoichiometric compounds between CH_4 and water at low temperatures and high pressures, they may be represented by the following general relation:

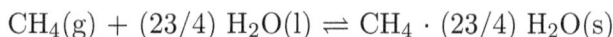

$$CH_4(g) + (23/4)\, H_2O(l) \rightleftharpoons CH_4 \cdot (23/4)\, H_2O(s)$$

whose $\Delta_r H^\ominus$ is -60.67 kJ/mol, while the equilibrium constants of the reaction at different temperatures are indicated in following table.

T/K	255	273	285
$K(\text{bar}^{-1})$	0.0769	0.0434	0.0125

Below 273 K, the phases in equilibrium are the solid clathrate, ice, and a vapor rich in methane. By considering that in the arctic region the temperature increases with the depth at a rate of 0.03 K/m, and the mean temperature at the surface can be taken as 261 K, and assuming that the ground is a formation of porous rocks able to support an hydrostatic pressure gradient of 9 kPa/m.

a) Draw the (p, T) phase diagram and discuss if it can have a point where the four phases coexist in equilibrium.

b) Calculate at what depth the clathrates are formed.

Answers: a) Yes, the coexisting phases are ice, clathrate, aqueous solution, and gas; b) approximately below 200 m

Chapter 9

Processes With Charge Transfer

9.1 Introduction

The characteristic properties of salts dissolved in polar solvents, particularly in water, led Arrhenius to postulate the hypothesis of total dissociation of the salts upon dissolution. Shortly after, it established that the strong electrolytes, that is, those which are fully dissociated in solution, exhibit deviation to the ideal behavior, even at low concentrations. Also, it was empirically known that, in dilute solutions, such deviations varied linearly with the square root of the ionic force (eqn. (6.12)).

9.2 Debye-Hückel Model

The Model

In 1923, Peter Debye and Erich Hückel proposed a very successful model for ionic solutions of strong electrolytes that allowed calculating the deviations of electrolytes from the ideal behavior. An important feature of the Debye-Hückel model is that it looks at the problem from the perspective of an ion in the solution.

Let's assume that we can measure, with a microelectrometer of atomic dimensions, the average charge in different volume elements of an electrolyte solution. If the microelectrometer is located on a solvent molecule, the measured instantaneous charge will fluctuate but, on average, will be zero. However, if it is in the same position of an ion, which we call central ion, the charge will be observed to fluctuate but the average charge detected by microelectrometer will not be zero, indicating the presence of more charges of opposite sign to that of the central ion. That is, on average, each ion is surrounded by a charge of opposite sign distributed throughout the solution. This is due to the existence of an ionic atmosphere surrounding the central ion, the key idea of the Debye-Hückel model schematized in figure 6.5, where it can be seen that close to the central ion there are more ions of opposite charge than ions of the same charge as the central one. That is, while the solvent molecules are homogeneously distributed around the central ion, oppositely charged ions accumulate preferentially around it, forming an ionic atmosphere having an opposite charge.

This model assumes that the only important interactions are the electrostatic ones, and the solvent only plays a role through its dielectric constant, ϵ_D, screening the interactions of the charges according to the Coulomb's law. The Debye-Hückel model applied to transport properties (like diffusion, electrical conductivity, etc.) also includes viscosity as a property of the solvent, which determines the friction that ions feel when moving through the liquid solvent. Consequently, it is a model that does not account for the molecular aspects of the solvent, and it only considers its macroscopic properties that modify the interactions among ions. For this reason, the Debye-Hückel model is called a primitive model.

Calculation of the Electrostatic Potential Around the Central Ion

A key point of the Debye-Hückel model is to calculate the ionic distribution surrounding the central ion. In chapter 2, it was shown that the probability of founding a system in an element of volume of the phase space, $d\omega$, is given by the Boltzmann distribution (eqn. (2.2)). In this case, we are interested in the probability of founding ions at a given distance from the central ion. Thus, if the system has N particles, the probability that these particles are located between r_1 and $r_1 + dr_1$, r_2 and $r_2 + dr_2$, $\ldots, r_N$ y $r_N + dr_N$, would be

$$P_N dr_1, dr_2 \ldots dr_N \propto \exp[-\beta V_N(r_1 \ldots r_N)]dr_1, dr_2, \ldots, dr_N$$

where $V_N(r_1 \ldots r_N)$ is the potential energy of the system, which in this case only considers electrostatic interactions, and $\beta = 1/kT$. As it was mentioned, the solvent is treated as a continuum medium without molecular details, hence without considering positions the its molecules; what we calculate is just the probability, P_{12}, that the ions 1 and 2 are separated by a distance r_{12}, independent of the position of the rest of the molecules. Thus,

$$P_{12} \propto \int \ldots \int \exp[-\beta V_N(\mathbf{r_N})]dr_3, \ldots, dr_N \equiv \exp(-\beta w_{12})$$

The quantity w_{12} is the mean force potential; this is the work necessary to bring particles 1 and 2 from the infinite, where they do not interact until they come to a distance r_{12}. Thus, the work w_{12} is related to the chemical potential of the ions.

According to the Debye-Hückel model, for a central ion j, the work to situate an ion i at the distance r_{ij} is given by $w_{ij} = z_i e \Psi_j(r_{ij})$, where $\Psi_j(r_{ij})$ is the electrostatic potential at the distance r_{ij} from the central ion.

Therefore, the concentration of i ions at a distance r_{ij} from the j central ion will be

$$n_{ij} = n_i^0 \exp[-\beta z_i e \Psi_j(r_{ij})] \tag{9.1}$$

where n_i^0 is the concentration of i ions expressed as particles per m^3, since we use SI units.

The relation between the electrostatic potential and the charge density of the system, Poisson's equation, is the basis for the solution of the problem. Poisson's equation indicates that

$$\nabla^2 \Psi_j(r_{ij}) = -\frac{1}{\epsilon_0 \epsilon_D} \rho_j \tag{9.2}$$

ρ_j being the charge density of j ions, and ∇^2 the Laplacian operator in polar coordinates and when the symmetry is spherical, as in this case–that is, when the function only depends of the polar distance–is

$$\nabla^2 = \frac{\partial^2}{\partial r^2} + \frac{2}{r} \frac{\partial}{\partial r}$$

On the other hand, ϵ_0 is the permitivity of vacuum,[1] and ϵ_D is the solvent's dielectric constant.

The charge density around the central ion can be expressed in terms of the local ionic concentrations, $\rho_j = \sum_i z_i e n_{ij}$, and this yields the Poisson-Boltzmann equation

$$\nabla^2 \Psi_j(r_{ij}) = \frac{1}{r} \frac{\mathrm{d}^2(r\Psi_j)}{\mathrm{d}r^2} = -\frac{1}{\epsilon_0 \epsilon_D} \sum_i z_i e n_i^0 \exp(-\beta z_i e \Psi_j) \tag{9.3}$$

Debye and Hückel assumed that the electrostatic energy is, on the average, lower than the thermal energy, that is, $\beta z_i e \Psi_j < 1$; hence, if the exponential in eqn. (9.3) is developed in series, only the two first terms need to be kept,

$$\sum_i z_i e n_i^0 (-1 + \beta z_i e \Psi_j)$$

The first term of the previous equation is zero due to the electroneutrality of the solution, so the Poisson-Boltzmann equations becomes

---

[1] The permitivity of vacuum is a universal constant, whose value is 8.854185×10^{-12} C/(V m) in the International System (SI) units.

$$\frac{\mathrm{d}^2(r\Psi_j)}{\mathrm{d}r^2} = \kappa^2(r\Psi_j) \tag{9.4}$$

with

$$\kappa^2 = \frac{e^2}{\epsilon_0 \epsilon_D kT} \sum_i z_i^2 n_i^0 \tag{9.5}$$

The integration of eqn. (9.4) gives the solution

$$r\Psi_j = C_1 \exp(-\kappa r) + C_2 \exp(\kappa r)$$

$C_2 = 0$, because the electrostatic potential at infinite distance from the central ion must be zero. Moreover, since the ions cannot penetrate each other, there should be a minimum distance of separation, or maximum approximation, between opposite charges that we will call a. That means that, inside a sphere of radius a around the central ion, there are no other ions and, therefore, there are no charges other than that of the central ion, that is, $\rho_i(r < a) = 0$; thus, equation (9.4) simplifies significantly, and the solution is

$$\Psi_j = C_3 + \frac{C_4}{r}$$

At shorter distances from the center of ion j, the potential will be determined by the charge of that ion and therefore $C_4 = z_j e/(4\pi\epsilon_0\epsilon_D)$. Thus, we finally get the following equations:

$$\Psi_j = C_1 \frac{\exp(-\kappa r)}{r} \qquad r \geq a \tag{9.6}$$

$$\Psi_j = C_3 + \frac{z_j e}{4\pi\epsilon_0\epsilon_D r} \qquad r \leq a \tag{9.7}$$

It is necessary that Ψ_j has the same value at $r = a$ as that calculated by any of the former two equations; it is also required that its two first derivatives be equal at $r = a$. By using these conditions we obtain

$$\Psi_j = -\frac{z_j e}{4\pi\epsilon_0\epsilon_D} \frac{\exp(\kappa a)}{(1+\kappa a)} \frac{\exp(-\kappa r)}{r} \qquad r \geq a \tag{9.8}$$

$$\Psi_j = -\frac{z_j e}{4\pi\epsilon_0\epsilon_D} \frac{\kappa}{(1+\kappa a)} + \frac{z_j e}{4\pi\epsilon_0\epsilon_D r} \qquad r \leq a \tag{9.9}$$

Figure 9.1 illustrates the change of the excess of anions, Δn_-, with the distance from a central cation having $z_j = 1$ in water at room temperature for a concentration around 0.01 M. The energy of the anions closest to the central cation ($r = a$) can be calculated with equation (9.8) or (9.9). For ions with size between 0.3 and 0.4 nm, the electrostatic energy at contact is approximately twice the value of the thermal energy.

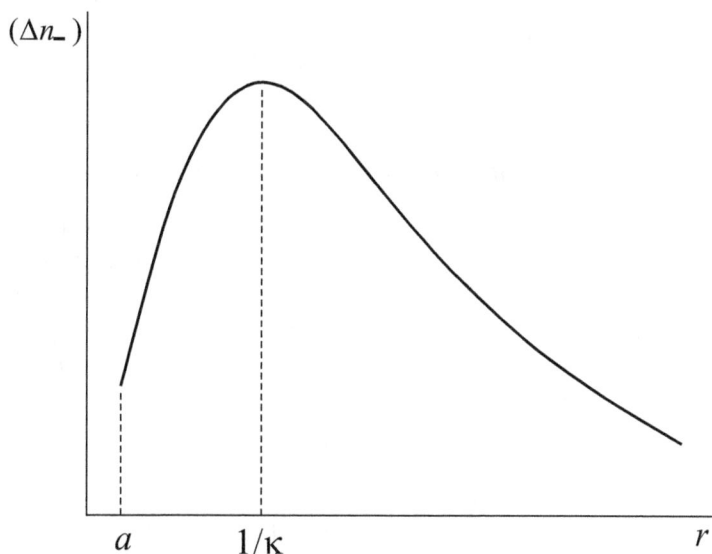

Figure 9.1: Anion excess concentration (Δn_-) as a function of the distance to the cation, corresponding to the scheme in figure 6.5.

This does not invalidate the Debye-Hückel approximation because, according to the figure, in that region there are few excess of ions. What is important is that the approximation be valid at the distance where most of the excess of opposite charge is located at $r = \kappa^{-1}$, that is, the distance known as Debye distance, or the radius of the ionic atmosphere. In fact, most of the phenomena produced by the presence of electrolyte in solution are equivalent to those produced by the presence of a charge opposite to that of the central ion located on an spherical arrangement at a distance κ^{-1} from the central ion. However, it should be noted the continuous nature of the distribution of the excess charge around the central ion.

The expressions derived for Ψ_j are only valid for relatively dilute solutions and account for the already mentioned observation about the important effect of the ionic strength in determining the deviation from ideality of the strong electrolyte solutions. Starting from eqn. (9.5), and recalling that a Faraday is $F = N_A e$, we obtain

$$\kappa^2 = \frac{1000F^2}{\epsilon_0 \epsilon_D RT} \sum_i z_i^2 c_i \tag{9.10}$$

where c_i is the molarity (mole dm^{-3}). The relationship between molality and molarity is, for dilute solutions, $c_i = \rho m_i$, where ρ is the solvent density. Thus, for density measured in g cm^{-3}

$$\kappa^2 = \frac{2000F^2\rho}{\epsilon_0 \epsilon_D RT} I \tag{9.11}$$

The Calculation of the Salt's Excess Chemical Potential

The excess chemical potential of the central ion is given by the electrostatic work necessary to charge it in the presence of the ionic atmosphere. Thus, the electrostatic potential that really matters is that due to the ionic atmosphere acting on the surface of the central ion ($r = a$). This potential is calculated with eqns. (9.8) or (9.9), discounting the electrostatic potential due to the charge of the central ion given by the second term of the right side in eqn. (9.9). Then, the potential due to the ionic atmosphere on the central ion with charge $q = z_j e$ is

$$\Psi_j(a) = -\frac{z_j e}{4\pi\epsilon_0\epsilon_D} \frac{\kappa}{(1 + \kappa a)} \tag{9.12}$$

The work needed to increase in dq the charge of the central ion in the presence of the ionic atmosphere, that is, the work required to bring dq from infinity to the surface of the ion, will be $\delta w = -\Psi_j(a, q)\mathrm{d}q$. The total work for charging the ion, or excess chemical potential of ion j, is obtained by integration between zero charge and $z_j e$,

$$\frac{\mu_j^{\text{ex}}}{kT} = \ln \gamma_j = -\frac{z_j^2 e^2}{8\pi\epsilon_0\epsilon_D kT} \frac{\kappa}{1 + \kappa a} \tag{9.13}$$

As discussed in chapter 6, for electrolyte solutions, the only experimentally accessible quantity is its mean activity, so it is necessary to obtain an expression for the mean activity coefficient using the Debye-Hückel model. If the electrolyte $C_{\nu_+}A_{\nu_-}$ is strong, one mole of it will generate ν_+ moles of cations C^{z+} with charge z_+, and ν_- moles of anions A^{z-} with charge z_-. In this case, the electroneutrality condition is

$$\nu_+ z_+ + \nu_- z_- = 0$$

According to eqn. (6.7)

$$\nu \ln \gamma_\pm = \nu_+ \ln \gamma_+ + \nu_- \ln \gamma_-$$

and replacing in the former equation with the expression by $\ln \gamma_i$, we obtain the expression for the mean activity coefficient

$$\ln \gamma_{\pm} = -\frac{z_+ \mid z_- \mid e^2}{8\pi\epsilon_0\epsilon_D kT} \frac{\kappa}{1+\kappa a} \qquad (9.14)$$

This is the Debye-Hückel equation for the mean activity coefficient of an electrolyte, already written [see eqn. (6.11)] as

$$\ln \gamma_{\pm} = -\frac{Az_+ \mid z_- \mid I^{1/2}}{1 + BaI^{1/2}}$$

In chapter 6, we have discussed the equations for the limit of very low electrolyte concentration. Also, we mentioned models that allow to extend the equation beyond the range covered by the Debye-Hückel theory.

9.3 The Conductivity of Electrolytes

The most distinctive property of electrolyte solutions is their capacity to conduct electrical current. Historically, this was an important evidence supporting Arrhenius theory, which maintains that strong electrolytes are fully dissociated in solution. The electrical conductivity is a consequence of the movement of the ions in the presence of an applied external electrical field, so electrical conductivity also implies mass transport. This subject will be treated in chapter 10, dealing with lineal processes out of equilibrium, including charge transport. However, there are two important reasons to dedicate a brief section to the conductivity of ionic solution in this chapter. On one hand, the Debye-Hückel model inspired Lars Onsager for modeling the effect of salt concentration on the variation of the electrical conductivity, and derive the corresponding limiting law. On the other hand, the electrolyte conductivity illustrates quite well the effect of the ionic atmosphere on the ions. Moreover, the determination of the electrical conductivity is one of the experimental methodologies used to obtain the dissociation constant of weak and moderately dissociated electrolytes, including the ion-pair formation or ionic association constant, discussed in chapter 6.

We have seen that the Debye-Hückel model considers the solvent as a continuum medium with dielectric constant ϵ_D. The extension of this primitive model to the field of transport processes requires the introduction of the solvent viscosity, η, which is related to the friction force exerted by the liquid solvent on the particles that move inside it.

Figure 9.2a is a schematic representation of a central cation surrounded by its negatively charged ionic atmosphere at equilibrium. When an external electric field is applied in the solution, a displacement of the central cation takes place in the direction of the external field but, at the same time, the ionic atmosphere moves in the opposite direction because of its opposite charge. This situation, depicted in figure 9.2b, triggers two effects that modify the mobility of the central ion in relation to that occurring at infinite dilution, that is, if the ionic atmosphere has infinite radius.

In a first approximation, when an spherical ion of radius r_i moves in a liquid, its velocity will be given by the Stokes law, which establishes that the friction force, f_i, exerted by the liquid on the ion is proportional to its velocity, v_i

$$f_i = A'\eta v_i r_i$$

$$\vec{E} = 0 \qquad\qquad \text{a}$$

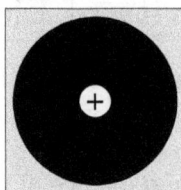

$$\vec{E} \qquad\qquad \text{b}$$

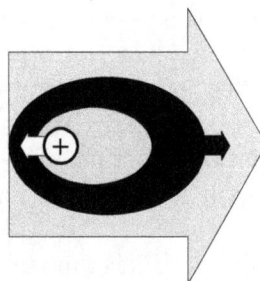

■ Ionic atmosphere   ▢ Solvent

Figure 9.2: Scheme of the negatively charged ionic atmosphere surrounding a central cation a) at equilibrium, and b) in the presence of an external electric field.

If the ionic solution is not infinitely diluted, the ion will be surrounded by its ionic atmosphere having a finite radius, and the electric field makes the ion move in one direction and the ionic atmosphere in the opposite one. The movement of the ions forming the ionic atmosphere is transmitted through collisions with the solvent molecules, which finally move in the opposite direction in relation to the central ion, as illustrated in figure 9.2b. That is, the central ion moves in a liquid media that moves itself in the opposite direction, and then the ion velocity with respect to laboratory coordinates will be slower at higher electrolyte concentration, when the radius of the ionic atmosphere is smaller. This slow-down mechanism is known as electrophoretic effect, and its contribution can be estimated by using the Stokes law approach to calculate the velocity of a sphere of radius $1/\kappa$, corresponding to the size of the ionic atmosphere. Thus, the ion change in velocity due to the electrophoretic effect will be proportional to the inverse of the radius of the ionic atmosphere,

$$\Delta v_{\text{el}} \propto \frac{\kappa}{\eta}$$

The second effect, also generated by the shift of the ionic atmosphere in relation to the central ion, is known as the relaxation effect and it is illustrated as well in figure 9.2b. In the steady state reached when an external electric field is applied, the center of the ionic atmosphere does not coincide with the position of the central ion. This relative shift induces a local electric field, opposite to the external one, which reduces the electrostatic force acting on the central ion and, consequently, the velocity of the central ion is reduced.

It can be consider that the relaxation effect provokes a variation in the ion velocity proportional to the relative velocity between cations and anions. In turn, this velocity change is proportional to the infinite dilution conductivity of the electrolyte times the ratio between the electrostatic energy, corresponding to the interaction of the central ion with the ionic atmosphere, and the thermal energy. Thus,

$$\Delta v_{\text{rel}} \propto \frac{z_+ \mid z_- \mid e^2 \kappa \Lambda^\infty}{\epsilon_0 \epsilon_D kT}$$

Then,

$$v_i = v_i^\infty - \Delta v_{\text{el}} - \Delta v_{\text{rel}}$$

and finally,

$$\Lambda = \Lambda^\infty - (\alpha \Lambda^\infty + \beta)\kappa = \Lambda^\infty - SI^{1/2}$$

In this expression the conductivity, Λ, and the ionic strength, I, correspond to those ions that are free, that is, the ions that do not form ionic pairs.

Figure 9.3 shows the molar conductivities of NaCl and $CuSO_4$ aqueous solutions at room temperature as a function of the electrolyte concentration. Also plotted are the limiting slopes given by the Onsager expression. NaCl represents the case of a typical strong electrolyte; the conductivity decreases as the concentration rises, but it is always higher than the corresponding limiting law conductivity. In the case of $CuSO_4$, the molar conductivity decreases with increasing concentration more abruptly than the prediction of the limiting law because of the ionic association, which leads to ion pairs that does not contribute to the solution conductivity. The analysis of the conductivity of $CuSO_4$ brings to the same estimation of the ionic association as the determination of activity coefficients.

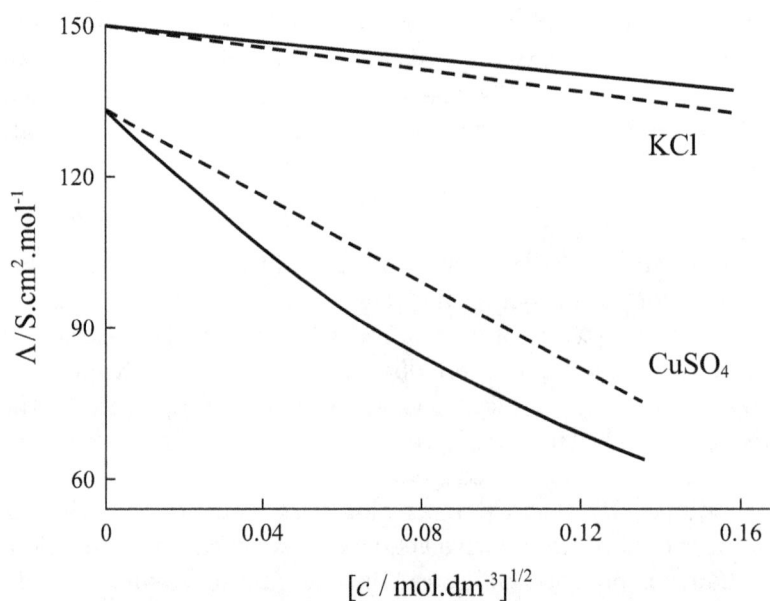

Figure 9.3: Conductivity of aqueous electrolytes at 298 K as a function of $c^{1/2}$: a) NaCl, and b) $CuSO_4$. Full lines: Experimental curve. Dashed lines: Onsager's limit law.

9.4   Charged Interfaces. The Electrochemical Potential

If we are interested in analyzing the charge transfer between two phases having charges, either due to ions or electrons, which are in equilibrium, the contribution of the electric work to the Gibbs energy of the phases has to be taken into account.

Thus, we will write for phase α

$$\mathrm{d}G^\alpha = -S^\alpha \mathrm{d}T + V^\alpha \mathrm{d}p + \sum_{i=1}^{C} \mu_i^\alpha \mathrm{d}n_i + \sum_{i=1}^{C} \Phi^\alpha \mathrm{d}q_i \qquad (9.15)$$

where Φ^α is the electrochemical potential in the phase α, and q_i is the charge of specie i. Then, since $\mathrm{d}q_i = z_i F \mathrm{d}n_i$, the former expression becomes

$$\mathrm{d}G^\alpha = -S^\alpha \mathrm{d}T + V^\alpha \mathrm{d}p + \sum_{i=1}^{C} \tilde{\mu}_i^\alpha \mathrm{d}n_i \qquad (9.16)$$

where the electrochemical potential of species i, $\tilde{\mu}_i$, is defined as

$$\tilde{\mu}_i^\alpha = \mu_i^\alpha + z_i F \Phi^\alpha \qquad (9.17)$$

Eqn. (9.16) is identical to eqn. (1.25) if in the former we replace the chemical potential by the electrochemical one. Thus for charged phases, $\tilde{\mu}_i$ plays the important role that μ_i has in processes among phases without net charge change. Consequently, an equation equivalent to eqn. (1.28) should exist and holds for the transfer of charged species from the phase α to the phase β, but now in terms of the electrochemical potentials, it can be written

$$\sum_{i}^{C} (\tilde{\mu}_i^\beta - \tilde{\mu}_i^\alpha) \mathrm{d}n_i^\beta \leq 0 \qquad (9.18)$$

The inequality implies a spontaneous process and the equality a process at equilibrium; in the second case, $\tilde{\mu}_i^\alpha = \tilde{\mu}_i^\beta = \tilde{\mu}_i$.

It is worth noting that splitting the electrochemical potential in a chemical term and an electrical term (eqn. (9.17)) is arbitrary and it does not exist a way to determine both terms separately without ambiguity; this is only possible when one of the terms is zero. The most important examples of that situation are the following two:

1. Transfer of ions or electrons between two phases of *identical materials*, but which are at different electrical potentials; for instance, two pieces of copper. In this case the chemical potential of both phases are the same and we will have

$$\tilde{\mu}_i^\beta - \tilde{\mu}_i^\alpha = z_i F (\Phi^\beta - \Phi^\alpha)$$

For this system, the difference of electrochemical potential, a quantity that is always measurable, is directly related to the difference of electrical potentials between both phases.

2. The second situation refers to processes between different phases, in such a way that the transfer of charged species does not affect the state of charge of the phases. Equal numbers of positive and negative charges can be transferred among phases, or the same number of charges (of each sign) of the phase β to the phase α than that transferred from the phase α to the phase β (processes illustrated in figure 6.2). In these situations the differences of electrochemical potentials will be equal to the difference of chemical potentials.

$$\sum_{i}'(\tilde{\mu}_i^{\beta} - \tilde{\mu}_i^{\alpha}) = \sum_{i}'(\mu_i^{\beta} - \mu_i^{\alpha})$$

where $\sum'$ means the sum over all the species transferred in allowed processes (see figure 6.2).

This analysis allows understanding, in a rigorous form, why the transfer of charged species among phases that do not involve changes in the charge state are still governed by the chemical potentials of the species.

Now, let's imagine a solid surface immersed in an ionic solution in equilibrium with it. If the state of charge of the solid phase is modified, the electrostatic potential of this phase will return to the equilibrium value if, for instance, a redox reaction takes place, which implies there will be a transfer of electrons between the phases (it is the case of processes in reversible electrodes. However, there are systems where, under certain conditions, it is not possible that a redox process occurs; then, the modification of the state of charge of the difference of electrostatic potential between the charged solid and the solution will remain–this is the case of ideally polarized electrodes. This change will alter the distribution of ions near the electrode surface, similar to the already studied case of the distribution of the ionic atmosphere surrounded by the central ion in the Debye-Hückel model. If the charge of the solid phase in contact with the solution is positive, the anions in solution will mainly approach the surface, while the cations will separate from it. This charge distribution is known as electrical double layer, and it is a universal phenomenon that occurs in charged phases surrounded by electrolytes and plays a fundamental role in surface phenomena (see chapter 7). In this case, it is assumed that the charged phase is on a plane in order to facilitate the calculation of the charge distribution in the solution, because there is just one relevant coordinate (the distance, x, to the interface).

The Poisson-Boltzmann equation (9.2), can also be used to solve this problem, but in this case it is a simpler one-dimension problem. If there is a single electrolyte in solution,

$$\frac{\mathrm{d}^2\Psi}{\mathrm{d}x^2} = -\frac{1}{\epsilon_0\epsilon_D}[z_+e\,n_+^0\exp(-\beta z_+e\Psi) + z_-e\,n_-^0\exp(-\beta z_-e\Psi)]$$

For the simple case where $z_+ = -z_- = z$, and $\nu = 2$, we have

$$\frac{\mathrm{d}^2\Psi}{\mathrm{d}x^2} = \frac{2ze\,n^0}{\epsilon_0\epsilon_D}\sinh(\beta ze\Psi)$$

Linearizing this expression, which is here even more justifiable than in the case of ions in solution, we obtain eqn. (9.5)

$$\frac{\mathrm{d}^2\Psi}{\mathrm{d}x^2} = \frac{2F^2\rho}{\epsilon_0\epsilon_D RT}I\Psi = \kappa^2\Psi$$

Again, the reciprocal of the radius of the ionic atmosphere, κ, is the central parameter determining the ionic distribution and, for the case of an interface, it is more appropriate to consider it as the thickness of the ionic atmosphere.

It is common the case of metals, as charged phases, immersed in electrolytes. Mercury in an aqueous solution containing simple electrolytes is an example of a metal that is not easily discharged by charge transfer processes; this is the archetype of an ideally polarized electrode. If the interface is charged, it is necessary to take into account the electrical work when writing the thermodynamic equations. Thus, eqn. (7.11) becomes

$$\mathrm{d}\gamma = -s^s\mathrm{d}T + \frac{V^s}{\mathcal{A}}\mathrm{d}p - \sum_{i=1}^{C}\Gamma_i\mathrm{d}(\mu_i + z_iF\Phi^s) \tag{9.19}$$

where a term of electrical work has been incorporated. This expression indicates that at constant p, T, and composition, a change in the electrostatic potential of the interface leads to a change in the interfacial tension (resulting in the phenomenon known as electrocapilarity). The variation of the interfacial tension with the electrostatic potential will be given by

$$\left(\frac{\partial\gamma}{\partial\Phi}\right)_{p,T\mu_i} = \sum_{i=1}^{C}\Gamma_i z_i F = q^{\mathrm{m}} = -q^{\mathrm{s}} \tag{9.20}$$

where q^{m} and q^{s} are the charge per unit area in the metal and the solution, respectively.

An increase of the interface area requires a surface work, as it was shown in chapter 7. In the case of charged interfaces, as the area of the interface increases, the charge density (charge per unit area) decreases, and the corresponding increase of surface work is in part compensated by diminishing the electrical work; the variation of Gibbs energy is the result of a compromise between both effects. That is, we have two phenomena with opposite contribution upon increasing the interfacial area.

In chapter 7, we discussed the difficulty for nucleating a drop or a crystal because the surface Gibbs energy is very large when the nucleus is very small. Now, we see that an electrical charge would help the nucleation phenomenon. This is employed in practice in relation to the formation of bubbles and crystals. With the purpose of determining the presence of ionizing radiation or nuclear particles that are able to ionize molecules, cloud chambers (e.g., Wilson chambers) and bubble chambers are used. In both cases, a fluid is supersaturated in water vapor in the cloud chamber, or in liquid hydrogen in the bubble chamber; however, the formation of ice crystals or gas bubbles, respectively, are not observed, except when there are charges induced by radiation, or charges that are intrinsic to nuclear particles. In those cases, a *string* of ice crystallites, or bubbles, follows the path of ionizing radiation because in spite of the fact that the fluids are not supersaturated enough to nucleate homogeneously, the presence of charges enhances the nucleation process.

9.5   Galvanic Cells

We have analyzed the case of charged interfaces when the processes of oxidation-reduction are not possible, that is, for ideally polarized electrodes. Now, we must analyze the opposite situation, a piece of metal immersed in an electrolyte solution when it is possible to transfer electrons between the metal (electrode) and some of the species in solution. These systems, which are the base of the galvanic cells, have a paramount importance because they provide a direct method to determine mean activity coefficients of electrolytes, but also because they are the basis of the electrochemical processes used in batteries to generate electrical work. In batteries, these processes occur completely out of equilibrium, since in order to deliver electrical work, it is necessary to extract electrical charge in a reasonable time, that is, the electrical current is not zero. On the contrary; in the galvanic cells, the processes occur very near to equilibrium conditions, that is, the current is null and the conditions of membrane or osmotic equilibrium among the phases are maintained (cf. chapter 1), which requires that some substances can exist in the both phases in equilibrium. If we have the following galvanic cell

$$(-)\ Cu,\ Tl\ |\ TlNO_3\ ||\ AgNO_3\ |\ Ag,\ Cu\ (+)$$

where the oxidation ($Tl \rightarrow Tl^+ + e$) occurs in the anode, and the reduction ($Ag^+ + e \rightarrow Ag$) occurs in the cathode, the global reaction will be

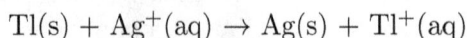

$$Tl(s) + Ag^+(aq) \rightarrow Ag(s) + Tl^+(aq)$$

Let's examine for this case which are the species that can exist in the two phases forming the interface and which can be transferred between them. The transfer of electrons is possible between the metallic phases Cu|Ag and Cu|Tl, while Ag^+ and Tl^+ ions can be transferred between the metals and the corresponding solutions, because we can assume the following equilibrium within the metals:

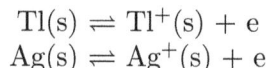

$$Tl(s) \rightleftharpoons Tl^+(s) + e$$
$$Ag(s) \rightleftharpoons Ag^+(s) + e$$

Now, it is possible to apply the equilibrium conditions for charged phases to each electrode. The analysis is applicable for the cathode: (1) in the Cu|Ag interface the condition is $\tilde{\mu}(e, Cu) = \tilde{\mu}(e, Ag)$; (2) the equilibrium among atoms, ions, and electrons in the metallic silver implies $\tilde{\mu}(Ag, Ag) = \tilde{\mu}(Ag^+, Ag) + \tilde{\mu}(e, Ag)$; (3) in the Ag/AgNO$_3$ solution interface, the Ag^+ ions guarantee the equilibrium condition $\tilde{\mu}(Ag^+, Ag) = \tilde{\mu}(Ag^+, aq)$. The summary of all the equilibrium conditions gives the following result for the cathode

$$\tilde{\mu}(e, Cu) = \tilde{\mu}(Ag) - \tilde{\mu}(Ag^+, aq)$$

Taking into account the analog equation for the anode, the definition of electrochemical potentials (eqn. (9.17)) and the charges of all the species involved, we get finally[2]

$$\Phi[e, Cu(+)] - \Phi[e, Cu(-)] =$$

$$= \frac{\mu^{\ominus}(Tl) - \mu^{\ominus}(Ag) + \mu^{\ominus}(Ag^+, aq) - \mu^{\ominus}(Tl^+, aq)}{F} + \frac{RT}{F} \ln \frac{a(Ag^+, aq)}{a(Tl^+, aq)}$$

This is the Nernst's equation that describes the potential, E, for the analyzed cell

$$E = E^{\ominus} + \frac{RT}{F} \ln \frac{a(Ag^+, aq)}{a(Tl^+, aq)}$$

where $E^{\ominus}$ is the difference of standard potentials of the Ag^+/Ag and Tl^+/Tl electrodes, defined as:

$$E^{\ominus}(Tl^+, aq/Tl, s) = \mu^{\ominus}(Tl^+, aq) - \mu^{\ominus}(Tl, s)$$

---

[2]We have assumed that the AgNO$_3$ and the TlNO$_3$ solutions are in the same phase and, consequently, have the same Φ value. This is not strictly true in the discussed example because neither the Ag^+ ion should enter in contact with the cathode, nor the Tl^+ ion with the anode to avoid direct chemical reactions. However, the assumption is not conceptually erroneous; the most general case will be discussed in the problem 3 of the next chapter.

and

$$E^{\ominus}(\text{Ag}^+, \text{aq}/\text{Ag}, \text{s}) = \mu^{\ominus}(\text{Ag}^+, \text{aq}) - \mu^{\ominus}(\text{Ag}, \text{s})$$

In this analysis, we have used the conditions detailed: Between the two metal copper pieces the difference of electrochemical potential will be given by the difference of electrostatic potentials because they are a unique substance; in the aqueous solutions the differences of electrochemical potentials of the Ag$^+$ and Tl$^+$ ions will be equal to the differences of chemical potentials, because in the aqueous phase the electrostatic potential is the same and because both ions have the same charge, so their contribution cancels out.

This argument can be extended to different types of electrodes, the conventional ones or ion selective electrodes, including the glass electrodes. It can be also applied to galvanic cells with transport.

9.6   Determination of Mean Activity Coefficients and Equilibrium Constants

We will consider here the determination of mean activity coefficients by means of the measurement of the electromotive force (emf) of galvanic cells. We will introduce only a cell, commonly used by H. S. Harned, which served as a basis to establish the pH scale. The cell has a hydrogen electrode that acts as anode, and silver/silver chloride electrode that acts as cathode. Both electrodes are of great precision and reversibility, so they have been largely used for the study of the properties of electrolyte solutions. The cell containing HCl solution is represented as

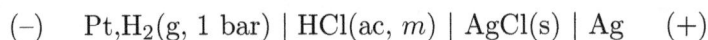

$$(-)\quad \text{Pt,H}_2(\text{g, 1 bar}) \mid \text{HCl(ac}, m) \mid \text{AgCl(s)} \mid \text{Ag}\quad (+)$$

The electrochemical reaction is

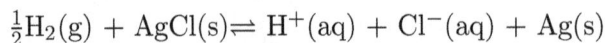

$$\tfrac{1}{2}\text{H}_2(\text{g}) + \text{AgCl(s)} \rightleftharpoons \text{H}^+(\text{aq}) + \text{Cl}^-(\text{aq}) + \text{Ag(s)}$$

The emf of the cell will be

$$E = E^{\ominus} - \frac{RT}{F} \ln \frac{p^{\ominus\,1/2} a(\text{H}^+)\, a(\text{Cl}^-)}{p^{1/2}\,(m^{\ominus})^2}$$

Considering unit hydrogen pressure and writing the ionic activities as the product of molalities times activity coefficients, we obtain

$$E = E^{\ominus} - \frac{2RT}{F} \ln m - \frac{2RT}{F} \ln \gamma_{\pm} \tag{9.21}$$

Moving all the known parameters to the lhs member, we obtain

$$Y \equiv E + \frac{2RT}{F} \ln m = E^{\ominus} - \frac{2RT}{F} \ln \gamma_{\pm}$$

Thus, by plotting Y as a function of $m^{1/2}$ we can determine $E^{\ominus}$ from the ordinate at the origin and calculate $\gamma_{\pm}$ for each concentration. This method is illustrated in figure 9.4 for the cell

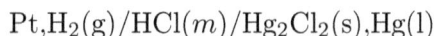

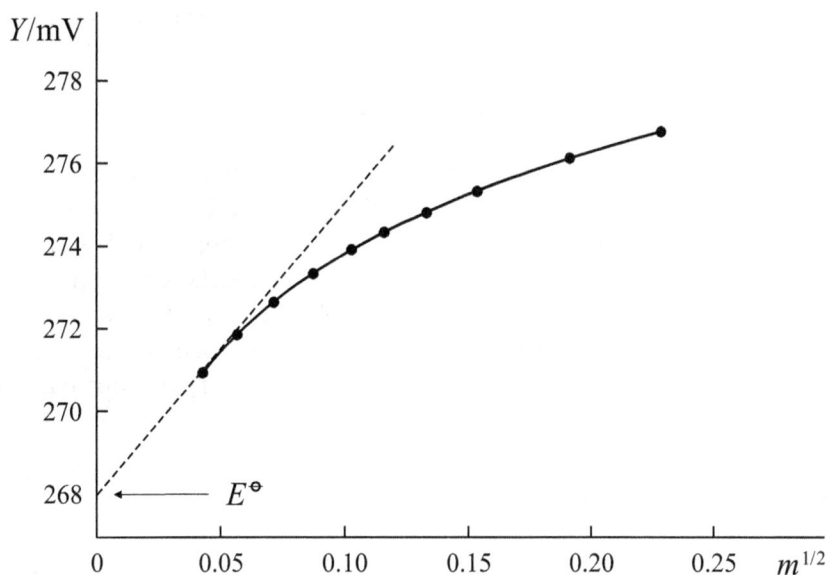

$$\text{Pt,H}_2(\text{g})/\text{HCl}(m)/\text{Hg}_2\text{Cl}_2(\text{s}),\text{Hg}(\text{l})$$

Figure 9.4: Harned's method to determine standard potentials, $E^{\ominus}$, and mean activity coefficients, $\ln \gamma_{\pm}$.

Another procedure frequently employed consists in expressing $\ln \gamma_{\pm}$ using some theoretical equation, and then obtaining the unknown parameters of the equation. For instance, using eqn. (6.13),

$$Y' \equiv Y + \frac{2RT}{F} \frac{A \mid z_- \mid z_+ I^{1/2}}{1 + I^{1/2}} = E^{\ominus} - CI$$

Now, we will describe the case where the same galvanic cell is used but the solution contains a weak acid HA, its salt NaA, and NaCl in concentrations m_1, m_2, and m_s, respectively. The emf of this cell is also given by eqn. (9.21), where $a(H^+)$ can be expressed in terms of the dissociation equilibrium constant of the weak acid, as

$$K = \frac{a(H^+)\, a(A^-)}{a(HA)\, m^\ominus}$$

Replacing in eqn. (9.21) we have

$$E = E^\ominus - \frac{RT}{F} \ln \frac{m_1 m_s}{m_2 m^\ominus} - \frac{RT}{F} \ln K - \frac{RT}{F} \ln \frac{\gamma_{Cl^-}}{\gamma_{A^-}}$$

where we have assumed that the activity of the non-dissociated acid is equal to m_1. Again, regrouping all the known quantities in the Y term, we have

$$Y = E + \frac{RT}{F} \ln \frac{m_1 m_s}{m_2 m^\ominus} = E^\ominus - \frac{RT}{F} \ln K - \frac{RT}{F} \ln \frac{\gamma_{Cl^-}}{\gamma_{A^-}}$$

By plotting Y as a function of the solution's ionic strength, we obtain as the ordinate at origin, $E^\ominus - (RT/F) \ln K$; the value of $E^\ominus$ can be determined as described in the first example, using a strong electrolyte; in that way, this cell allows us to determine the dissociation constant, K. From the curve, it is possible to get information on the ratio of the activity coefficients of the two anions, a quantity that is measurable because the process can be represented as the introduction of one of them in solution and the extraction of the other, without changing the state of charge of the solution.

9.7   Colloidal Systems

Colloids are frequently present in natural or industrial systems. They consist in aggregates of atoms or molecules suspended in a fluid media. These material systems are formed by particles of sizes that can be observed with a microscope, typically in the range of 0.1 to 100 μm; that is, they are small particles, but with a size much larger than those of simple atoms and molecules. Each colloidal particle can be formed by aggregates of simple molecules or macromolecules, crystalline material, etc. These particles, when suspended in a homogeneous media, gas or liquid, form a colloid that can be stable; that is, its particles remain dispersed without forming deposits. In this section, we will analyze the case of colloids suspended in liquids, to see if their behavior can be described on the basis of molecular interactions among their particles.

A characteristic of all the colloidal systems is that they disperse the light that goes through them (Tyndall effect) in such a way that if we look at the system in a direction perpendicular to the direction of the beam incidence, the light ray can be observed. When the size of the colloidal particles are adequate and their concentration is high, the light dispersion gives the liquid a cloudy aspect. In all cases, the dispersion of the visible light is due to the fact that the system contains particles that are commensurable with the wave length of the light; that is, particles in suspension are around 0.5 μm in size.

The microscopic aggregates that form colloidal particles contain atomic groups that are charged or are susceptible to develop charges by ionization or other chemical reaction. For instance, they have groups -COOH, -OH, -NH$_2$, etc., or have strongly adsorbed ions on their surface, as is the case of metals or metal oxides, or have anionic groups in their particles and alkaline or alkaline-earth ions that neutralize their charge (this is the case of clays and other natural silicates). Thus, the colloidal particles interact with each other through their charges which, in principle, can be explained by resorting to the description of the electric double layer for ideally polarized electrodes given in section 9.4. However, an important difference exists with the case discussed in that section. Ideally polarized electrodes have a charge introduced from the outside, and the electric double layer is formed by electrolytes that are electrically neutral. On the contrary, in the case of colloids, the particles have fixed charges that are neutralized by the corresponding counterion, for instance Na$^+$ ions if the colloid has -COO$^-$ groups. That is, the behavior of the colloidal systems can be described using as a model a charged particle that behaves as a kinetic unit moving as a whole, with counterions that form the ionic atmosphere or double layer. Moreover, the colloidal systems contain other electrolytes that will modify the interactions among the colloidal particles by modifying their ionic atmosphere. The effect of adding salts is quite important for establishing the state of the colloidal system. As quoted, the electrostatic interactions are fundamental for describing the behavior of colloidal systems, but dispersion, or van der Waals forces, originated by Lennard-Jones interactions among the molecules or the individual atoms that compose them (see section 2.10), are also present. Although colloidal systems can be stable during a long time without particle aggregation or deposition, it is observed that the particles could coalesce and flocculate, Moreover, when their state of charge changes due to changes in the pH, or by addition of electrolytes that screen their charges, a certain concentration is reached where the colloid particles coalesce or flocculate. Which are the interactions that dominate among particles in a stable colloid and what destabilizes them by adding an electrolyte?

The description of the interactions among colloidal particles is known as DLVO theory, due to the scientists who help to develop it: Derjaguin, Landau, Verwey, and Overbeek. This model is the simplest one to account for the behavior of colloids on the basis of the interactions among their particles. It uses the potential energy due to electrostatic interactions that is essentially given by the expression of the electric double layer (cf. section 9.4), and the attraction energy between non-polar particles (dispersion forces, also called van der Walls interaction, (cf. figure 2.9). Taking into account these two types of interactions, the following equation for the potential energy between two colloidal particles of diameter d, separated by a distance R, is obtained,

$$w(R) = A_{\text{elec}} kTd \frac{\exp(-\kappa R)}{\kappa} - \frac{Bd}{6R} \tag{9.22}$$

where κ is the inverse of the radius of the ionic atmosphere (eqn. (9.10)). Figure 9.5 shows the potential energy curves between particles, $w(R)$, as a function of the separation distance.

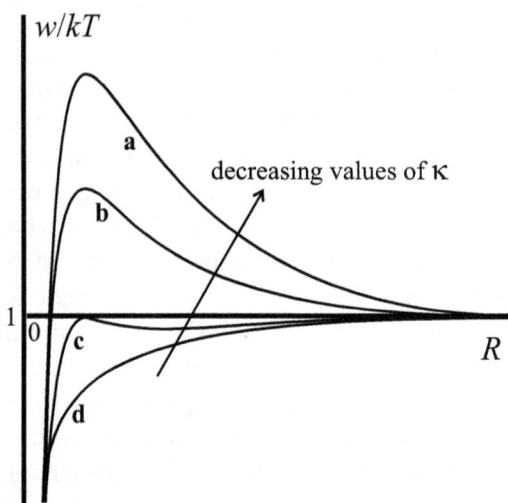

Figure 9.5: Potential energy between two colloidal particles, $w(R)/kT$, as a function of the separation distance, for different values of κ.

There are two terms in the former equation; the first one refers to the interaction between the ionic atmospheres of the two particles and it is a repulsive contribution because they have the same charge. The figure illustrates that the particles cannot approach each other for solutions at low ionic strength, that is, for small κ values, because the thermal energy is much lower than the repulsive electrostatic barrier between them. As more electrolyte is added, the double layers are thinner, being confined to a space closer to the particle, and the repulsive barrier becomes lower. For a given salt concentration, the height of the repulsive barrier is zero, as illustrated in figure 9.5. In this case, there is no impediment for the particles to approach each other, and the coagulation concentration is reached. For higher concentrations, the attractive van de Waals interactions dominate, leading to aggregation of the colloidal particles and to the coalescence of the system.

Donnan Effect and Membrane Potential

The use of the dialysis method to eliminate small ionic or non-ionic solutes that can contaminate solutions containing neutral or charged macromolecules, such as polyelectrolytes, proteins, or colloids is frequent. The method consists of putting in contact the solution containing the macromolecules with pure water through a membrane that is only permeable to small molecular solutes; that is, the membrane impedes the transfer of macromolecules or colloids to the pure water. We will analyze the case of macromolecules or colloids having electrical charge (P^{z-}), having counterions (M^{+}). Figure 9.6 schematizes the process for a macroion with negative charge z and the salt that contains M^{+} and X^{-} ions. We want to know what the situation will be when the system reaches equilibrium.

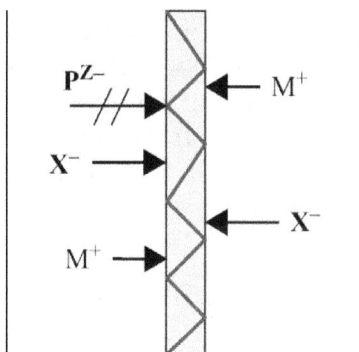

Figure 9.6: Donnan effect generated by the presence of a membrane impermeable to the P^{z-} species.

According to the treatment used in previous chapters, particularly in chapter 6, the equilibrium implies the equality of the chemical potentials of the species that can be transferred between the two compartments through the membrane that separates them. The permeable components are, in this case, H_2O, M^{+}, and X^{-}. For the component H_2O, an osmotic pressure is generated that will equilibrate the value of $\mu(H_2O)$ on both sides of the membrane. For the ions, the equilibrium condition will be the equality of the mean chemical potential of the MX electrolyte on both sides of the membrane, which leads to the equality of the mean activities in the left, $a_\pm^l(MX)$, and right, $a_\pm^r(MX)$, compartments. Assuming ideal solutions we have,

$$m^l(M^+)m^l(X^-) = m^r(M^+)m^r(X^-)$$

In the right compartment, where there is only MX, it will hold that $m^r(M^+) = m^r(X^-)$, because the solution must be electrically neutral. For the compartment at the left, we have MX and the macroion with negative charge z, and counterions M^+, and the electroneutrality condition implies that $m^l(M^+) = zm_p^l + m^l(X^-)$, where m_p is the concentration of the macroion. With these relationships, it is possible to derive the expressions

$$\frac{m^r(X^-)}{m^l(X^-)} = \sqrt{1 + \frac{zm_p}{m^l(X^-)}} > 1$$

and

$$\frac{m^r(M^+)}{m^l(M^+)} = \sqrt{1 - \frac{zm_p}{m^l(M^+)}} < 1$$

Thus, the equilibrium condition leads to a difference of concentrations of M^+ and X^- ions on both sides of the membrane. This phenomenon is known as the Donnan effect, and it is clear from the previous analysis that it is a fully ideal phenomenon. If the system illustrated in figure 9.6 is in equilibrium and we introduce in both compartments electrodes sensitive (reversible) to the same ion, for instance to the ion M^+, we have built the following galvanic cell

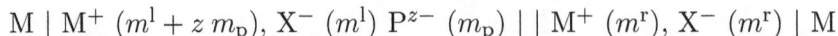

$$M \mid M^+ \ (m^l + z \, m_p), \ X^- \ (m^l) \ P^{z-} \ (m_p) \mid\mid M^+ \ (m^r), \ X^- \ (m^r) \mid M$$

If we measure the emf of this cell, it will be found that it is zero, as corresponds to a system in equilibrium. This is confirmed by writing the corresponding Nernst's equation

$$\Delta E = \frac{RT}{F} \ln \frac{a_{(+)}^l a_{(-)}^l}{a_{(+)}^r a_{(-)}^r}$$

and, since the activities of M^+ and X^- are equal on both sides of the membrane as has been already noted. If instead of electrodes sensitive to M^+ we place in each compartment a reference electrode, such the saturated calomel electrode [Hg | Hg$_2$Cl$_2$, KCl (saturated), the emf of this cell will not be zero, but *approximately*,

$$\Delta E = \frac{RT}{F} \ln \frac{m^l(M^+)}{m^r(M^+)}$$

that is, by definition, the membrane potential.[3] Comparing this expression with the former one, it is easy to see that the ratio of activity coefficients of M^+ at each side of the membrane is equal to the membrane potential. This fact is a consequence of the arbitrary division of the electrochemical potential in its chemical and electrostatic components, and it is also related to the liquid junction potential that appears in some galvanic cells.

Problems

Problem 1

The potential energy between two drops of dodecane suspended in water varies with the distance, R between them, according to

$$w(R) = d \left[\pi \epsilon_0 \epsilon_D \psi_0^2 \exp(-\kappa R) - \frac{B}{24R} \right]$$

This equation is equal to eqn. (9.22), but it includes the surface potential, ψ_0, of the drops. At 298 K, $\psi_0 = -50$ mV, the Hamaker constant $B = 0.5 \times 10^{-20}$ J, and $\epsilon_0 = 78$.[4]

Calculate the distance where $w(R)$ is maximum, and the corresponding value of $w(R)$, in kT units, for drops 0.6 μm in diameter, when:

a) There is no electrolyte added (κ^{-1} is estimated to be 50 nm in this case)

b) The ionic strength is 0.01 mole/kg

---

[3]It is usual to define the membrane potential in terms of a ratio of ionic activities; although, in practice, there are no appreciable differences between both definitions, the only thermodynamically rigorous definition is that given here.

[4]R. M. Pashley, "Effect of Degassing on the Formation and Stability of Surfactant-Free Emulsions and Fine Teflon Dispersions", *J. Phys. Chem. B*, 107 (2003): 1714-1720.

Answers: a) $R_{max} = 1.41$ nm, $(w/kT)_{max} = 750$; b) $R_{max} = 0.36$ nm, $(w/kT)_{max} = 607$

Problem 2

a) Write the hemireactions and the global reaction corresponding to the following galvanic cell

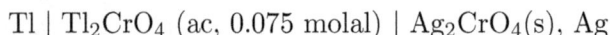

$$Tl \mid Tl_2CrO_4 \text{ (ac, 0.075 molal)} \mid Ag_2CrO_4(s), Ag$$

b) Calculate the emf of the cell at 298 K.

c) If the solution contains, in addition, 0.5 molal KNO_3, its emf decreases or increases? Justify your answer.

Data: $E^{\ominus}$ (Tl^+/ Tl) $= -0.336$ V; $E^{\ominus}$ (Ag^+/ Ag) $= 0.800$ V, and $K_{sp}(Ag_2CrO_4) = 1.12 \times 10^{-12}$

Use the equation $\log \gamma_{\pm} = -0.5|z_+z_-|[\sqrt{I}(1+\sqrt{I})^{-1} - 0.3I]$

Answers: b) $E = 0.888$ V; c) it decreases slightly

Problem 3

We have the following galvanic cell, at $T = 298.0$ K,

$$Pt(s) \mid H_2 \text{ (g, } p = 2 \text{ bar)} \mid HI \text{ (ac, 0.1 molal)} \mid Ag(s)$$

a) Write the equation corresponding to the global process and calculate the cell emf.

b) Determine the emf of the same cell at $T = 330$ K.

c) Determine $\Delta_R G^{\ominus}(298$ K) per mole of HI produced, and the corresponding dissociation constant.

Data: $E^{\ominus}$ (AgI/Ag) $= -0.1518$ V; $\Delta_f H^{\ominus}$ [$AgI(s)$] $= -61.84$ kJ/mole; $\Delta_f H^{\ominus}$ [$I^-(aq)$] $= -55.2$ kJ/mole; $\Delta_f H^{\ominus}$ [$H^+(aq)$] $= 0$ kJ/mole.

$$\ln \gamma_{\pm} = -A(T) |z_+z_-|[\sqrt{I}(1+\sqrt{I})^{-1} - 0.3I] \quad \text{and} \quad A(T) = 1.17(T/298)^{-3/2}$$

in the usual units

Answers: a) -0.0124 V; b) -0.00743 V; c) $\Delta_R G^{\ominus} = 14.65$ kJ/mole, $K = 2.70 \ 10^{-3}$

Problem 4

Calculate the potential energy between the ions K^+ and Br^- separated by 0.7 nm at 298 K, when:

a) The ions are in vacuum

b) The ions are dissolved in water (ϵ_D) at $I = 0$

c) The ions are in water at $I = 0.01$ (assume $a = 0.30$ nm)

d) Compare the former energies with the mean thermal energy

Answers: a) 3.28×10^{-19} J; b) 4.19×10^{-21} J; c) 2.29×10^{-21} J

Problem 5

A fuel cell of the PEM (proton exchange membrane) type has two electrodes of platinum finely dispersed on carbon separated by a membrane of poly(styrenesulphonate) (Nafion). This membrane contains–SO_3^- groups and conducts the electrical current by transporting H^+ ions. The cell is fed with H_2 at a pressure of 5 bar in the anode, and O_2 (from the air) at a partial pressure of 0.2 bar in the cathode. The reactions taking place on the electrodes are

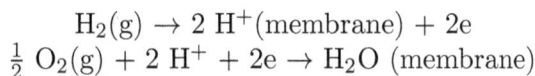

$$H_2(g) \rightarrow 2\ H^+(\text{membrane}) + 2e$$
$$\tfrac{1}{2}\ O_2(g) + 2\ H^+ + 2e \rightarrow H_2O\ (\text{membrane})$$

in such a way that during the spontaneous process the cell delivers energy and produces water that, partially, is retained in the membrane. At 298 K, the measured reversible potential (without current circulation) of this cell was 1.244 V. Calculate the water activity in the Nafion membrane and discuss the approximations made to solve the problem.

Answer: $a(H_2O) = 0.72$

Chapter 10

Non-Equilibrium Thermodynamics and Transport Processes

10.1   Introduction

So far, we have treated closed systems, where entropy changes are only due to internal processes, such as chemical reactions, mass transfer among phases, or inhomogeneities within a phase. In this chapter, we will deal with systems where mass and/or energy flows, from or toward the environment, are possible and the irreversible processes that take place as a consequence of them will be described. The main characteristic of the processes that occur out of equilibrium is the production of entropy. By resorting to eqn. (1.2), we obtain a general expression for the time dependence of the entropy during an irreversible process, valid for closed and open systems

$$\frac{\mathrm{d}S}{\mathrm{d}t} = \frac{\mathrm{d_I}S}{\mathrm{d}t} + \frac{\mathrm{d_e}S}{\mathrm{d}t} \tag{10.1}$$

In all cases, the signature of an irreversible process is that $\mathrm{d}_iS > 0$ (eqn. (1.2)). If the system is adiabatically isolated

$$\frac{\mathrm{d_I}S}{\mathrm{d}t} = \frac{\mathrm{d}S}{\mathrm{d}t} \geq 0$$

On the other hand, if entropy flux exists between the system and the environment, and there are not internal irreversible processes, we have

$$\frac{\mathrm{d}S}{\mathrm{d}t} = \frac{\mathrm{d_e}S}{\mathrm{d}t}$$

and the flux of entropy in the system will be equal to the entropy flowing from (toward) the environment. In the cases where there are internal processes, and mass and/or energy external exchanges, the production of entropy, $P \equiv (d_i S/dt)$, which is a measure of the process spontaneity, is given by

$$P = \frac{dS}{dt} - \frac{d_e S}{dt} \geq 0 \tag{10.2}$$

For macroscopic systems the entropy flux from or toward the environment can have any sign, meanwhile the production of entropy is always positive (see chapter 1 for other cases). In this chapter, two types of systems will be analyzed: i) heterogeneous (discontinuous) and ii) inhomogeneous and continuous.

Heterogeneous or Discontinuous Systems   These systems have at least two phases, and the thermodynamics properties are constant inside each phase (see figure 10.1a). A particular case is a system in which the phases are separated by an interface, or a membrane, where a change in the thermodynamic properties occur (figure 10.1b). It should be noted that, unlike the discussion on the interface in chapter 7, here the fluxes are not described from the molecular point of view, but macroscopically. Thus, figure 10.1a depicts the change of a property X in a discontinuous or heterogeneous system without describing the structure of the interface, while figure 10.1b shows the macroscopic change of X inside a membrane that separates two phases, without considering what occurs in the two interfaces separating the membrane and the system.

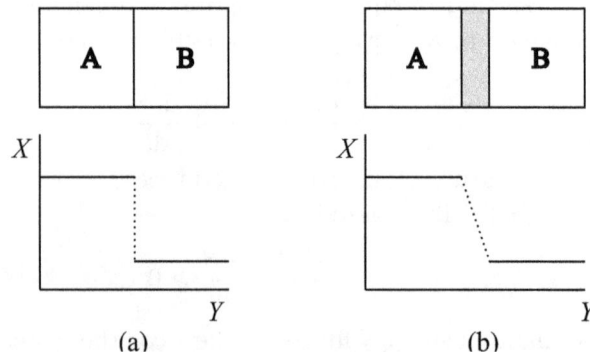

Figure 10.1: Extensive property X as a function of the position Y in a system: (a) discontinuous, and (b) discontinuous with a membrane.

Inhomogeneous and Continuous Systems   These are systems composed by an unique phase, where the values of (p, T, ρ, x_i) depend on the spatial coordinates in a continuum way, as illustrated in figures 10.2a and 10.2b. Figure 10.2a shows the case of a Inhomogeneous system where X changes with Y; that is, there is a gradient of X. A particular case is a continuous system in which there are no gradients of (p, T, ρ, x_i), as indicated in figure 10.2b, and the only internal processes that can take place in such a system are chemical reactions, under ideal stirring, to avoid concentration gradients. The chemical reactions that occur in the bulk of a phase are scalar or isotropic processes; that is, they do not depend on a particular direction, as is the case of vectorial processes.

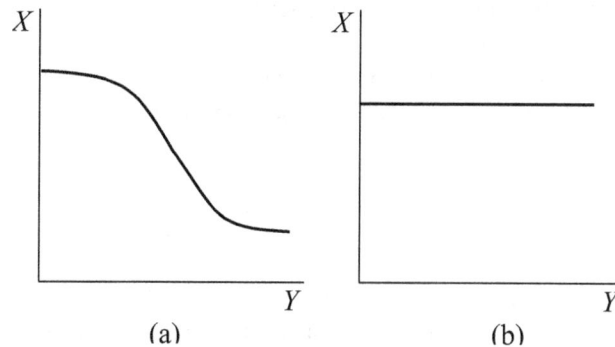

Figure 10.2: Extensive property X as a function of the position Y in a system: (a) continuous with gradient, and (b) continuous and isotropic.

10.2   The Local Equilibrium Hypothesis

In order to extend the laws of the equilibrium thermodynamics to processes that evolve out of equilibrium, it is necessary to introduce the local equilibrium hypothesis. It states that the relationships among the local values of the thermodynamics properties of the perturbed system (out of equilibrium) are the same that apply to the system under equilibrium conditions. From the microscopic point of view, this hypothesis is valid when the time required for achieving a stable distribution of local kinetic energy and densities, through molecular collisions, is much shorter than the characteristic local evolution time of the system. In this case, an equation of state can be applied to each region of the system, considering its local density and temperature.

Nevertheless, for processes where the fluxes of energy, matter, charge, and other quantities are proportional to the forces that originate them (the so-called linear irreversible processes), the use of thermodynamics does not require the local equilibrium hypothesis, as we will see next, because the thermodynamic quantities can be developed as a series, truncated in the first term around its equilibrium state. In his treatment of irreversible processes, Lars Onsager formulated a new principle or axiom based on a microscopic vision of such processes occurring in material systems, which cannot be derived from thermodynamics.

10.3   The Continuity Equation

In the case of discontinuous systems with homogeneous phases, the equations of conservation of mass, moment, and energy, have simple forms. For instance, the case of a system formed by two phases (α and β), where there are not chemical reactions, was already analyzed in chapter 1 to establish the spontaneity conditions. Now, we will analyze the flow of matter between both phases, leading to the following expression for the mass balance

$$\frac{\mathrm{d}n_k^\alpha}{\mathrm{d}t} + \frac{\mathrm{d}n_k^\beta}{\mathrm{d}t} = 0 \qquad (10.3)$$

where k is one of the chemical species of the system. In an open system, $\mathrm{d}n_k$ can include a contribution due to the matter exchange with the environment, $\mathrm{d}_e n_k$, and a contribution, $\mathrm{d}_i n_k$, due to the matter exchange between the system phases, or due to chemical reactions in these phases.

In the case of a inhomogeneous system, the mass balance requires the continuity equation, whose general expression for any extensive property X, as electric charge, mass, thermal energy, etc. is given by

$$\frac{\mathrm{d}X}{\mathrm{d}t} = -\nabla \mathbf{J}_X + q(X) \qquad (10.4)$$

where the first term on the rhs is the divergence of the X flux[1] exiting a volume element, and $q(X)$ is the rate of production (or consumption) of X within the system, due to chemical reactions.

If X is the mass, the mass flux equals the product, ρv, being ρ the density of the system, and v the mass flow velocity. If there are not chemical reactions, $q(X) = 0$,

---

[1]Flux is a vectorial quantity defined as the change of an extensive function per unit of time and area. In equations containing vectors the use of the operator $\nabla = (\partial/\partial x)\mathbf{i} + (\partial/\partial y)\mathbf{j} + (\partial/\partial z)\mathbf{k}$ is common. When applied to a scalar function X, the gradient of X vector is obtained as $\nabla X = (\partial X/\partial x)\mathbf{i} + (\partial X/\partial y)\mathbf{j} + (\partial X/\partial z)\mathbf{k}$. If the operator is applied to a vector $\mathbf{X}$, it yields the scalar product divergence of $\mathbf{X}$, which can be written as $\nabla \mathbf{X} = (\partial X_x/\partial x) + (\partial X_y/\partial y) + (\partial X_z/\partial z)$, which is a scalar.

and the continuity equation expressing the balance of mass in terms of the molar concentration, for a given k-component of the system, is

$$\frac{dc_k}{dt} = -c_k \nabla \mathbf{v} - \nabla \mathbf{J}_k \qquad (10.5)$$

In this equation, the contribution to the flux of mass due to the velocity of displacement of the global system ($\mathbf{v}$) is separated from the contribution due to the relative flow of the k-component. Thus, the flux of k is given by $\mathbf{J}_k = c_k(\mathbf{v}_k - \mathbf{v})$, $\mathbf{v}$ being the reference velocity (for instance, the velocity of the center of mass of the system or, in the case of a solution, the velocity of the solvent). This relationship is the result of defining the flux as the amount of moles of the k-component that cross an area $d\mathcal{A}$ in a time dt. As indicated in figure 10.3, the molecules that cross a virtual transversal plane during the interval dt are those located in the cylinder of length $(\mathbf{v}_k - \mathbf{v})dt$ and area $d\mathcal{A}$. If c_k is the molar concentration of k, the number of moles crossing the virtual plane is $c_k(\mathbf{v}_k - \mathbf{v})\, dt\, d\mathcal{A}$, then the flux of k is $\mathbf{J}_k = c_k(\mathbf{v}_k - \mathbf{v})$.

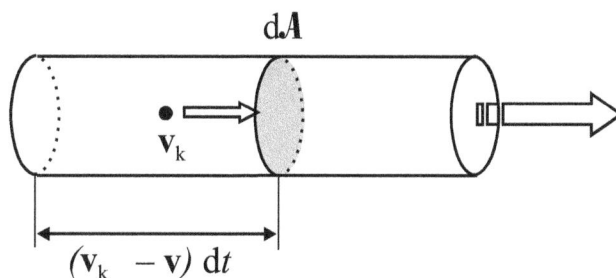

Figure 10.3: Flux of molecules moving in relation to a reference system, with a velocity $\mathbf{v}$.

Equation (10.5) is known as the diffusion equation, where the first term represents the convective part of the mass (density) change with time, while the second term represents the diffusive part.

Without considering in detail the hydrodynamic of the system, for a turbulent flux there are velocity gradients of the fluid in all directions, resulting in a much better mixing than for a laminar flux. Let us assume a fluid, for instance, water, inside a cylindrical tube flowing in laminar regime, in such a way that the velocity along the cylinder will be only a function of the distance to the central axis. If a drop of ink is introduced in the tube at a given point, the ink will start to flow and diffuse simultaneously. Thus, the ink spot will become larger and disperse due to the diffusive process that results from the random molecular movements. Moreover, the spot will move due to the viscous effect, following the water flux. If the flux becomes turbulent, or objects that acts as barrier for the water flux are introduced in the tube, the spot will disperse more quickly due to the macroscopic mixing that occurs under this condition, leading to the complete homogenization of the system.

The first Fick's law corresponds to the following linear phenomenological equation relating the flux of k with its concentration gradient, according to

$$\mathbf{J}_k = -D_k \nabla c_k \tag{10.6}$$

D_k being the diffusion coefficient of k. If the system is stationary, that is, there is not convection, the continuity equation becomes identical to the second Fick's law

$$\frac{\mathrm{d}c_k}{\mathrm{d}t} = D_k \nabla^2 c_k \tag{10.7}$$

as can be verified replacing $\mathbf{J}_k$ by eqn. (10.6) in the continuity equation (10.5).

For the case of linear infinite diffusion in the direction x, assuming that at $t = 0$ all the mass of k is at the origin ($x = 0$), the solution of equation (10.7) is given by

$$c_k(x,t) = \frac{m_k}{(2\pi D_k t)^{1/2}} \exp\left(-\frac{x^2}{4D_k t}\right) \tag{10.8}$$

The shape of the distribution of concentration of k, $c_k(x,t)$ at two different times is indicated in figure 10.4. The curves have the classical form of a Gaussian function describing a random process.

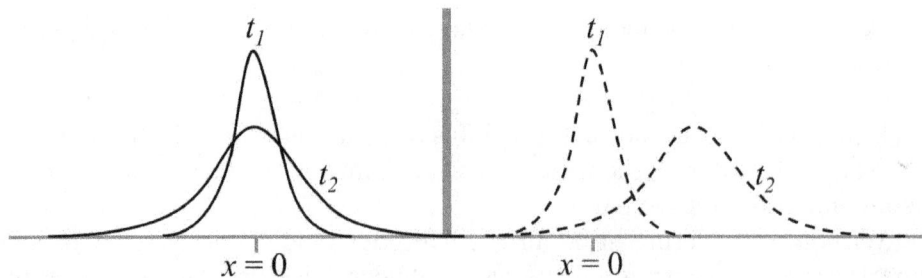

Figure 10.4: Concentration profiles of a diffusing species in a stationary liquid (full lines), and in a liquid moving along the x-axis (dashed lines).

A simple limiting case is that of fluid moving under laminar regime; that is, each volume element of the fluid at equal distance of the center of the tube, moves with the same velocity, $\mathbf{v}$, in the direction of the flux. In this case, the distribution given by eqn. (10.8) is still valid, with the simple change of variable $x' = x - vt$. The concentration profile of the diffusing species is the same, except that it shifts along the tube axis, as indicated in figure 10.4. The half-width of the distribution curve is related to the mean quadratic displacement, $\langle x_k^2 \rangle$, of the diffusing molecules, and it can be derived from eqn. (10.8) that

$$\langle x_k^2 \rangle = 2D_k t \tag{10.9}$$

This expression can be considered as the definition of the coefficient of diffusion.

A Microscopic Vision of the Transport Process

An alternative way of calculating the mean quadratic displacement of a particle of mass m, under Brownian motion in a fluid, is by resorting to the Langevin equation

$$m\frac{dv}{dt} = -\xi v + f(t)$$

where v is the particle velocity, ξ is the friction coefficient of the particle with the solvent, and $f(t)$ is a random force that constantly changes the direction of the particle.

Having into account the relationship

$$mx\frac{dv}{dt} = m\left[\frac{d(xv)}{dt} - v^2\right]$$

and by multiplying this equation by x, and calculating its mean value in time, indicated by the symbol $\langle \cdots \rangle$, it results

$$m\left\langle \frac{d(xv)}{dt} \right\rangle = -\xi \langle xv \rangle + m \langle v^2 \rangle$$

The average on $xf(t)$ is null because $f(t)$ is random. Since the kinetic energy of the particle, $m \langle v^2 \rangle /2$, is equal to $kT/2$, according to the equipartition principle, the solution to the former equation is

$$\langle xv \rangle = Ae^{-\xi t/m} + \frac{kT}{\xi}$$

The integration constant can be calculated from the boundary condition, $x = 0$ for $t = 0$, giving $A = -kT/\xi$. Taking that into account, it results

$$\langle xv \rangle = \frac{1}{2}\frac{d\langle x^2 \rangle}{dt}$$

and, on integrating this equation, we obtain the following expression for $\langle x^2 \rangle$

$$\langle x^2 \rangle = \frac{2kT}{\xi} \left[t - \frac{m}{\xi} \left(1 - e^{-\xi t/m} \right) \right]$$

For long periods compared to m/ξ (the system's relaxation time), the equation reduces to

$$\langle x^2 \rangle = \frac{2kTt}{\xi}$$

Comparing the previous equation with eqn. (10.9), we obtain

$$D = \frac{kT}{\xi}$$

That is, the diffusion coefficient of the particle is linked to its friction coefficient, in the solvent where it moves. One of the simpler models to calculate this friction coefficient is that developed by Stokes, which considers a spherical particle with radius r, moving in a continuous fluid with a dynamic viscosity, η. The expression for the friction coefficient in this model depends on the boundary conditions for the interaction between the particle and the solvent, being $\xi = 6\pi r \eta$ for the solvent sticking the particle. Combining this expression for the friction coefficient with the experimental diffusion coefficient, the hydrodynamic radius of the molecule can be estimated.

Energy Balance

The enthalpy (per volume unit) balance takes a simple form in absence of forces acting on the system, or chemical reactions, leading to

$$\rho \frac{\mathrm{d}H}{\mathrm{d}t} = -\nabla \mathbf{J}_q - \nabla \sum_{i=1}^{n} H_i \mathbf{J}_i \qquad (10.10)$$

The sum in the equation is over the n components of the system, and $\mathbf{J}_q$ is the heat flux, phenomenologically related through Fourier's law, to the temperature gradient in the system, ∇T, according to the expression

$$\mathbf{J}_q = -\lambda \nabla T \qquad (10.11)$$

where λ is the system's thermal conductivity. The first term on the right side in eqn. (10.10) is the thermal energy (heat) that enters or exits the system by conduction, due to the temperature difference with the environment. The second term involves the energy changes inside the system, due to the diffusive flux of their components. When a species diffuses from one point of the system to another, its partial molar enthalpy, H_i, changes due to the variable composition and temperature of the system, and therefore the energy content of the system also varies. It should be noted again that in eqn. (10.10) the enthalpy changes due to chemical reactions have not been considered.

10.4   The Entropy Production

An equation similar to the balance of enthalpy (eqn. (10.10)) can be written for the balance of entropy, per unit of volume, leading to

$$\rho \frac{\mathrm{d}S}{\mathrm{d}t} = -\nabla \mathbf{J}_S + \sigma \tag{10.12}$$

where $\mathbf{J}_S$ represents the flux of entropy (that can be negative or positive) and, consequently, the first term on the rhs of eqn. (10.12) is the entropy change due to exchange with the environment, $\mathrm{d}_e S/\mathrm{d}t$, while the second term represents the internal entropy production per unit of volume, σ, that is always positive.

In a closed system with a single phase, considering isothermal and isobaric processes, it results from eqn. (1.25) that

$$\mathrm{d}G = \frac{1}{\rho} \sum_{i=1}^{n} \mu_i \mathrm{d}c_i \tag{10.13}$$

and then

$$\rho T \frac{\mathrm{d}S}{\mathrm{d}t} = \rho \frac{\mathrm{d}H}{\mathrm{d}t} - \sum_{i=1}^{n} \mu_i \frac{\mathrm{d}c_i}{\mathrm{d}t} \tag{10.14}$$

By using eqns. (10.4) and (10.10), and considering that the system is stationary, we obtain

$$\rho T \frac{\mathrm{d}S}{\mathrm{d}t} = -\nabla \mathbf{J}_q - \nabla \sum_{i=1}^{n} H_i \mathbf{J}_i + \sum_{i=1}^{n} \mu_i (\nabla \mathbf{J}_i) \tag{10.15}$$

Having into account that

$$\nabla (\mu_i \mathbf{J}_i) = \nabla \left[(H_i - TS_i) \mathbf{J}_i \right] = \mu_i (\nabla \mathbf{J}_i) + \mathbf{J}_i \nabla \mu_i \tag{10.16}$$

it is possible to replace the last term on the rhs of eqn. (10.15), and after some algebra it results

$$\rho \frac{\mathrm{d}S}{\mathrm{d}t} = -\nabla \left[\frac{\mathbf{J}_q}{T} + \sum_{i=1}^{n} S_i \mathbf{J}_i \right] - \frac{1}{T} \sum_{i=1}^{n} \mathbf{J}_i \nabla \mu_i \tag{10.17}$$

The first term on the rhs represents the flux of entropy by heat conduction, and the second one is the flux of entropy by diffusion. By comparing this expression with the balance of enthalpy (eqn. (10.12)), it is clear that the flux of entropy is given by

$$\mathbf{J}_S = \frac{\mathbf{J}_q}{T} - \sum_{i=1}^{n} S_i \, \mathbf{J}_i \tag{10.18}$$

That is, $\mathbf{J}_S$ comprises the heat exchange with the ambient, as well as the entropy changes, due to mass exchange fluxes with the ambient. The entropy internal production is given by

$$\sigma = -\frac{1}{T} \sum_{i=1}^{n} \mathbf{J}_i \, \nabla \mu_i \tag{10.19}$$

which can be extended to generalized forces ($\mathbf{X}_i$) and fluxes ($\mathbf{J}_i$), resulting in

$$\sigma = \frac{1}{T} \sum_{i=1}^{n-1} \mathbf{J}_i \, \mathbf{X}_i > 0 \tag{10.20}$$

where the fluxes are defined with respect to a plane fixed to solvent. For the case of mass transport, $\mathbf{X}_i$ is given by

$$\mathbf{X}_i = -\sum_{j=1}^{n-1} \left(\delta_{ij} + \frac{c_j}{c_n} \right) \nabla \mu_j \tag{10.21}$$

being $\delta_{ij} = 0$ if $i \neq j$, and $\delta_{ij} = 1$ if $i = j$. Equation (10.20) indicates that the internal entropy production is the sum of products of fluxes, $\mathbf{J}_i$, times the generalized forces, $\mathbf{X}_i$. For the isothermal and isobaric case analyzed here, these forces are the gradients of chemical potentials, $\nabla \mu_i$, of the components of the system.

10.5 Transport Processes

Transport processes are irreversible processes where, by action of a driving force (gradient of some property), a flow of mass, electric charge, energy, moment, etc., is induced. There are phenomenological laws that describe these processes when they are caused by only one type of driving force or gradient. Two of these laws have been already described: Fick's law in eqn. (10.6), which holds for the mass transport, and Fourier's law in eqn. (10.11), describing the heat transport.

Another well-known phenomenological equation is Ohm's law, $\mathbf{i} = -\kappa \nabla \Phi$, where the current density (or flux of charge), $\mathbf{i}$, is linearly related to the potential gradient, $\nabla \Phi$ (or electric field, $\mathbf{E} = -\nabla \Phi$), through the specific conductivity, κ, of the system.

The treatment of transport processes, according to the formalism developed in the previous section, allows a generalization of thermodynamics to irreversible processes that are valid for linear transport processes, which are those where the fluxes are linearly related to the generalized forces; that is,

$$\mathbf{J}_i = \sum_{k=1}^{n} L_{ik} \mathbf{X}_k \qquad (10.22)$$

By comparing this expression with the phenomenological laws, such as the laws of Fick, Fourier, or Ohm, summarized in table 10.1, it is observed that the gradients of concentration, temperature, and electrical potential appear as generalized forces for the transport of mass, heat, and charge, respectively. The coefficients L_{ii} corresponding to the fluxes that are associated with the conjugated forces–flux of mass for the gradient of chemical potential (concentration), flux of heat for the gradient of temperature, and flux of charge for the gradient of electrical potential–are the transport coefficients known as diffusion coefficient, thermal conductivity, and electrical conductivity, respectively, which are always positive quantities.

Table 10.1: Fluxes and forces associated

Transport	Flux	Force
Mass	$\mathbf{J}_m$	$\mathbf{X}_m = -\nabla\mu$
Charge	i	$\mathbf{X}_c = -zF\nabla\phi$
Thermal energy (heat)	$\mathbf{J}_q$	$\mathbf{X}_q = -\nabla T/T$

Equation (10.22) establishes that if more than one driving force is acting on the system, there ise flux coupling due to the action of different forces. Thus, there will be contribution to the flux $\mathbf{J}_i$ from all the forces $\mathbf{X}_k$, with $k \neq i$. The transport coefficients, L_{ik}, associated to these fluxes satisfy Onsager's reciprocal relationship (ORR), given by

$$L_{ik} = L_{ki} \qquad (10.23)$$

This is the principle, or axiom, of reciprocity proposed by Onsager, who derived it from the principle of microscopic reversibility that establishes the invariability of the movement laws by a temporal inversion, that is, by the transformation of t into $-t$, in the dynamics of the particles.

For *non-linear* irreversible processes, that is, those that take place in states far from equilibrium, the fluxes can vary by transforming t into $-t$. Then, it is not possible to apply the principle of microscopic reversibility to these processes because on time inversion there could be transitions that do not allow the system to return to the initial state. This subject will not be treated in this chapter.

As we will see later, the reciprocity relationships allow us correlate different transport coefficients, apparently independent from each other.

Alternative Description of Chemical Reactions

In the case of a homogeneous system that is isotropic (figure 10.2b), where there are not gradients of temperature, pressure, electrical potential, etc., the entropy production is only due to chemical reactions. We saw in chapter 8 that in these systems (eqn. (8.6)) holds, leading to

$$\frac{d_i S}{dt} = \frac{\mathbb{A}}{T}\frac{d\xi}{dt} = -\frac{1}{T}\left(\sum_i \nu_i \mu_i\right)\frac{d\xi}{dt}$$

where $d\xi/dt$ is the reaction rate. We will perform an analysis similar to that in the discussion subject of page 247, but in this case focused on the Onsager formulation. The former equation can also be generalized for the case of several reactions, as

$$T\frac{d_i S}{dt} = \sum_r \mathbb{A}_r v_r \tag{10.24}$$

Here, the association of the chemical affinity to a driving force, and the reaction rate with a flux, is evident. Let's assume now that in the system occurs two reactions (subindexes 1 and 2), corresponding to chemical processes not far from equilibrium. The expressions for the chemical rates, according to eqn. (10.22), will be

$$v_1 = L_{11}\mathbb{A}_1 + L_{12}\mathbb{A}_2$$

$$v_2 = L_{21}\mathbb{A}_1 + L_{22}\mathbb{A}_2$$

where the second term on the rhs for the rate of reaction 1 represents the effect of reaction 2 on the rate of reaction 1. Similarly, the first term on the rhs for the rate of reaction 2 represents the contribution of reaction 1 on the rate of reaction 2. The coefficient $L_{12} = L_{21}$ is a measure of the coupling between both reactions.

In this particular example, the reaction rate is a function of $p, T, c_i,...,$ c_n, as well as of the affinity that is shifted from its equilibrium value ($\mathbb{A}$

$= 0$). As a result, $v_r = f_r(p, T, c_i, ..., c_n, \mathbb{A})$, and the equilibrium condition corresponds to $v_r = 0$, for $\mathbb{A}_r = 0$, and the rest of the variables in their equilibrium values. Then, the reaction rate can be developed in powers of $\mathbb{A}$ around the equilibrium

$$v_r = \left(\frac{\partial f_r}{\partial \mathbb{A}_r} \right)_{\xi_{4eq,r}} \mathbb{A}_r + \frac{1}{2} \left(\frac{\partial^2 f_r}{\partial \mathbb{A}_r^2} \right)_{\xi_{eq,r}} \mathbb{A}_r^2 + ...$$

This expression is equivalent to that derived on page 249, but is now referred to the reaction affinity. Close to the equilibrium state, we can ignore the higher terms, keeping only the linear variation, resulting in $L_{ii} = (\partial f/\partial \mathbb{A})_{eq}$. The validity of eqn. (10.22) is limited to systems where v_r depends linearly on $\mathbb{A}_r(\xi)$.

10.6 Generalization of Le Chatelier's (or Moderation) Principle

In chapter 8, we showed how the entropy production, σ, equals the product of a force ($\mathbb{A}/T$) and a flux ($d\xi/dt$), a particular case of eqn. (10.20). On page 250, we also derived the expression

$$\frac{d_i S}{dt} = \frac{v_+}{2R} \left(\frac{\mathbb{A}}{T} \right)^2 \geq 0$$

for small departures of equilibrium. In this equation, all the factors are positive, even if $\mathbb{A} = -\Delta_R G < 0$; if this is the case, the reaction proceeds in the inverse direction and more reactants will be formed, but v_+ will be always positive.

In the linear regime, that is, when the fluxes are proportional to the forces, it can be considered that the equilibrium thermodynamics we deal with in the previous chapters is applicable to the irreversible processes, provided that the conditions for using the local equilibrium hypothesis are satisfied. Under these conditions, it is commonly observed that irreversible processes lead the system to stationary states, which are those where, for a fixed position in the system, the intensive properties do not change with time, even when fluxes are present.

Let's imagine a tube at constant pressure with a liquid inside. If the temperature is different in each extreme of the tube, the temperature gradient is a force, $\mathbf{X}_q$, driving the heat flux, $\mathbf{J}_q$, but at the same time a flux of matter will be induced, $\mathbf{J}_m$.

In this case, the entropy production is

$$T\frac{d_iS}{dt} = \mathbf{J}_q\mathbf{X}_q + \mathbf{J}_m\mathbf{X}_m$$

If the temperature gradient between the ends of the tube remains constant, a stationary state will be reached, where there will be flux of heat, but the flux of matter will be absent. Using the Onsager formalism (eqn. (10.22)) the expression for the entropy production becomes

$$T\frac{d_iS}{dt} = L_{qq}\mathbf{X}_q^2 + L_{qm}\mathbf{X}_m\mathbf{X}_q + L_{mq}\mathbf{X}_q\mathbf{X}_m + L_{mm}\mathbf{X}_m^2$$

To evaluate the effect of changing the force $\mathbf{X}_m$ in the entropy production equation, we can take the derivative with respect to $\mathbf{X}_m$, and using the ORR, that is, $L_{qm} = L_{mq}$, it yields

$$T\frac{\partial}{\partial\mathbf{X}_m}\left(\frac{d_iS}{dt}\right)_{\mathbf{X}_q} = 2L_{qm}\mathbf{X}_q + 2L_{mm}\mathbf{X}_m = 2\mathbf{J}_m$$

Once the system reaches the stationary state $\mathbf{J}_m = 0$ and the entropy production does not change with the force $\mathbf{X}_m$, reaching the system an extreme point (the entropy production is a minimum at the stationary state). For linear irreversible processes, this represents an important correlate with the Le Chatelier's principle, analyzed in chapter 8. Some authors call moderation principle to the extended Le Chatelier's principle, which is a consequence of thermodynamic stability of the equilibrium states (and the states close to equilibrium in linear regime), even if irreversible processes are taking place.

10.7   Transport in Homogeneous Systems

In this section, we will analyze the application of eqns. (10.22) and (10.23) to some particular systems to obtain the relationship between the Onsager transport coefficients, L_{ij}, and the corresponding phenomenological coefficients. Before that, it is convenient to review the form of the generalized forces associated to the various fluxes we have seen so far, summarized in table 10.1.

It is interesting to note that the force associated to the flux of mass is not the concentration gradient, but the gradient of chemical potential, such as indicated in eqn. (10.21). Therefore, it is obvious that the diffusion coefficient, D, defined by Fick's law (eqn. (10.6)) is different to the coefficient L_{mm} obtained from eqn. (10.22), with $\mathbf{X}_m = -\nabla\mu$.

In a continuous system, where there are only gradients of chemical potentials, just mass fluxes of the different components can be observed. Thus, for a binary system formed by a solvent (1) and a solute (2), we expect fluxes of both components. However, we have seen that fluxes are always measured in relation to a reference velocity, for instance the velocity of the solvent, and in this case $\mathbf{J}_1 = 0$ and $\mathbf{J}_2 = c_2(\mathbf{v}_2 - \mathbf{v}_1)$. Thus, in a binary system, there is only one diffusion coefficient, $D = D_{12} = D_{21}$, which expresses the relative flux of one component with respect to the other, and it is called inter-diffusion coefficient. Recalling eqn. (10.9), it is clear that, even in the case of a system formed by one component, that component could diffuse through the molecules of the same sort. We call that type of process self-diffusion. The self-diffusion coefficient can be determined if we add to the system a tracer of the same chemical component in the form of an isotope. In this case, we will observe, after a time, that the isotope is distributed homogeneously all over the system. It is evident that when we add the isotope of the component, we generate a gradient of chemical potential of this species but, since the major species and its isotope are chemically identical, this gradient becomes an entropy gradient that drives the mixing of both constituents.

Figure 10.5 could clarify these concepts. Drawn there, are the inter-diffusion coefficient and the tracer diffusion coefficients of each component, for a binary mixture of benzene and cyclohexane. The tracer diffusion is determined using isotopes in mixtures of variable composition.

Since benzene and the isotope-labeled benzene are chemically equivalent, the tracer diffusion of the labeled benzene in cyclohexane is equal to the interdiffusion of benzene, at infinite dilution of cyclohexane. The same is observed fot the tracer diffusion of cyclohexane. The curves corresponding to the diffusion of tracers start at the values of the self-diffusion coefficient of each component.

In general, for a system of n components we will have $(n-1)$ independent fluxes, and it is easy to demonstrate, using the RRO (eqn. (10.23)), that the number of independent diffusion coefficients is $n(n-1)/2$.

One of the non-isothermal processes of interest that occurs in homogeneous systems is the thermal conduction described by Fourier's law. The thermal conductivity, λ, is related to the corresponding Onsager coefficient by

$$\mathbf{J}_q = -\lambda \nabla T = -\frac{L_{11}}{T} \nabla T \qquad (10.25)$$

Thermodiffusive Effects

These processes imply the simultaneous existence of temperature and chemical potential gradients, that, in addition to the known processes of thermal conduction

$Dx10^5 cm^2/s$

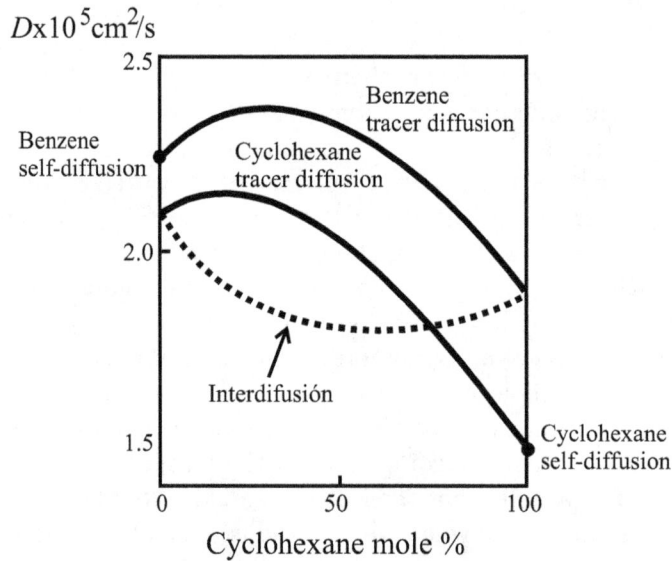

Figure 10.5: Scheme of the composition dependence of interdiffusion, and tracer diffusion of cyclohexane and benzene, in a binary mixture of these two solvents (adapted from figure 7.5-1 in Cussler's book, cited in appendix C)

and diffusion, induce cross effects, such as Soret effect (thermal diffusion due to temperature gradients), and Dufour effect (heat transport due to gradients of concentration). In a binary system, the fluxes of mass and heat can be expressed in the following way

$$\mathbf{J}_m = -L_{11}\nabla\mu_2 - \frac{L_{12}}{T}\nabla T \tag{10.26}$$

$$\mathbf{J}_q = -L_{21}\nabla\mu_2 - \frac{L_{22}}{T}\nabla T \tag{10.27}$$

where L_{11} and L_{22} are clearly related to the diffusion coefficient (D) and the thermal conductivity (λ), respectively, while L_{12} is the thermo-diffusion coefficient which, according to the RRO, is equal to the Dufour coefficient L_{21}.

Thermo-diffusion is a very common phenomenon, taking place when an homogenous system is warmed up, leading to the formation of concentration gradients. For instance, if the end of a column of an aqueous NaCl solution is heated, the salt concentration increases in the cold end and decreases in the hot one. In the same way, if we have two bulbs containing a mixture of He and Ar, joined through a capillary tube, and we increase the temperature of one of the bulbs by transferring heat, the mole fraction of He increases in the hot bulb, and decreases in the cold one. The microscopic treatment of this process that permits understanding why the lighter gas is concentrated in the hot region is very complex, but it is clear that the effect can be used for the separation of gaseous mixtures.

10.8   Transport in Heterogeneous Systems

In a discontinuous system, the separation between two phases can be achieved using a membrane, a capillary, or a porous plug. In some cases, there is no physical separation between the phases and the transport process occurs directly, for instance, through the interface between two metals, an electrode and a solution, or two immiscible liquids, etc. When the phases contain charged species, the allowed charge transport processes are those schematized in figure 6.2.

Thermoelectric Effects in Solids

Thermoelectric effects in fluids are complex to treat because temperature gradients induce concentration gradients, that in turn generate mass transport. This is not the case in solids, where temperature gradients induce electrical potentials at the interfaces, and the charges are transported by electrons or defects in the crystals' structure.

The processes that can take place are i) electrical conduction (due to the electrical potential gradient), ii) thermal conduction (due to the temperature gradient), iii) charge flux (due to the temperature gradient, called Seebeck effect), and iv) heat flux (due to the potential gradient, called Peltier effect). The cross effects [(iii) and (iv)] are observed when two metallic conductors from different metals (A and B) get in contact forming a discontinuous system, as shown in figure 10.6, for a thermocouple.

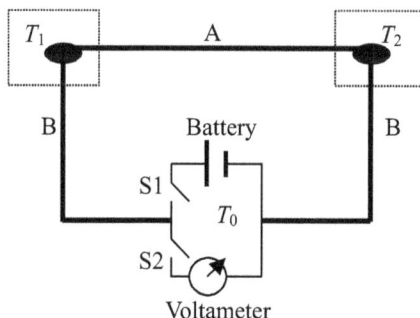

Figure 10.6: Scheme of two metallic conductors A and B to study the Seebeck (thermocouple) and the Peltier effects. In the case of a thermocouple, T_1 and T_2 represent the sample and the reference temperatures, respectively.

In this case, $T_1 \neq T_2$, and the switch S2 can be closed to connect a micro-voltameter in the circuit, for measuring the potential generated by the temperature difference. In the case of the Peltier effect, the temperatures of A and B are initially the same ($T_1 = T_2 = T_0$), and when the switch S1 is closed and the battery is connected to the circuit, the potential gradient generates heat transport from one metal to the other (if the junctures A-B are not thermostatized, one of them will get hot and the other will get cold).

For the general situation, where there are gradients of temperature as well as electrical potential, the flux of heat is

$$\mathbf{J}_q = -L_{11}\left(\frac{\nabla T}{T}\right) - L_{21}\nabla\phi \tag{10.28}$$

and the currect that circulates through the circuit is

$$\mathbf{I} = -L_{21}\left(\frac{\nabla T}{T}\right) - L_{22}\nabla\phi \tag{10.29}$$

For the Seebeck effect, the condition $\mathbf{I} = 0$, implying a stationary state, leads to

$$-\left(\frac{d\phi}{dT}\right)_{I=0} = \frac{L_{21}}{TL_{22}} \tag{10.30}$$

that is, an expression that relates the Onsager's coefficients with the magnitude of the thermoelectric effect of the metallic couple.

Electrokinetics Effects

The electrokinetic effects are generated when two phases exhibit differences of pressure and electrical potential. The common case is that of phases separated by a membrane with thickness δ, and area $\mathcal{A}$, as indicated in figure 10.7. We will observe fluxes of volume (permeation) and charge (current) through the membrane.

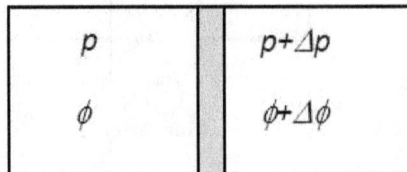

Figure 10.7: Scheme of a discontinuous system formed by two phases separated by a membrane, with pressure and electrical potential gradients.

In the simplest case, when only a driving force is present, the corresponding transport coefficients are identified with the permeation process (flux of volume, in the absence of electric field), and the electrical conductivity (flux of charge, in absence of pressure gradient). If both forces are present, crossed fluxes will appear: The electroosmotic flux due to the potential gradient, and the flux of charge due to the gradient of pressure through the membrane.

The flux of volume and the current are proportional to the area of the membrane, and to the inverse of its thickness, being

$$\mathbf{J}_V = \mathcal{A}\left(a\frac{\Delta p}{\delta} + b\frac{\Delta \phi}{\delta}\right) \tag{10.31}$$

$$\mathbf{I} = \mathcal{A}\left(b\frac{\Delta p}{\delta} + \kappa\frac{\Delta \phi}{\delta}\right) \tag{10.32}$$

We have assumed here that the gradients through the membrane are linear, and consequently the gradient of the corresponding property Z, ∇Z, can be replaced by $\Delta Z/\delta$. Under this assumption, the changes in the properties are limited to the membrane, and they remain constant within the phases.

The coefficient a is the permeability of the membrane, associated to $-\delta L_{11}/\mathcal{A}$, and κ is the specific electrical conductivity, associated to $-\delta L_{22}/\mathcal{A}$. The coefficient b represents both the electroosmotic permeability and the stream conductivity, because $b = -L_{12}/\mathcal{A} = -L_{21}/\mathcal{A}$.

The electrokinetic effects are common in clay-like materials in contact with aqueous solutions, as is the case of soils. In this type of system, an electroosmotic flux is observed when an electrical potential is applied between two regions. On the other hand, if a flow of ionic solution is forced through the pores of the system, an electrical current (stream current) is generated in the direction of the volume flux, even in absence of an external electric field.

Other Transport Processes Through Membranes

Several processes of chemical, physical, and biochemical interest occur in the presence of a membrane that separates two phases. In order to make the treatment simpler, we will consider the system as isothermal, but pressure, electrical potential, and composition gradients, are present through the membrane, as indicated in figure 10.8. The particular case where $\Delta c = 0$ (c is the molar concentration of solute) was discussed in the previous section, and now we will deal with processes with concentration gradients.

For the system illustrated in figure 10.8, the flux of volume and the electrical current (when charged species are present) will be influenced by the diffusion of species

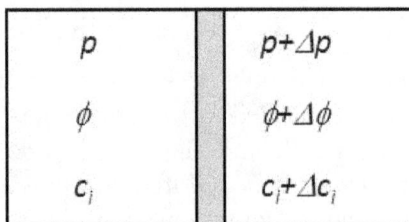

Figure 10.8: Scheme of a discontinuous system formed by two phases separated by a membrane, under pressure, electrical potential, and solute concentration gradients.

that can permeate through the membrane. It is the case of electrolyte solutions with different concentrations, separated by a membrane, in the presence of different pressure and electrical potentials on each side. The simplest situations involve permeation ($\Delta\phi = 0$, $\Delta c_k = 0$), electrical conductivity ($\Delta p = 0$, $\Delta c_k = 0$), and electroosmotic permeability ($\Delta c_k = 0$), already analyzed in a previous example. The new processes that can occur in this case are related to the concentration gradients through the membrane ($\Delta c_k \neq 0$). If $\Delta\phi = 0$ and $\Delta p = 0$, we obtain the following expression for the mass flux

$$\mathbf{J}_k = -L_{kk}\left(\frac{\partial\mu_k}{\partial c_k}\right)_{T,p}\frac{\Delta c_k}{\delta} \tag{10.33}$$

The coefficient $P_k = -L_{kk}(\partial\mu_k/\partial c_k)$ is similar to the permeability coefficient, a, for the case of flux of volume (eqn. (10.31)), but in this case the permeation is driven by a gradient of concentration (chemical potential) and this parameter is called coefficient of osmotic permeability.

By comparing eqn. (10.33) with Fick's law

$$\mathbf{J}_k = -D_k\frac{\Delta c_k'}{\delta} \tag{10.34}$$

where $\Delta c_k'/\delta$ is the concentration gradient inside the membrane (see figure 10.9). It is obvious that, if we define the partition coefficient of the solute between the membrane and the solution ($K_k = c_k'/c_k$), the relation between the diffusion coefficient and the osmotic permeability coefficient of the solute in the membrane is

$$P_k = D_k K_k \tag{10.35}$$

The membrane potential, $\Delta\phi_M$, is the potential drop between both sides of the membrane, when $\Delta p = 0$ and $I = 0$ (see page 289). This potential appears, for instance, in electrochemical cells with liquid junction, as in the cell

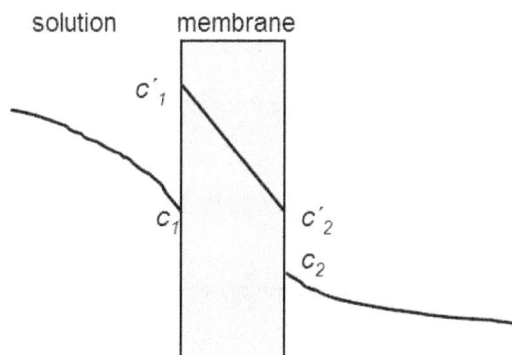

Figure 10.9: Concentration gradients of a solute in the membrane and the solutions on both sides, for a case with $K_k > 1$.

$$\text{Cu, Ag} \mid \text{AgNO}_3 \| \text{TlNO}_3 \mid \text{Tl, Cu}$$

studied in section 9.5. If the aqueous solutions are not dilute, the potential of this cell contains a membrane potential term due to the activity difference of the solutions. This contribution, called the liquid junction potential, is negligible in dilute solutions, and for this reason was not included in section 9.7.

Controlled Drug Delivery

The diffusion of solutes through membranes can be used with several purposes, generally related to purification and separation processes. One of the most interesting applications of this transport process is the controlled delivery of biomolecules or pharmaceutical compounds. In this case, the goal is to achieve a constant level of concentration of the substance in the medium toward which it is permeating, in such a way that the level of toxicity is never reached, while the concentration is still above the level of efficiency, as indicated in figure 10.10.

When a pharmaceutical compound is consumed in the form of a pill, its dissolution in the biological fluid controls its concentration in the organism. As it is shown in figure 10.10, its concentration can rapidly reach a high value, eventually higher than the toxicity level, and then it may rapidly decay below the limit of effectiveness.

There are several ways of keeping the drug concentration almost constant over long periods. One of them is to disperse the drug into a capsule

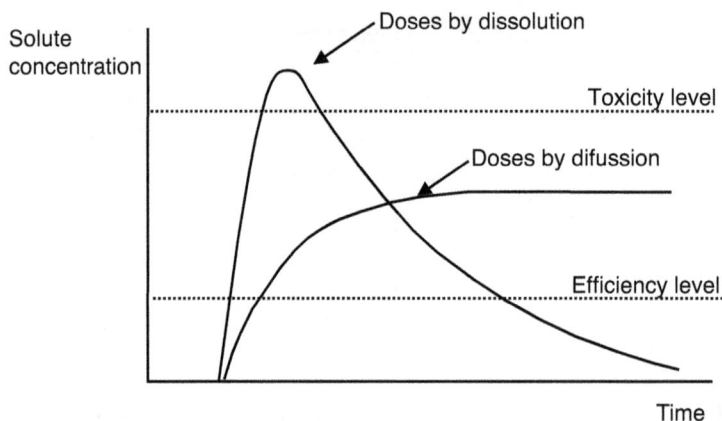

Figure 10.10: Drug concentration profile as a function of time for different types of drug delivery (adapted from figure 18.0-1 of Cussler's book, cited in appendix C).

of polymer swollen with water, which is then dispersed in the biological fluid. In this case, the transport conditions are similar to those of the linear semi-infinite diffusion described in section 10.3, and the concentration of the solute will be a function of the distance to the capsule surface, with a temporal dependence given by eqn. (10.8). If the initial concentration of solute in the capsule is c_o, the amount of solute, M, delivered to the medium in contact with the capsule, at a time t, is obtained by integration of the concentration along the diffusion direction

$$M(t) = \int_0^\infty c(x,t)\mathcal{A}\mathrm{d}x$$

The result for short times being

$$M = \sqrt{\frac{4Dt}{\pi}}\mathcal{A}c_o$$

indicates that the drug concentration increases as $t^{1/2}$ in the beginning, and at longer times as its concentration reaches a constant value.

Even more flexible is a method that delivers the drug by controlled diffusion by means of a reservoir, having a wall or membrane, that permits the osmotic permeation of the solute (drug), but not of the solvent (figure 10.11).

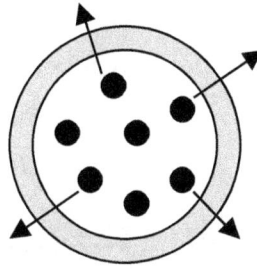

Figure 10.11: Delivery of a drug contained in a reservoir by permeation through the membrane.

In this case, the problem can be solved if the osmotic permeability, P, of the solute through the membrane is known. According to eqn. (10.34), and assuming that the concentration of the solute inside the reservoir, c_o, is much higher than that in the external medium, the amount of solute delivered in a time t is

$$M = J\mathcal{A}t = \frac{Pc_o}{\delta}\mathcal{A}t$$

where $\mathcal{A}$ represents the reservoir area, and δ is the membrane thickness. Therefore, we conclude that the concentration of a solute in the external medium increases with time to reach a final value given by the moles delivered (calculated from M) and the volume of the medium, provided that the latter is much larger than the volume of the reservoir.

Since P is the product of the diffusion coefficient of the solute in the membrane times the partition coefficient of the solute (eqn. (10.35)), it would be possible to select the adequate membrane to control the delivery rate of the drug to the medium.

Problems

Problem 1

a) Using the condition $\Delta T = 0$ for the Peltier effect, and the Onsager reciprocal relationships, show that the Peltier coefficient, defined as $\Pi = (\mathbf{J}_q/T)_{\Delta T=0}$, is related to the thermoelectric effect by the Kelvin relationship

$$\Pi = -T\frac{\mathrm{d}\phi}{\mathrm{d}T}$$

b) For the couple Cu–Ni, $(\mathrm{d}\phi/\mathrm{d}T) = 22.3$ μV/K at 298 K. Assuming that between the Cu–Ni junctures in the scheme of figure 10.6 a potential of 0.1 mV is applied, what temperature is achieved in the juncture that releases heat, if the juncture that absorbs heat is at 295 K? Indicate what does limit the application of higher potentials to obtain higher temperature differences? (the specific electrical resistivity, at 298 K, is 1.71×10^{-6} ohm.cm for Cu, and 7.12×10^{-6} ohm.cm for Ni).

c) The frigistors are couples formed by conveniently doped BiTe semiconductors (type p and n). For these pairs $(\mathrm{d}\phi/\mathrm{d}T) = 33$ μV/K, over a wide range of temperature. Calculate ΔT when a potential of 2 Volts is applied between the terminals, keeping the cold juncture at 295 K.

Answer: b) $T = 290.5$ K; c) $\Delta T = 60.6$ K

Problem 2

The flux conductivity or electroosmotic permeability, b, is determined by measuring the flux potential, $\Delta\phi$, at zero current. Demonstrate that

$$\left(\frac{\Delta\phi}{\Delta p}\right)_{I=0} = -\frac{b}{\kappa}$$

where κ is the specific electrical conductivity.

Problem 3

The electrical mobility of an ion is defined by $u_i = \mathbf{v}_i/\mathbf{E}$, $\mathbf{E}$ being the electric field that drives the charge flux, $\mathbf{i} = z_i e F c_i u_i \mathbf{E}$, where z_i is the charge of the ion i, and c_i its concentration.

In dilute solutions, there is a relationship between the electrical mobility of the ion and its diffusion coefficient, known as the Nernst-Einstein law, given by

$$u_i = \frac{z_i e}{kT} D_i$$

Derive an expression for the electrical potential generated in the liquid junction between two NaCl aqueous solutions, with concentrations c_1 and c_2.

Hint: Write down the flux equations through the liquid junction for both ions and assume that, in the stationary state, the net current density through the junction, $i = \mathbf{J}_+ - \mathbf{J}_-$, is zero

Answer: $\Delta\phi = \frac{RT}{F} \times \frac{(D_- - D_+)\nabla c}{D_- c_- + D_+ c_+}$

Problem 4

The ionic conductivity of the tetraethylammonium ion in ethanol at infinite dilution is 29.4 S cm^2/mole, at 298 K. Calculate:

a) The friction coefficient of the ion

b) The displacement velocity of the ion under an external electric field of 100 V/m

c) Calculate the ratio between the kinetic energy of the ion due to its migration to the mean kinetic energy in that medium

Data: Ethanol viscosity, at 298 K, is 1.089 mPa.s

Answer: a) $\xi = 5.26 \times 10^{-12}$ N s/m; b) $v = 3.05 \times 10^{-6}$ m/s; c) 4.1×10^{-16}

Problem 5

Calculate the diffusion coefficient of hemoglobin in water at 37 ^{o}C. Hemoglobin is a globular protein whose diameter is 5.5 nm, and the water viscosity at that temperature is 0.694 mPa s.

Answer: $D = 5.95 \times 10^{-7}$ cm^2/s

Problem 6

The permeability of oxygen through polyethylene, at 25 ^{o}C, is 2.2×10^{-13} cm^3 (NCPT) cm^{-1} s^{-1} Pa^{-1}.

A polyethylene membrane with 100 μm in thickness and 10 cm^2 of area, protects the oxidation of a substance packed under vacuum. If the oxidation requires 2×10^{-3} moles of oxygen, for total deterioration, estimate the time required for oxidizing 10 % of the packed substance in contact with air, at 25 ^{o}C (consider that the partial pressure of oxygen in air is 0.2 bar).

Answer: 1,420 hours

Problem 7

The diffusion coefficient of glucose in a 0.02 molal aqueous solution, at 298 K, is 0.673×10^{-10} m^2/s. If the diffusion of glucose occurs through a porous membrane, 1 mm in thickness, that separates a 0.03 molal glucose solution from a 0.01 molal glucose solution, calculate (assuming that the solutions are ideal):

a) The friction coefficient of glucose in water

b) The velocity of the glucose molecule

c) Calculate the ratio between the translational energy of the glucose molecule to the mean kinetic energy in that medium

Answer: a) $\xi = 6.107 \times 10^{-12}$N s/m; b) $v = 6.73 \times 10^{-13}$ m/s; c) 3.3×10^{-23}

Problem 8

In order to avoid the presence of some insects, it is common to attract them sexually by using pheromones. The method consists in placing pheromones sources in such a way to keep them away. To get an efficient and perdurable effect, the delivery of pheromones must be controlled.

Considering that it is necessary to dose 50 mg per day of pheromones, using spherical capsules formed by a membrane 100 μm in thickness, containing inside a saturated[2] pheromone solution, 0.85 mg.dm^3, and knowing that the permeability of the pheromone in the membrane is 1.5×10^{-4} cm^2/s, calculate the needed total area of the capsules.

Answer: 45.3 cm^2

---

[2]An excess of solid drug is used in equilibrium with the solution, to assure a prolonged delivery.

Chapter 11

Future Developments

11.1 Behavior of Matter in Temporal and Spatial Micro- and Nanoscopic Scales

In this section, we will discuss the effect of changes in the time scale and in the spatial scale and how they affect the behavior of the systems. These days physical chemistry is covering the description of material systems in more fields, some of them connected with other disciplines. It is particularly interested in covering smaller scales of time and dimensions of systems and processes, and obtaining information about their behavior. It frequently deals with biomolecules, and also with clusters having a small number of simpler molecules, as well as the reactions occurring in these systems in fractions of picoseconds involving forces of the order of pN.

What can be said about the behavior of systems under those conditions? Is it possible to use the same relations we have obtained and applied to macroscopic systems in the different chapters of this book?

We will consider first the Brownian movement that affects the molecules in fluids. Experiments show that this phenomenon can be observed by the random movement imposed to larger macroscopic particles suspended in a fluid, like particles of pollen having a size between a tens or some hundreds of micrometers. The name given to this phenomenon relays in R. Brown, who observed it for the first time with a microscope. If a grain of pollen is suspended in water, we see that it is animated by a random movement that in seconds will displace the pollen particle in a given direction, then in another direction, and so on, in successive moments. This occurs in spite of the fact that the fluid is in an equilibrium state and that no external forces are applied on the system.

What happens when a small force is applied to the same pollen grain? In that case, we would say that a spontaneous process will occur, and the grain will adopt a given direction when that force is applied.

In 1905, A. Einstein wrote in the introduction of his first article about Brownian motion: "If the movement described here can be observed, then the classical thermodynamic cannot be considered applicable with precision, not even to (those) bodies of dimension which can be distinguished with a microscope[1] ..." This statement refers to the fact that in a macroscopic system at equilibrium no changes are observed with time, but when they are observed in a microscopic scale and during short time intervals, it is confirmed that the particles that constitute the system are in constant random motion. Einstein's observation was, at the beginning of the twentieth century an interesting academic question; nowadays that observation has become also a practical question due to the increasing interest on micro- and nano-devices in numerous practical applications, as well as to understand the behavior of matter at small scales of time and space. This requires a careful consideration of the capacity of the relations presented throughout this book to describe these very small scales of time and size.

In this chapter, a few articles will be presented and discussed, and all of them focalize their interest in those space and time scales. Although they are only examples, they indicate clearly an area of future development, which will very probably be central for studies on matter and molecules in the near future.

Many relevant processes occurring at cellular and supramolecular levels, have been used to operate devices and sensors. The use of modified electrodes, involving for example reactions occurring at the protein' membranes, which are sensitive to the presence of electric fields, formation of crystal nuclei, or processing of materials at microscopic level, are topics of increasing importance. All of them are related to phenomena that occur at very small spatial and time scales.

All these processes take place at finite rates and, traditionally, their description would require a detailed knowledge of the system's features of the environment, and of the rates of the processes. This differs from the equilibrium states, which are completely defined by the values of some thermodynamic variables. Consequently, the possibility of relating equilibrium thermodynamic functions with the observations made on systems that are not in equilibrium, is very important.

When a system is in equilibrium, will the fluctuation of the properties' values, originated in the Brownian motion, force the system to evolve in one direction or in the opposite one with equal probability, according to the principle of microreversibility discussed in chapter 10? When an external small force is applied to the system, it will tend to evolve in the direction of the applied force, but also it will have a

---

[1]A. Einstein, "Investigations on the Theory of the Brownian Movement". New York: Dover Publications, 1956.

small probability, small but not zero, to evolve in the opposite direction. Hence, if the force is small and it is applied for a short time, there is a chance that the process evolves in a different direction to that of the spontaneous one. The probability of moving along a trajectory against that associated to the spontaneous evolution cannot be neglected in many cases, as we will see. For doing this, the system will consume entropy (entropy reduction). When the work produced by an engine undergoing a cyclic path is of the order of kT (per degree of freedom), it can be expected that the engine has the possibility of working in a way that consumes entropy, for a small period. Thus, thermal energy from the environment can be transformed into usable work, resulting in the engine working in opposite direction to that of the spontaneous evolution. If, on a macroscopic scale, these type of situations were to occur, they would violate the second principle of thermodynamics, because it would imply that entropy is consumed during the process, and the entropy would change in a negative direction. It has always resulted a paradox to accept the principle of microreversibility (chapter 10) and, at the same time, the fact that the spontaneous evolution of macroscopic systems has a unique direction: That in which systems evolve spontaneously producing entropy, and that this is not reversible.

In the years close to the twenty-first century, the study of fluctuations happening in fluid systems, like those responsible of the Brownian motion, have allowed the derivation of relations with which it is possible to calculate the probability of process taking place in the opposite direction than that corresponding to the spontaneous evolution (i.e., decrease of the overall entropy production), as it is required by the second principle of thermodynamics. Those relations indicate that the violations of the second principle are measurable and appreciable in small systems, and for short time intervals. A relation exists between the ratio of both probabilities of entropy production (P): The spontaneous trajectories that will be positive $(P_+ > 0$, spontaneous processes), with respect to those which consume entropy $(P_- < 0$, non-spontaneous processes). A frequently used equation relating the two types of trajectories makes use of the exponential of the ratio of P_+ to P_- (i.e., the ratio of the entropy produced by the conventional trajectories and the opposite ones). Since entropy is an extensive property, as the size of the system grows, the trajectories leading to a decrease in entropy will be less probable. The same happens when the observation time increases. In both situations, the systems return to a behavior that agrees with the second principle; in the same fashion that Brownian motion leads to equilibrium when the system's extension and/or the time of observation increases.

Recently,[2] the validity of this relation, using colloidal particles having a diameter

---

[2]G. M. Wang, E. M. Sevick, E. Mittag, D. J. Searles, and D. J. Evans , "Experimental Demonstration of Violations of the Second Law of Thermodynamics for Small Systems and Short Time Scales", *Phys. Rev. Lett.*, 89 (2002): 050601.

of 6.3 μm suspended on water, was verified experimentally. A particle was pushed be means of optical tweezers,[3] which exerted over the particle a force between 1.0 and 0.01 pN. Figure 11.1 is a schematic representation of the fact that, when the force was small, ($\mathbf{F}_1$), some of its displacements are in the opposite direction to $\mathbf{F}_1$. Instead, when the force increases to $\mathbf{F}_2$ values, there are almost no displacement in a direction different from that of $\mathbf{F}_2$.

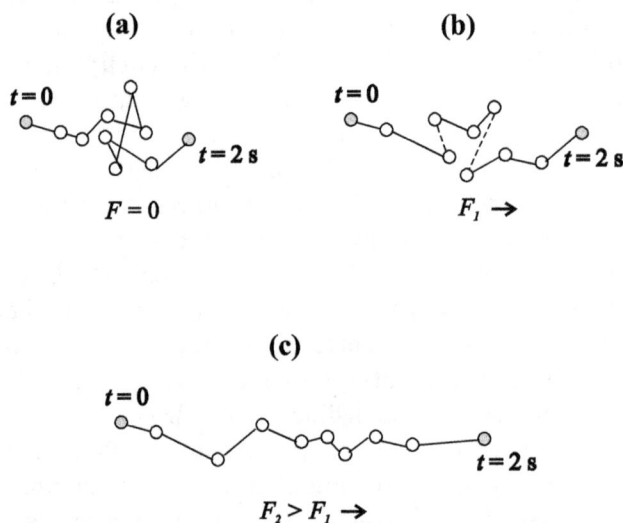

Figure 11.1: Scheme of molecular trajectories in a 2 s interval when applying a force of value: (a) zero, (b) small force, and (c) appreciable force. In (b), the dashed line indicates a displacement in a contrary direction to the applied force.

In this study, different trajectories of micrometer particles were found where entropy was consumed, when the observation time intervals were in the order of seconds. These results are important to design and use nanomechanical and molecular devices. The important conclusion is that the smaller the devices and the shorter their displacements, the larger the probability that they may work contrary to what is established according to the second principle applied to macroscopic systems. Nevertheless, a relation exists that links the probability of non-spontaneous trajectories with the equilibrium thermodynamics quantities, in this case ΔS.

Figure 11.2 shows histograms of the results of two experiments, one for 0.01 s and the other for 2 s duration. In the first case, the larger quantity of events are those that

---

[3]Optical tweezers are described in detail by A. Ashkin, *Proc. Nat. Acad. Science*, 94 (1997): 4853-4860.

Number of trajectories

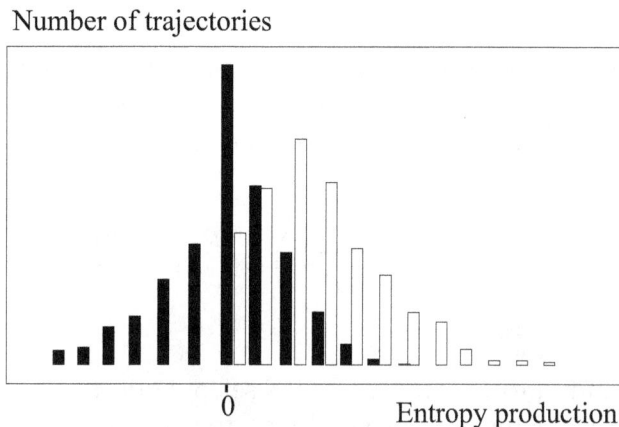

Figure 11.2: Histograms of the number of trajectories as a function of the production of entropy. Black bars, 0.01 s experiment; white bars, 2 s experiment.

produce no entropy change; the other events, in a larger number, produce changes in entropy (either producing or consuming entropy). The histogram corresponding to the 2 s observation shows no trajectory that consumes entropy.

There is another very useful relationship connecting non-reversible events with equilibrium quantities, derived by C. Jarsisnsky.[4] Like in the first example, this expression is applicable to microscopic systems and processes taking short times to occur. When a system evolves between the states A and B by following a *quasi-static* isothermal trajectory,[5] which indicates that the reversible isothermal work W is equal to the change of Helmholtz energy necessary to take the system from A to B [see eqn. (1.22)], according to

$$W = \Delta A_{\mathrm{A} \to \mathrm{B}} \qquad (11.1)$$

if the process occurs at a finite rate, over an irreversible path (i.e., when the system does not take a quasi-static path), the work done over (by) the system will be larger (smaller) than $\Delta A_{\mathrm{A} \to \mathrm{B}}$, then

$$W \geq \Delta A_{\mathrm{A} \to \mathrm{B}} \qquad (11.2)$$

---

[4]C. Jarzynski, "Non-equilibrium Equality for Free Energy Differences", *Phys. Rev. Lett.*, 78 (1997): 2690-2693.

[5]Strictly, no process can be carried out through equilibrium states. Nevertheless, a process can be accomplished by separating it slightly from the equilibrium system state. This type of trajectory is denoted quasi-static.

Jarzynski derived a relation between $\Delta A_{A \to B}$ and the work necessary to take the system from A to B at a finite rate, that is, when the process its not carried out in a quasi-static path. In that case, the relation is

$$\exp\left(-\frac{\Delta A_{A \to B}}{kT}\right) = \left\langle \exp\left(-\frac{W}{kT}\right) \right\rangle \tag{11.3}$$

where $\langle (...) \rangle$ indicates an average over a set of trajectories. This means that the experiment must be repeated several times, to take the system from the initial state to the final one by different paths, and this implies a change in the speed with which the process is carried out. This relation tells us that it is possible to obtain information about the value of the quantity $\Delta A_{A \to B}$ from a set of measurements, corresponding to the process out of equilibrium, and that this is independent of the particular trajectory that takes A to B and of the speed with which the variables that determine the state of the system change.

Studies have been published showing that the expression given in eqn. (11.3) is correct. Bustamante and colleagues[6] measured the work necessary to fold and unfold an RNA molecule (as if it were a spring); for that purpose, they bound the ends of the RNA molecule to two spheres of polystyrene having diameters of 2 to 3 μm. One of the spheres was attached to a delicate piezoelectric dinamometer, and the other was kept fixed at a given position with an optical tweezer. The RNA molecule can be unfolded by increasing the distance between the two spheres, using the optical tweezer (as depicted in figure 11.3, when $F \neq 0$), and making again the RNA molecule fold back (i.e., decreasing the distance between the two spheres). This situation is indicated in figure 11.3 by $F = 0$. The result of measuring this very delicate elastic work, done over different trajectories and with different forces, was averaged according to eqn. (11.3), and the authors verified the correctness of the Jarzynski relation. Moreover, the forces applied in this case could not be larger that kT, not because eqn. (11.3) is limited to that type of work responsible for the energy exchange, but because the number of runs that need to be performed to obtain a sufficiently precise average increases significantly with the amount of energy being exchanged.

---

[6] J. Liphardt, S. Dumont, S. B. Smith, I. Tinoco Jr., and C. Bustamante, "Equilibrium Information from Non-equilibrium Measurements in an Experimental Test of Jarzynski's Equality", *Science*, 296 (2002): 1832-1835.

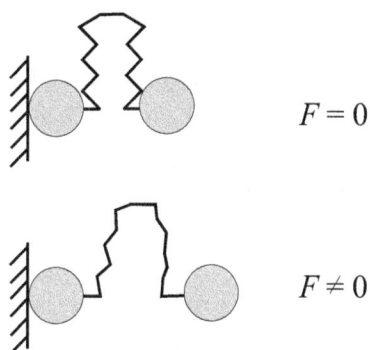

Figure 11.3: Folding and unfolding of an RNA molecule.

Figure 11.4 shows unfolding (U) and refolding (R) trajectories of the RNA molecule involving different forces. When the rate of the stretching force change is between 2 and 5 pN/s, the processes U and R are reversible. When the speed is larger, the process becomes irreversible (see that, for 52 pN/s the trajectory marked U in the figure is different from that marked with R).

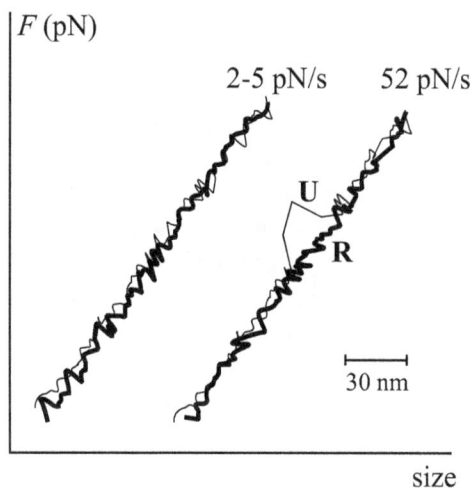

Figure 11.4: Force applied to unfold and fold an RNA molecule using two different rates for changing the force.

The studies discussed so far in this chapter illustrate clearly that when using biomimetic systems, in which the relevant time and space scales are small, the results obtained can be different to that expected for macroscopic systems. The possibility that the system may evolve in a direction that consumes entropy cannot be neglected, and the probability of this to occur can be small, but not zero. On the other hand, the possibility of obtaining equilibrium values of thermodynamic quantities from work being carried out through non-quasi-static paths is very valuable. Also it is important to be able to predict the behavior (or the trajectory) of these small systems on the basis of equilibrium thermodynamic quantities.

11.2   Behavior of Systems Which Follow the Laws of Thermodynamics When Symmetry is Broken

This second section of chapter 11 will be dedicated to presenting processes occurring in two systems that seem to defy the second principle of thermodynamics, but are within the realm of thermodynamics. We'll give a short qualitative account of the evaluation of the systems' evolution and point out how these apparent abnormalities are really due to a change in the systems' symmetry (this term covers also problems of rugosity), which again are very important issues for the present and near-future application of thermodynamics. The detailed explanations are given in the original references, which will be given.

First, it is important to mention Curie's postulate, which establishes that scalar systems cannot produce vectorial or tensorial fluxes. This is easily demonstrated by using Onsager's reciprocity postulate, presented in chapter 10, section 10.5. Let us assume that in a homogeneous medium, it could be a temperature gradient and also a chemical reaction going on. Now, we display the linear flux equations, as described in chapter 10, and then we assume that the system is at constant temperature (i.e., the chemical reaction proceeds isothermally). Under these conditions, it is not possible for this reaction to produce a flux, which is a vectorial quantity. We have already analyzed a system of that kind in chapter 1 when presenting the active transport which, as shown, generates a flux of Na^+ ions starting from the same concentration of these atoms in both compartments. There, the symmetry is broken by the hydrophobic membrane containing monensine that separates both compartments. Thus, since the system's symmetry is changed from two homogeneous bulk phases to a situation where the separating membrane has a preference for the sodium ion, its concentration change in one of the compartments is possible as the spontaneous process. Now, we will look into two more complex systems.

Drops of Hydrophobic Liquids Find Their Way to Exit From a Maze

The process in which hydrophobic liquid drops move in a maze containing an alkaline aqueous solution has been analyzed in detail in a recent article by Lagzi et al.[7] The drops have a volume of approximately 1 mm^3 and they are either motor oil (MO) drops or CH_2Cl_2, both hydrophobic liquids; the maze was prepared by photolithography and filled with 240 mm^3 of KOH (pH $\approx$ 12.0). At the exit of the maze there is a gel of agarose soaked in HCl (pH $\approx$ 1.2), and the width of the maze's channels are 1.4 mm. The whole system is kept at constant ambient temperature. If a drop of one of the hydrophobic liquids is added at the entrance, nothing happens, but upon addition of 2-hexyldecanoic acid (HAD, a surfactant) to the drops, they start moving from the entrance of the maze toward its exit (cf. figure 11.5).

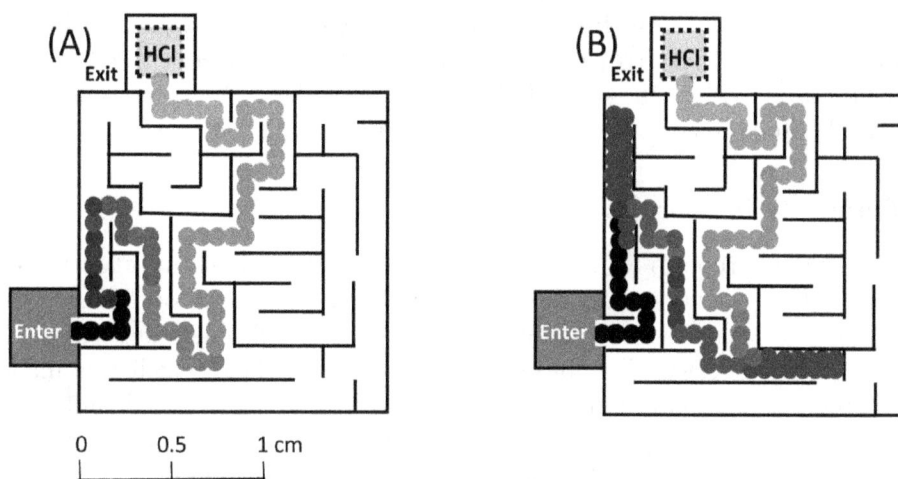

Figure 11.5: Hydrophobic (A, motor oil; B, CH_2Cl_2) drops in a maze. This figure is an adaptation of the figure 1 in the original paper. Black paths, initial movement as the drops go toward the exit; pink paths, successful movements to exit the maze.

This occurs due to a break of symmetry induced by HDA, since this molecule is ionized at the back of the drop (alkaline region) and protonated at the front (acid region). The protonated HDA has a much larger surface tension than the ionic base, and this produces an interfacial tension (air-liquid) gradient that breaks the isotropic symmetry of the drops. Consequently, it is the interfacial tension gradient that is responsible to draw the drops toward the maze's exit. Figure 11.5 is a schematic drawing of the paths followed by the drops (interesting videos were provided by

---

[7]I. Lagzi, S. Soh, P. J. Wesson. K. P. Browne, and B. A. Grzybowski, "Maze Solving by Chemotactic Droplets", *J. Amer. Chem. Soc.*, **132** (2010): 1198-1199.

the authors in the supporting information to their paper). Now, when the drop is formed by motor oil, (A) in the figure, moves more slowly (1 mm/s) than when constituted by CH_2Cl_2, (B) in the figure, having a larger velocity (10 mm/s). Since the latter drop moves faster, the drop's inertia takes the drop not always by the shortest path toward the exit, although it will finally hit the exit. In the case of MO drops, they advance more slowly and go directly to the exit. This illustrates some of our discussions regarding to the memory of the previous position (non-Markovian path). The fuel for these phenomenona to occur is the neutralization of the HDA.

The interpretation of the collected evidence is schematically shown in figure 11.5. In A) the MO drop moves directly to the end of the maze (lower speed), and in B) the CH_2Cl_2 drop moves faster, eventually missing the path toward the acid exit, to finally take the right trajectory, shown on the right panel of the figure. The effect of inertia is evident in some of the snapshots, where the larger momentum of the CH_2Cl_2 drops make it choose the wrong path.

Trampolining Drops

This example reports a short account of the paper by Schtzuis et al.,[8] where all the details (they are very many details about the process) have been carefully described by the authors. The system consists of 2 mm water droplets deposited over a rigid, hydrophobic surface. The basis of this surface is silica, which was made strongly hydrophobic by chemical fluorination. It results in a highly rough and heterogeneous surface (cf. inset in figure 11.6), having a hierarchical structure from inside out, upon which water drops were deposited. The question of symmetry is in this case very clearly broken and complicated.

The system is at a constant temperature, and the vapor pressure of water in the environment is about 0.01 bar. Now, when a water droplet is over this hierarchically heterogeneous surface, which means that the it sits over the narrower channels on top of the surface (cf. inset in panel A of figure 11.6), they are capped by the droplet without allowing the water vapor to escape from under it. The drop's bottom, capping the top smaller channels on the surface of the rigid structure, will have a vapor pressure that exceeds that in the enclosed system, and it will emit a jet of water vapor toward the surface. This will levitate the drop from the surface and make it jump, due to an increase in the mechanical momentum of the drop. The authors have shown that the force of the jet is larger than the adhesive force between drops and the solid material. When the droplet falls down, it is observed that it rebounds on the surface, even increasing the height of its following jumps (as seen in panel

[8]T. M. Schutzius, S. Jung, T. Maitra, G. Graeber, M. Köhme, D. Poulikakos, "Spontaneous Droplet Trampolining on Rigid Superhydrophobic Surfaces", *Nature*, 527 (2015): 82-85.

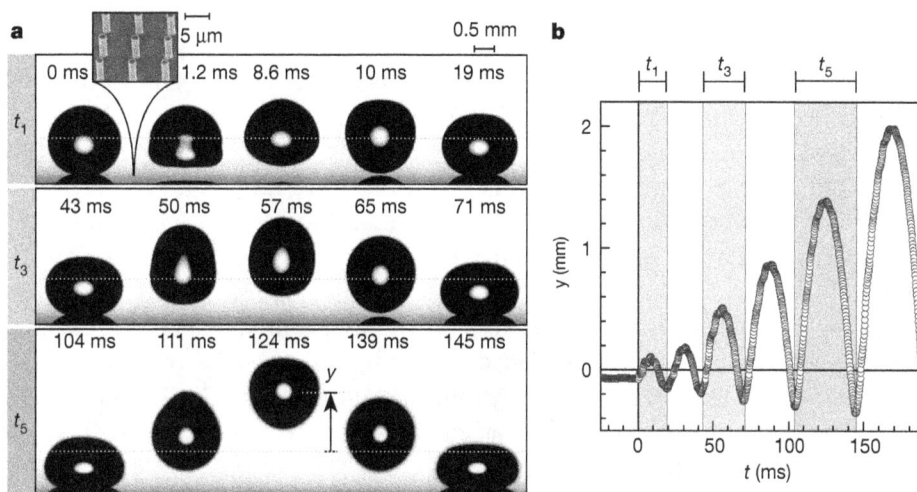

Figure 11.6: Trampolining drops. See footnote 8 on page 330.

B of figure 11.6). At the same time, the drop suffers various phase changes which provide the energy of the process to occur (there are changes in temperature that even induce crystallization of water inside the drop). Summarizing the experiment, in the small capillaries at the top of the surface, the drop evaporates, projecting a jet of vapor toward the bottom of the surface (into the channels) that pulls it up and also cools it down. Where is the energy necessary for this increase coming from, and how can this behavior be explained?

Clearly, the second principle of thermodynamics is not violated in the experiment because part of the water of the drop is evaporated (consuming an amount of the enthalpy of evaporation) and there are recalescent processes[9] that also contribute to provide the necessary mechanical momentum for the drop's jump.

The authors have described the various aspects of how the surface structure influences the trampolining effect. They are related to the narrowness of the space between the columns of material in the outer surface, and the concomitant effects due to it; also, the elastic energy when rebounding is transformed in gravitational energy that helps further levitation. Under the experimental conditions, the drop moves up and down, rebounding several times. This process goes on until there is

---

[9]Increase in temperature when the rate of heat liberation during a phase transformation exceeds the rate of heat dissipation required to cool down the system. Since in this case liquid and solid water are both dense phases, there is very little change of entropy, and the whole process is due to enthalpic changes (in this case, solidification of ice is an exothermic process that increases the temperature and the momentum of the droplet).

no more water on the surface (cf. figure 11.6). The length of the bouncing may be altered by changing the features of the hierarchical surface, as the authors discuss.

This phenomenon is related to that described in 1756 by Leidenfrost: Small water drops falling on a hot metal plate do not evaporate immediately; rather, they start a jerky movement over the metal plate until they evaporate (cf. figure 11.6). This is due to the formation of a vapor cushion under the drops, which isolate them from the plate.

When the water droplet rebounces on the hydrophobic surface, reaching an even greater height upon levitation and due to its resilience, the resulting elastic energy is converted in an increase of the gravitational energy. Consequently, the process described does not violate the second principle of thermodynamics. The process will continue until the droplet fully evaporates.

All the examples discussed in this final chapter show the importance of handling thermodynamics for systems and processes having different space and temporal scales as the usual ones in a lab, and also the great influence of the symmetry of the system/process to describe the outcome of many phenomena.

Appendix A

Mathematical Tools

In this appendix, we give more elements for using the mathematical properties of the thermodynamic functions and obtain, in this way, many relations among different properties which are of importance. In section 1.3, we have included an abridged presentation of that aspect but, in order to avoid discontinuities in the discussion of the systems' behavior, we prefer to leave a more detailed presentation of some mathematical consequences, to be described in the appendix. It is only necessary to remember that the differentials of the thermodynamic functions are exact, which implies that their integrals do not depend on the particular path for a process, but only the limits of the integration, which are given by the initial and the final states of a process. This means that for $z = f(x, y)$ we have

$$\mathrm{d}z = \left(\frac{\partial z}{\partial x}\right)_y \mathrm{d}x + \left(\frac{\partial z}{\partial y}\right)_x \mathrm{d}y \tag{A.1}$$

Also, it should be remembered that the second cross derivatives are equal

$$\frac{\partial}{\partial x}\left(\frac{\partial z}{\partial y}\right) = \frac{\partial}{\partial y}\left(\frac{\partial z}{\partial x}\right) \tag{A.2}$$

which generate the known Maxwell's relations. This presentation is based essentially on thermodynamic transformations of a system of a single component, when there is only mechanical work (volume changes). Nevertheless, the material given is easily extended to systems having more thermodynamic degrees of freedom, either because they have more than one component, or because the process includes other types of work than the mechanical one. At the end of this appendix, we give an example of the mathematical way this type of problem can be handled.

Using the definitions of the Helmholtz and Gibbs functions,

$$A = U - TS, \qquad G = H - TS \tag{A.3}$$

333

for reversible processes that only involve volume work, dA and dG are given in eqns. (1.10) and (1.11). From these expressions, it is clear that the entropy is related with the temperature derivatives of both thermodynamic quantities, but following different paths. Thus, they yield

$$S = -\left(\frac{\partial A}{\partial T}\right)_V = -\left(\frac{\partial G}{\partial T}\right)_p \tag{A.4}$$

In eqns. (1.16) and (1.17), the relations between heat capacities with U and H (using the first principle of thermodynamics) are given. If the reader starts with eqns. (A.3) and calculates the second derivatives with respect to temperature, at constant volume and pressure, respectively, and takes into account eqn. (A.4), it results that the heat capacities at constant volume and at constant pressure are related to entropy variations with T, given by

$$C_V = T\left(\frac{\partial S}{\partial T}\right)_V \qquad C_p = T\left(\frac{\partial S}{\partial T}\right)_p \tag{A.5}$$

Other very useful relations to which one frequently recurs in this book are those relating U and H, respectively, with the derivatives of A and G with respect of temperature. We will deduce the second relation, and the reader should do it with the first one. The task implies the calculation of the derivative

$$\left(\frac{\partial G/T}{\partial T}\right)_p = \left[\frac{\partial}{\partial T}\left(\frac{H - TS}{T}\right)\right]_p$$

Using eqns. (1.17) and (A.5) one gets

$$\left(\frac{\partial G/T}{\partial T}\right)_p = -\frac{H}{T^2} \tag{A.6}$$

and also,

$$\left(\frac{\partial G/T}{\partial 1/T}\right)_p = H \tag{A.7}$$

On the other hand, the application of the Maxwell's relations to the second derivatives of A and G, obtained starting from eqns. (1.10) and (1.11), yields the following interesting relations

$$\left(\frac{\partial p}{\partial T}\right)_V = \left(\frac{\partial S}{\partial V}\right)_T \qquad \text{and} \qquad \left(\frac{\partial V}{\partial T}\right)_p = -\left(\frac{\partial S}{\partial p}\right)_T \tag{A.8}$$

which are very useful.

It is also interesting to find the Maxwell relation resulting from dU (cf. eqn. (1.6)) for a system of one component

$$\left(\frac{\partial T}{\partial V}\right)_S = -\left(\frac{\partial p}{\partial S}\right)_V \tag{A.9}$$

and compare it with the similar relation found to the right of eqn. (A.8); try to relate the adiabatic thermal expansion with the isobaric one, and discuss the meaning of the difference.

All the operations that have been described are simple ones, and can be resolved using a mnemonic scheme presented in the following frame.

Mnemonic Scheme to Calculate the Derivatives of the Derivatives of Thermodynamic Fundamental Functions

One way to summarize briefly the thermodynamic relations for one-component systems, where only mechanical work is done, can be carried out using the quadrilateral that follows (cf. Callen's book, among the suggested books for consultation). It consists in obtaining them in the following manner:

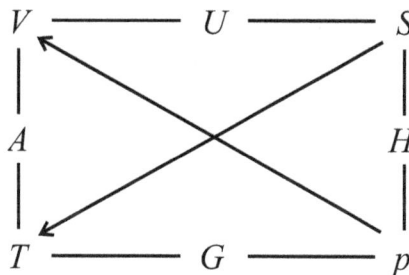

On the middle of each side of the quadrilateral, the fundamental thermodynamic functions are written; the two canonical variables on which they depend are written on the vertices at each side of the quadrilateral, embracing the corresponding fundamental property–for example, $G(T, p)$ or $H(S, p)$. The procedure to calculate a derivative (e.g. $(\partial G/\partial T)_p$) is as follows:

- We move from the thermodynamic function we want to differentiate (G) toward the differentiation variable (T).

- The variable located at the opposite vertex, on the same side, (p) remains constant.

- The derivative will be equal to the property found on the opposite side of the corresponding diagonal (S).

- If the arrow on the diagonal points in the direction of the path that has been followed, the sign of the derivative will be positive; instead, if the arrow points in the opposite sense, the derivative will be negative (in this example, $(\partial G/\partial T)_p = -S$).

In order to calculate a Maxwell relation, using as example $(\partial T/\partial V)_p$, we follow the following procedure:

- The Maxwell relation connects i) the derivative of one variable placed in a vertex (T) with the variable that lies on the other vertex of the same side of the quadrilateral (V), keeping constant the variable on the diagonal (p) (that is, $(\partial T/\partial V)_p$), with ii) the derivative on the opposite side of the quadrilateral (i.e., $(\partial p/\partial S)_T$).

- Both derivatives are equal when the two arrows point toward the derivatives, otherwise the two derivatives are equal but one of them has a negative sign, as in the present case where $(\partial T/\partial V)_p = -(\partial p/\partial S)_T$.

What we have illustrated for the case of fundamental functions that only depend on two variables (i.e., for pure components with a single kind of work), it could be extended to any number of variables. However, due to the difficulty of the representation in more than two dimensions, it is not recommended to use this scheme in such cases, except when one is only interested in the variation of two variables, even having a multicomponent system or more than one type of work.

In chapter 1, we used the so-called equations of state, eqns. (1.14) and (1.15), and it was possible to obtain the first derivative of these. The reader can derive the second equation of state that relates the variables (p, V, T) with the change of enthalpy with pressure, at constant temperature. These two thermodynamic equations of state were not deduced using a particular molecular model, which highlights their importance and practical value (i.e., they can be applied to any molecular model that one desires to use). As an example, let us consider a gas where the molecules have an attractive intermolecular effect given by a constant value, corresponding to the average potential energy between molecules. Under this assumption, the energy of every molecule will be proportional to the number of particles present, per unit volume in the gas (i.e., to the density $\rho = n/V$, where n is the number of moles contained in the volume V). The average energy of interaction will be equal to the excess internal energy, that is the total internal energy minus that of the system considered ideal, $U^{ex} = U - U_{id} = a\rho$. Now, calculating the change of U with the molar volume, we get the expression

$$\left(\frac{\partial U}{\partial V_m}\right)_T = -\frac{a}{V_m^2}$$

because the ideal part of U only depends on temperature. Hence, using eqn. (1.14), we get

$$p = -\frac{a}{V_m^2} + \frac{RT}{V_m - b}$$

The last term of the rhs member was calculated assuming that $(\partial p/\partial T)_V$ is that of an ideal gas, but the free volume that may be occupied by molecules excludes the volume they already occupy, b, usually denoted as co-volume. The previous equation of state is the well-known van der Waals equation, where we have denoted the molar volume as $V_m = (V/n)$, to avoid confusion.

One property obeyed in general by every thermodynamic function is that relating the partial derivative with respect to the different variable. Assume, for instance, the thermodynamic function $z = f(x, y)$, the total differential of which, dz, is given by eqn. (A.1); making a process at constant z, that is $dz = 0$, we have

$$\left(\frac{\partial z}{\partial x}\right)_y (dx)_z + \left(\frac{\partial z}{\partial y}\right)_x (dy)_z = 0$$

In this expression, we have marked that the process is at $z = $ constant, by adding it as a subscript. Arranging the previous expression, the following equation is obtained

$$\left(\frac{\partial z}{\partial x}\right)_y \left(\frac{\partial y}{\partial z}\right)_x \left(\frac{\partial x}{\partial y}\right)_z = -1 \tag{A.10}$$

This relation and its analogous ones for a larger number of variables are very useful.

It is convenient to illustrate the relation between the variations of some thermodynamic quantities when different paths are followed. We want to calculate the change of the internal energy with temperature when the process is: i) isocoric (at constant volume), ii) isobaric (at constant pressure), and iii) keeping constant the phase coexistence (a process that will be denoted by the subindex σ). The first relation is given by eqn. (1.16), and in this case the derivative is C_V. The case (ii) is obtained starting from

$$dU = \left(\frac{\partial U}{\partial T}\right)_V dT + \left(\frac{\partial U}{\partial V}\right)_T dV$$

and calculating how U varies with T, at constant pressure, and by differentiation we get

$$\left(\frac{\partial U}{\partial T}\right)_p = \left(\frac{\partial U}{\partial T}\right)_V + \left(\frac{\partial U}{\partial V}\right)_T \left(\frac{\partial V}{\partial T}\right)_p$$

Finally, taking into account the first thermodynamic equation state, eqn. (1.14), we obtain the desired relation

$$\left(\frac{\partial U}{\partial T}\right)_p = C_V + \left(\frac{\partial V}{\partial T}\right)_p \left[T\left(\frac{\partial p}{\partial T}\right)_V - p\right]$$

The reader will be able to deduce the case (iii), remembering that $V = f(p, T)$; the final relation is

$$\left(\frac{\partial U}{\partial T}\right)_\sigma = C_V + \left[T\left(\frac{\partial p}{\partial T}\right)_V - p\right]\left[\left(\frac{\partial V}{\partial T}\right)_p + \left(\frac{\partial V}{\partial p}\right)_T \left(\frac{\partial p}{\partial T}\right)_\sigma\right]$$

The expression being used to define the functions H, A, and G from U are very useful, because in the new functions the independent variables of U are exchanged by their conjugated ones [see eqns. (1.9), (1.10), and (1.11)]. For example, while U has as independent variables S and V, H has S and p, where V and p are conjugated variables. The mathematical procedure employed to do this is known as Legendre transformation, and is used at any time that, due to the nature of the studied problem, a change of independent variable is more convenient. It is necessary to make clear that the Legendre transformation maintains all the information about the behavior of the studied system. For instance, if one wishes to describe the evolution of a system susceptible of mechanical work and also of elastic work, it seems preferable to operate using a function having as independent variables T and p. But which variable, the tension J or the length L, is more convenient to express the elastic work $J dL$? If we were interested in the evolution of a system formed by a metallic rod with a fixed length, it would be convenient to use L as independent variable. On the other hand, if the rod has a free end from which hangs a mass applying a determined tension, its evolution will be better described using J as the independent variable. In the first case, $(G)_{T,p,J}$ will be ≤ 0, and in the second case, $(dG^*) \leq 0$, where $G^* = G + JL$.

The Velocity of Propagation of Sound and Thermodynamic Propeties

Sound is produced by an acoustic wave transmitted through a given medium: Water, air, a metal, etc. The speed with which it propagates may look essentially as a kinetic property of the medium, but it is a thermodynamic property. In order to discuss this point in the simpler way, we will refer to the velocity of sound as it travels through a gaseous medium. The acoustic wave is sinusoidal, and it is produced by a periodic change of pressure in the medium, with regions at a larger pressure (maxima in the wave ondulatory movement) and regions having a lower pressure (valleys). Then, its propagation depends on how this periodic variation of pressure evolves with time, and this will depend on how the pressure changes when the gas density varies, in adiabatic conditions. As a consequence of the sound perturbation, the temperature of the system will moderately oscillate around its mean value. It is accepted that the pressure or the temperature oscillations are very small; for instance, the intensity of normal sound in air produces a change in pressure of ca. 0.02 Pa, and when the pressure variation reaches 200 Pa $= 2$ mbar, it becomes harmful.

The propagation of sound (u) is given by the equation

$$u^2 = \left(\frac{\partial p}{\partial \rho}\right)_S$$

The greater the change in pressure for a given change of density, the faster the sound signal will be propagated. The previous expression can also be written in terms of the adiabatic compressibility, $\kappa_S = (\partial \rho/\partial p)_S/\rho$, being

$$u^2 = (\rho\,\kappa_S)^{-1}$$

How is κ_S related with the isothermal compressibility κ_T? This is a good example to demonstrate the applications of the mathematical tools presented in this appendix. The gas density and its entropy will be functions of p and T, and their differentials are written as in the general eqn. (A.1). If we want to know how the density ρ varies with pressure, under adiabatic conditions, we have

$$\left(\frac{\partial \rho}{\partial p}\right)_S = \left(\frac{\partial \rho}{\partial p}\right)_T + \left(\frac{\partial \rho}{\partial T}\right)_p \left(\frac{\partial T}{\partial p}\right)_S \qquad (A.11)$$

The expression equivalent to eqn. (A.1) for $S(p,T)$ lets us obtain the equivalent expression to eqn. (A.10) for an adiabatic path (i.e., at constant S), given by

$$\left(\frac{\partial T}{\partial p}\right)_S = -\left(\frac{\partial S}{\partial p}\right)_T \Big/ \left(\frac{\partial S}{\partial T}\right)_p$$

Using this expression and one of the Maxwell's relations given in eqn. (A.8), we get

$$\left(\frac{\partial T}{\partial p}\right)_S = \left(\frac{\partial V}{\partial T}\right)_p \frac{T}{C_p}$$

Replacing it in the expression of the adiabatic change of density with pressure, eqn. (A.11), and remembering that $V = 1/\rho$, we arrive to

$$\left(\frac{\partial \rho}{\partial p}\right)_S = \left(\frac{\partial \rho}{\partial p}\right)_T - \left(\frac{\partial \rho}{\partial T}\right)_p^2 \frac{T}{C_p \rho^2}$$

Moreover, if one uses the equation given for $(C_p - C_V)$, we finally get the relation

$$\kappa_S = \kappa_T \frac{C_V}{C_p}$$

an expression that can be derived by the reader.

This example can be resolved in a more direct manner using jacobians (another very useful mathematical tool), but this method will not be described here; the interested reader can consult Callen's book for learning to work with them.

Appendix B

Tabulation and Calculation of Standard Thermodynamic Quantities

Only one of the fundamental thermodynamic functions, the entropy S, can be determined in an absolute way on the basis of the third principle of thermodynamics (cf. chapter 2). The other functions, namely U, H, A, and G, express energies, and hence it is only possible to determined differences of these properties and not their absolute values. Due to this reason, it is necessary to adopt reference values for all these functions of the substances in their standard state. This conventions are very important in the thermodynamic description of chemical reactions. It should be remembered that, as shown in this book, standard quantities only depend on temperature.

The reaction of formation of one substance in the standard state refers to the chemical process that produces such substance, starting from the elements that compose it, in their most stable aggregation state at $T = 298.2$ K and $p = 0.1$ MPa (which is equal to 1 bar). That is, the conventional base of all modern tables of thermodynamic properties, and it is necessary to have this point clear in order to use the tables in all calculations of chemical processes. In order to calculate changes in the standard thermodynamic properties due to changes in the temperature, it is necessary to know the heat capacities of the intervening substances in the given process or reaction. The heat capacities, like the entropy, can be determined in an absolute manner (cf. chapter 2); the more important relations involving the heat capacities are now summarized

$$C_V = \left(\frac{\partial U}{\partial T}\right)_V = T\left(\frac{\partial S}{\partial T}\right)_V$$

and

$$C_p = \left(\frac{\partial H}{\partial T}\right)_p = T\left(\frac{\partial S}{\partial T}\right)_p$$

When a given substance changes its aggregation state when temperature changes, it is necessary to add to the value of the property one wishes to calculate, the contribution due to the change of the state of aggregation of the substance. It is useful to describe it by means of an example. We will use the reaction

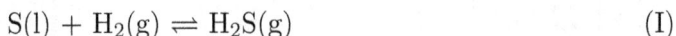

$$S(l) + H_2(g) \rightleftharpoons H_2S(g) \tag{I}$$

Let us assume that we want to calculate the values of the thermodynamic properties associated to the reaction (I), at $T = 473.2$ K and $p = 0.1$ MPa $= 1$ bar. As said, the data summarized in tables in general refers to a different temperature ($T^{\ominus} = 298.2$ K) and the same pressure.

The change of reaction enthalpy for (I) is given by

$$\Delta_r H^{\ominus}(T) = \Delta_f H^{\ominus}(H_2S, g, T) - \Delta_f H^{\ominus}(S, l, T) - \Delta_f H^{\ominus}(H_2, g, T) \tag{B.1}$$

where $\Delta_f H^{\ominus}$ are the enthalpies of formation of the substances, which are calculated on the basis of the tabulated values referring to the enthalpies of formation of the stable aggregation state, at $T^{\ominus}$, of the three substances involved in the process (I). That is, gas for H_2 and H_2S, and rhombic crystal for S. For the substances that do not undergo changes in their state of aggregation between $T^{\ominus}$ and T, we get

$$\Delta_f H^{\ominus}(T) = \Delta_f H^{\ominus}(T^{\ominus}) + \int_{T^{\ominus}}^{T} C_p(g)\mathrm{d}T$$

However, the sulfur is a rhombic crystal at $T^{\ominus}$, and undergoes a transition to a monoclinic crystal at $T_{tr1} = 386.1$ K, and a transition to the liquid phase at $T_{tr2} = 392.5$ K.[1] In order to calculate $\Delta_f H^{\ominus}(S, l, T)$, it is necessary to add the contributions due to both phase transitions, leading to

---

[1]These temperatures correspond to pressures where there is equilibrium between the two phases; the equilibrium pressures are really different from 0.1 MPa. For condensed phases, this pressure difference has a negligible effect in the transition temperature.

$$
\begin{aligned}
\Delta_f H^{\ominus}(\mathrm{S},\mathrm{l},T) \;=\;& \Delta_f H^{\ominus}(\mathrm{S},\mathrm{rhomb},T^{\ominus}) + \int_{T^{\ominus}}^{T_{\mathrm{tr1}}} C_p(\mathrm{S},\mathrm{rhomb})\mathrm{d}T \; + \\[2mm]
+\;& \Delta_{\mathrm{tr1}} H(\mathrm{S},386,1\mathrm{K}) + \int_{T^{\mathrm{tr1}}}^{T_{\mathrm{tr2}}} C_p(\mathrm{S},\mathrm{mon})\mathrm{d}T \; + \\[2mm]
+\;& \Delta_{\mathrm{tr2}} H(\mathrm{S},392,5\mathrm{K}) + \int_{T_{\mathrm{tr2}}}^{T} C_p(\mathrm{S},\mathrm{l})\mathrm{d}T
\end{aligned}
$$

Once the values of $\Delta_f H^{\ominus}(T)$ are obtained, we can get $\Delta_r H^{\ominus}(T)$ using eqn. (B.1). It is also possible to obtain $\Delta_r G^{\ominus}(T)$ and the equilibrium constant K from the definition $G = H - TS$ and the use of eqn. (8.11). That requires the calculation of the change of reaction entropy, given by

$$
\Delta_r S^{\ominus}(T) = S^{\ominus}(\mathrm{H}_2,\mathrm{S},\mathrm{g},T) - S^{\ominus}(\mathrm{S},\mathrm{l},T) - S^{\ominus}(\mathrm{H}_2,\mathrm{g},T)
$$

remembering that, if there are no phase transition, between $T^{\ominus}$ and T, one has

$$
S^{\ominus}(T) = S^{\ominus}(T^{\ominus}) + \int_{T^{\ominus}}^{T} C_p \mathrm{d}\ln T
$$

If there exist phase transitions, their contribution should also be added to the thermodynamic equations.

The terms added to $\Delta_r H^{\ominus}(T)$ and $\Delta_r S^{\ominus}(T)$, due to phase changes, cancel each other when calculating $\Delta_r G^{\ominus}(T)$, if the values of p and T are those corresponding to the phase equilibrium.

With regard to galvanic cells, also in this case, a convention is necessary with respect to the electrodes because, again, only differences of electric potential between both electrodes can be measured, and not at an absolute value. The convention consists in considering that at all temperatures the standard hydrogen electrode potential is equal to 0 V; that is, the Gibbs free energy for the half reaction (2 H$^+$ + 2 e $\rightarrow$ H$_2$) is zero at all temperatures.

Appendix C

Books Suggested for Consultation

- Callen, Herbert B. *Thermodynamics and an Introduction to Thermostatistics.* Hoboken, NJ: John Wiley & Sons, 1987.

- Chandler, David. *Introduction to Modern Statistical Mechanics.* Oxford, UK: University Press, 1987.

- Cussler, Edward L. *Diffusion: Mass Transfer in Fluid Systems.* Cambridge, UK: University Press, 1997.

- Guggenheim, Edward A. *Thermodynamics: An Advanced Treatment for Chemists and Physicists* 8th edition. North Holland, 1986.

- Hill, Terrell L. *An Introduction to Statistical Thermodynamics.* New York: Dover Publications, Inc., 1986.

- Israelachvili, Jacob N. *Intermolecular and Surface Forces.* USA: Academic Press, 1994.

- Prausnitz, John M., and Lichtenthaler, Ruediger N., and Gomes de Azevedo, E. *Molecular Thermodynamics of Fluid-Phase Equilibria.* New York: Prentice-Hall, 1986.

- Rowlinson, John S., and Swinton, F. L. *Liquids and Liquid Mixtures.* Kent: Butterworth-Heinemann, 1982.

Index

Credits

www.ingramcontent.com/pod-product-compliance
Lightning Source LLC
Chambersburg PA
CBHW082010190326
41458CB00010B/3140